Victor Papanek

For Constance Loadwick & Solomon Charles

Contents

Acknowledgments

The breadth of archival and oral history research, spanning several years, that underpins this monograph was made possible through the support of the Botstiber Institute for Austrian-American Studies, and the Fonds zur Förderung der wissenschaftlichen Forschung (FWF, Austria), for which I am enormously thankful. I gratefully acknowledge the support of The Graham Foundation in enabling the production of this book.

At various points my research took me to Northern Europe, North America, South America, and South Asia. This work depended as much on the oral histories and insights individuals shared with enthusiasm and critical perspective as it did on archival and visual documentation. Their experiences as colleagues, students, and friends of Victor Papanek, and their own involvement with design activism in the 1960s and 1970s, proved invaluable to the project and its critical framing. Some of those individuals preferred to remain anonymous; I thank them, and the following people: Olli Alho, Maria Benktzon, Pak Imam Buchori, Steen Juhler, Goroslav Keller, Barbro Kulvik, Satish Kumar, Gunilla Lindahl, Peter Mackeprang, Per Mollerup, Vuokko Nurmesniemi, Dieter Raffler, Mohd Azali Abd. Rahim, M. P. Ranjan, Parko Severi, Antti Siltavuori, Howard Smith, Liisa Tenkku, Lise Lotte Wiingaard.

Exceptional thanks are due to Mikko Pyhälä, author and former Finnish diplomat, who provided strong support for the project through his own personal perspective on Papanek and the 1960s in Finnish history and his tireless direction to appropriate resources and oral history contacts. Jim Hennessey, lifelong friend, collaborator, and coauthor with Papanek, along with Sara Hennessey were extremely generous with their time and sources. Yrjö Sotamaa, prominent design educator and former colleague of Papanek at Purdue University in the late 1960s, has been an abiding supporter of the research into the origins of social design. Investigating industrial design during this period, despite the number of women involved as students or tutors, is fraught with the methodological issue of representing women's largely undocumented presence. Sheila Levrant de Bretteville, a colleague and critic of Papanek during his deanship at CalArts (California Institute of the Arts), generously gave of her time discussing the pedagogic approach of design activism from a personal as well as historical feminist perspective. Gui Bonsiepe, leading design theorist and critic of Papanek's ideas around development in the Global South through the 1960s and 1970s, offered hours of interviews in his home in La Plata, Argentina, and proved to be a welcome ally in exploring the problematic dimensions of design for the "real world." Friends and family of Papanek provided detailed context and more intimate perspectives on his work and life: Thanks to Nicolette Papanek, Harlanne Roberts, Jennifer Satu, and Al Gowan.

The nuanced translation of key documents and transcripts from Finnish, Swedish, Danish, and Norwegian proved pivotal to the project. In this respect, I am particularly grateful to David Jones and Jyri Kokkonen. As well as providing translations, Robert Gassner shared his unbridled enthusiasm for primary research and his extensive knowledge made research in Denmark enormous fun. Thanks to Michael Pierazzi in San Francisco and Jennifer Kaufmann at Purdue University, as well as amateur genealogist Michael Foster, all of whom generously provided local research materials and insights. Bart Beaty shared nuanced insights from his extensive research on Fredric Wertham that proved vital to understanding the impact of 1940s social psychiatry on Papanek's anticonsumerist discourse. Fiona Raby, Jamer Hunt, Paola Antonelli, and John Thackara—through engagement with the materials in various contexts from exhibitions to symposia—have offered great insights. Historian Felicity D. Scott and anthropologist Arturo Escobar helped shape my thinking through their own work and early related discussions. Thanks also to Tanishka Kachru and Carey Hedlund. Undoubtedly, I have inadvertently omitted numerous other individuals integral to this project, and for this I apologize unreservedly.

The following institutions, archives, and libraries and their teams offered vital input to a research project that spanned several years. My thanks go to the Avery Architectural and Fine Arts Library, Columbia University, New York; CalArts Library, Special Collections and Institute Archives, Valencia, California; Frank Lloyd Wright Foundation, Scottsdale, Arizona; Föreningen Svensk Form, Historical Centre for Business History, Stockholm, Sweden; Archive and Library, Konstfack, Stockholm, Sweden; National Institute of Design, Archive and Library, Ahmedabad, India; Hagley Museum and Library, Wilmington, Delaware; University of Jyväskylä Archive and Library, Jyväskylä, Finland; International Institute of Social History, Amsterdam, Netherlands; Museum of the City of New York, New York; NC State University Libraries, Special Collections, Raleigh, North Carolina; Jane Kessler Memorial Archives, Penland School of Craft, Penland, North Carolina; Parsons New School of Design, Archive and Library, New York; Cooper Union, Archive and Library, New York; Purdue University Archives and Special Collections, West Lafayette, Indiana; Visual Resources and Special Collections, Dorothy H. Hoover Library, OCAD University Archives, Toronto; Schumacher College, Devon, UK: San Francisco Public Library, newspaper collection, San Francisco, California;

UC Berkeley Library, University Archives, Berkeley, California; United Nations International Development Organization (UNIDO), Vienna; University of Brighton Design Archives, UK; Volvo Museum, Gothenburg, Sweden.

The Victor J. Papanek Foundation archive at the University of Applied Arts Vienna formed the initial locus of my research. The collection was first painstakingly collated by architectural historian Bryleigh Morsink. My great thanks go to Bryleigh, an inspired researcher, with whom I spent many an hour in the early stages of research mapping out Papanek's US residences and various biographical timelines. Thanks to my colleagues Martina Gruenewald and Elana Shapira for sharing numerous fruitful research conversations, and to Michelle Jackson-Beckett for her invaluable insight. Heartfelt thanks to Maria Blyzinsky for her abiding friendship and to the Nenning family. I am grateful to Mateo Kries at Vitra Design Museum, Germany, and the curatorial team there, who enthusiastically embraced my suggestion to co-curate an international traveling exhibition exploring Papanek's ideas in relation to contemporary design practice, and the contradictions of design activism shaped by the historical research for this book.

Great thanks to the editorial team at MIT Press, and the three anonymous reviewers of the manuscript for their insightful comments. Special thanks to Doug for his patience as the research remit and timeframe expanded for the monograph.

Boundless love and appreciation (for enduring the tedium of having an obsessive researcher in their midst) to Constance, Solomon, Paul, and Hester. Love and thanks to my ever-knowledgeable father, John Michael Clarke, for calculating the travel time by train and ship from Vienna to the East Coast, United States, in 1939, among other incidental but crucial things, and to my mother for her abiding love and inspiration—who sadly never saw the final manuscript, but whose memory is woven into the travels and words that went into making its pages.

Introduction

There are professions more harmful than industrial design, but only a few of them. Never before in history have grown men sat down and seriously designed electric hairbrushes, rhinestone-covered file boxes, and mink carpeting for bathrooms, and then drawn up elaborate plans to make and sell these gadgets to millions of people. Today industrial design has put murder on a mass production basis. By designing criminally unsafe automobiles that kill and or maim nearly one million people around the world each year, by creating whole new species of permanent garbage to clutter up the landscape, and by choosing materials and processes that pollute the air we breathe, designers have become a dangerous breed.[1]

In 1971, an obscure book by an unknown author, focusing on the arcane topic of design's social usefulness to humankind, was translated from Swedish to English, and published as *Design for the Real World: Human Ecology and Social Change.* The 300-page, breathlessly delivered, hyperbolic treatise on the failings of the contemporary design profession quickly emerged as a bestseller, and the Austrian American designer behind it, author Victor Papanek, was invited to appear on a prime-time TV chat show. Once installed in the broadcasting studio, cameras rolling, he pulled out a prototype of his most recent design to use as a prop to demonstrate what a socially responsible design object might look like. Wielding a bright-blue plastic inflatable bedpan that could be pumped up with warm air beneath the patient, the social designer elaborated on how his observations of the inhumanity of the conventional stainless steel contraption during a recent visit to hospital had inspired its redesign: "I began to realize there was something inherently wrong with bedpans; because they are noisy, they clutter, they're difficult to move around, disgusting to wash out and clean—and they are unbelievably chilly, and they're expensive; they need a small room in each hospital floor just to be stored, and they cause back injuries."[2]

The inflatable plastic bedpan, one of a few select designs Papanek prototyped, never took off, failing to reach production stage despite being showcased on peak time American television. And yet that never really was the intention. Rather, the unlikely (and impractical) medical appliance formed part of a repertoire of product designs, from portable play environments for disabled children to dung-powered tin can radios for nonliterate communities in the Global South, devised as agitprop interventions in a social design revolution.

Formations

A young Jewish refugee who launched his own design practice, Design Clinic, in New York City shortly after World War II, Papanek was soon confronted by what he described as the "narrow market dialectics" that dictated the ethics of industrial design.[3] To someone who had escaped the horrors of Nazi Vienna to be met with the Futurama dreamscape of the New York World's Fair in 1939, at first design proffered an overwhelming optimistic vision of progressivism and democracy. Paradoxically, Papanek soon discovered that its principal role lay in fueling the all-consuming, undiscerning cacophony of American commercial culture that surrounded him, and with it some of the less desirable aspects of humanity's seemingly insatiable desire for stuff. And so began his search, as an outsider, for an alternative model of design that might heal society's ills, remedy social inequality, and empower its users: "The designer bears a responsibility for the way the products he designs are received at the market place. But this is still a narrow and parochial view," he later wrote, reflecting on his first foray into audio-furniture design for the famed New York Gimbels department store in the 1940s.[4] For, he continued, "The designer's responsibility must go far beyond these considerations. His social and moral judgment must be brought into play long *before* he begins to design, since he has to make a judgment, an a priori judgment at that, as to whether the products he is asked to design or redesign merit his attention at all. In other words, will his design be on the side of the social good or not?"[5] What began as a critique of the "sexed-up" frivolities of a design culture ramped up to meet the demands of unbridled postwar consumer culture had, by the 1960s, culminated in a full-blown global campaign against a profession that wreaked irreversible ecological damage, endorsed neocolonial development and perpetuated social inequality—with *Design for the Real World* its erstwhile manifesto. Translated into more than twenty languages and taken up by a generation of designers and design students desperately seeking an alternative politics of design, Papanek's work took on a life of its own—its clarion call weaponized in the dismantling of the beaux arts hierarchies and modernist teachings of European design schools, and the opening-out of design beyond the reductionist dualities of Western and non-Western. As a designer and critic of international renown, Papanek consciously pitted himself against the ideologies of his modernist émigré forebears who had viewed mass-produced, standardized industrial design through the lens of Western rationalism. Describing the Bauhaus style, in a piece of his earliest design criticism, as a "fascist negation of living that is now proven a lie," he proposed in its place a humane, indigenized design approach imbued with anthropological sensitivity to the local, the vernacular, and an understanding of the broader cultural nuances of design's power in undermining or solidifying social inclusion.[6] In challenging design's assumed role as the originator of frivolous fripperies in an age of overabundance, his ideas pivoted on the overarching theory that design was the key agent of social change, not merely a technocratic tool for stylization or aestheticization, or a driver of increased consumer consumption. As an integral part of the social design agenda, he advocated non-Western tropes of design—from material cultures of the Inuit to the Suku Bali, as holistic models of design whereby things are understood as inseparable from the social relations, customs, rituals, and histories in which they are embedded. The politics of design, in other words, relied on understanding the practice as a cultural rather than rational, problem-solving phenomenon.

Contexts

Design for the Real World, and the social remit it engendered, did not of course operate in a vacuum. In Europe, the radical Italian designers Superstudio, who famously advocated "life without objects" as part of their self-consciously disaffected stance toward late capitalism, exercised a similar contemporary fascination with the vernacular and Indigenous object.[7] "Objects and tools represent a particular field of investigation; they lend themselves much better to being used as keys in the interpretation of complex relationships: Objects are the direct witnesses of the creative drive," recounted Superstudio designer Alessandro Poli, in describing the design group's Extra-Urban Material Culture pedagogic initiative of 1973: a project aimed at collecting, recording, and analyzing handmade peasant tools prior to their destruction through unabated technological advance.[8] Yet, unlike the Italian design radicals whose work had been famously showcased in the 1972 Musem of Modern Art (MoMA), New York exhibition *Italy: The New Domestic Landscape*, with a number of their catalog contributions renouncing the practice of architecture and design altogether in faux flourishes of avant-garde gestures,[9] Papanek stood by an unerringly (and often unfashionably) pragmatic stance that earned him praise and criticism in equal measure: "In an environment that is screwed up visually, physically, and chemically, the best and simplest thing that architects, industrial designers, planners, etc., could do for humanity would be to stop working entirely," he wrote several years prior to Superstudio's high-profile rejection of architectural and design practice.[10] But *Design for the Real World*, which Papanek dedicated to his students "for what they have taught me," was deliberately conceived as an antidote to inaction, and what Papanek would later describe as the "romantic bourgeois Marxism" that had spread like an epidemic across design schools.[11] "In all pollution designers are implicated at least partially," he wrote toward the end of its preface, "but in this book I take a more affirmative view: It seems to me that we can go beyond not working at all, and work positively.

Design can and must become a way in which young people can participate in changing society."[12] Publication of the groundbreaking volume perfectly preempted, and coincided with, a maelstrom of interconnected events, interventions, and political crises that led to the foregrounding of environmentalism, postindustrial futures, postcolonialist thinking, and degrowth economics: all of which threw into relief the politics of industrial design, and its intertwinement with Western capitalism's expansionist and development policies. In 1972, *Only One Earth*, a report commissioned by the Secretary-General of the United Nations Conference on the Human Environment (UNCHE) was released as a precursor to a groundbreaking Stockholm Congress of the same name. Led by British economist Barbara Ward and French-born American environmentalist René Dubois, it identified a full-blown consumer revolution, described as "a twenty-five year boom among the industrialized nations after 1945," as the key environmental hazard of the twentieth century, "gobbling up resources and increasing requirements of materials and energy at an unprecedented rate."[13] This publication was one of a multitude that emerged from a newly forming canon of environmental critique. But its specific mention of consumer culture, and its conduit advertising, formalized the phenomenon of First World consumer culture as a central tenet of environmental discourse. This critique highlighted the vicissitudes of an industrial development agenda, which simultaneously promoted manufacturing and full-throttle consumer economies while bemoaning their proven detriment to environment and social structure.

The critique was further buoyed by counterculture rhetoric and alternative economic theories of the late 1960s and early 1970s, exemplified by the neo-Marxist Wolfgang Haug's *Critique of Commodity Aesthetics* (*Kritik der Warenästhetik*, 1971), which identified the profession of design specifically as a key driver of capitalism, accusing it of complicity in generating an insatiable desire for commodities and the resultant dissolution of authentic social relations. The publication of the *Critique of Commodity Aesthetics* coincided with that of *Design for the Real World*, and though they addressed vastly different audiences, they placed the role of design firmly on the agenda as a central object of critical examination in sketching out models of alternative postindustrial economies and social life.

Most significantly, the UNCHE of 1972 served as a springboard from which emerged a broader design for development discourse that sought to remedy political dilemma over local-global dichotomies of industrialization and its environmental consequences. This same dilemma Papanek had striven to address since the early 1960s, through his codesigns for socially excluded groups as well as the so-called developing countries. His projects focused on prototyping for low-cost, self-assembly items including a sustainable "appropriate technology" television for use in Africa (addressing the inequalities of global media access) and a self-build revolver (intended for vermin control) in poor rural communities of the United States. More specifically, the UNCHE laid the groundwork for future initiatives that affected the design profession: the collaboration of International Council of Societies of Industrial Design (ICSID) and United Nations Industrial Development Organization (UNIDO), and the signing of the Design and Development Ahmedabad Declaration in 1979. Both organizations were represented at the 1972 UNCHE, and their collaboration five years later, in which Papanek played a formative, critical role as a member of the design and development working group, would mark a turning point in the formalization of the design profession's role in development policy impacting on small-scale manufacture and craft economies in countries ranging from India to Mexico. Despite his involvement in these policy schemes, Papanek also became an avid critic of their unintended consequences: engaging fully with the early 1970s debate around decolonizing design, he challenged top-down solutions to local conditions and rejected the use of copyright and patents, advocating instead an open source design model.

Legacies

Design for the Real World quickly became a classic of alternative culture. On the radical student bookshelf, the volume was sandwiched between tomes such as Rachel Carson's warning of imminent ecological disaster, *Silent Spring* (1962), and Teresa Hayter's critique of the mechanisms of neocolonialism, *Aid as Imperialism* (1971). Papanek's iconoclastic designs, provocative journalism, and unique pedagogic projects rattled a complacent corporate design establishment, and the design avant-garde alike. His "real world" paradigm posited a model of design that served humanity—culturally and practically, addressing the needs of the socially excluded, rather than oiling the lucrative wheels of consumer corporations or stoking the egotism of individual designers: "Design and production talent have been wasted" bemoaned Papanek in a typically acerbic think piece in the design press, "on the concocting of such inane trivia as mink-covered toilet-seats, electronic fingernail polish dryers and baroque fly-swatters."[14]

But there were comparable titles, such as Ulm School of Design, Germany, industrial design professor Tomás Maldonado's *Design, Nature and Revolution: Toward a Critical Ecology* (published in Italian in 1970; English-American edition in 1972). Like *Design for the Real World*, Maldonado wrote this work at the peak of the countercultural and New Left political wave. He invoked triadic ecological relations among design, nature, and revolution, but with a broader theoretical focus on human environment rather than industrial design per se, urging that "design is part of a larger totality which it's possible for the designer to approach consciously, politically, critically."[15] Papanek clearly saw parallels between his own work and that of Maldonado. Despite having had no prior knowledge of the Ulm professor's ideas during the preparation of his manuscript for *Design for the Real World*, the first edition copy of *Design, Nature and Revolution* held in Papanek's personal library is the most extensively marked-up book of his entire collection. Maldonado's writings clearly echo those of Papanek's socially responsible design polemic, yet they aim at a distinctly more politicized as opposed to pragmatic stance toward the transformative potential of design: "Politically speaking, the revolutionary sense of dissent is really only attainable through design. Dissent that rejects hope in design is nothing but a subtle form of consent," postulated Maldonado. [16] Both design theorists' works stood, then, as discrete interventions into a shared contemporary discourse around the social and political purpose of design in a postindustrial world of environmental crisis and social inequality. Notably, though, Papanek distinguished himself as a public polemicist willing to court controversy by leaning toward populism to spread his message rather than confining his ideas to what he perceived as the elitism and self-congratulatory tone of the design establisment. Interestingly, the one facet that linked the two figures together, however, came by way of Gui Bonsiepe—Maldonado's German Ulm School of Design protégé—who visited Papanek in the United States in the mid-1960s, and who would later become the social designer's rival and most outspoken critic, lambasting his approach to the Global South and even accusing the designer of collusion with the US military and intelligence services. Decades on, Victor Papanek is undisputedly recognized as a pioneer of social design, and *Design for the Real World* (which has never fallen out of print) remains one of the most widely and globally read critiques of the skewed priorities of industrial design to this day. As social design, humanitarian design, speculative design, design anthropology, transition design, and decolonizing design movements have gained such prescience in the twenty-first century, this book offers a critical historical and biographical account of a figure and genre of design central to the understanding the complex origins of these burgeoning areas. Papanek's words, in the opening pages of his seminal work penned over half a century ago now, still hold enormous resonance in the present-day context: "As socially and morally involved designers, we must address ourselves to the needs of a world with its back to the wall while the hands on the clock point perpetually to one minute before twelve."[17]

Structure

Papanek's life and career spanned the twentieth century, and as such its trajectory incorporates the shifting terrain of industrial design practice during this period, the historiography of which is dominated by monographs of great, male designers. While the subject of this book is also premised on the life and works of a male industrial designer, the intention has not been to chart the career trajectory of another "great" or "pioneering" figure. Rather it critically examines, through extensive original archival and oral history research, the life and work of a designer who remains widely exalted as an antihero of design. In a twenty-first-century context, the narrative behind the oeuvre of social and humanitarian design that Papanek pioneered is layered with problematic resonance, from white saviorism to full-blown misogyny. Yet the notion of social design as being, by default, progressive persists today, even though the origins of the phenomenon have been left largely unexamined. This monograph provides critical historical contextualization around the discourse of social and humanitarian design, as a means of beginning to chart its often ambivalent legacy in contemporary multinational corporations, independent design studios and humanitarian policy-making institutions around the world.

Chapter 1, "Design Clinic: The Early Formation of a Social Designer," explores the formative aspects of Papanek's early life as an Austrian American émigré. Having escaped Nazi persecution as a youth in 1939, he began to educate and establish himself as a designer in New York City, opening his own studio there, Design Clinic, in 1946. As a young refugee hailing from a bourgeois Viennese family and now immersed in the alien setting of a downtown working-class and ethnically mixed neighbourhood, Papanek pieced together an unlikely set of cultural references and connections (both real and imagined) that would go on to form his self-identity as a social designer. Chapter 2, "Seduction of the Innocent: From Bondage Fetishism to Frank Lloyd Wright," follows Papanek's complex journey as he navigated life as an ambitious young designer, heavily influenced by the brash novelty of US popular culture, on the one hand, and the organic approach of America's leading architect, Frank Lloyd Wright, on the other. Wright would prove to be an abiding influence and point of reference throughout the designer's life, Papanek having claimed to apprentice and work with the architect at his Taliesin studios in the 1940s. This chapter traces the origins and evidence of that relationship, Papanek's negotiation of a fraught personal life, newly defined in the United States by his Jewish ethnicity. It reveals his extraordinary and surprising connection to one of America's most famous critics of popular culture, Fredric Wertham, and the Black progressive thinkers behind Wertham's celebrated social psychiatry clinic for African Americans in Harlem. Like many designers before him, Papanek's commercial ambitions inspired his journey across the so-called émigré "bridge" from the East Coast to the West Coast, leaving behind him the city that had been so formative to him as a crucible of design thinking and social insight. Chapter 3, "Studio 44: Mid-Century California Calling," explores the shift in Papanek's design practice to that of design critic, a career move he combined with the launch of a new hybrid gallery and consultancy, Studio 44, in downtown San Francisco. Teaching experimental design at both the Art League of California, San Francisco, and the Chouinard Art Institute, Los Angeles, in the 1950s, Papanek gradually carved out a profile as an innovative design pedagogue offering a contemporary transdisciplinary approach that challenged conventional, vocational models of industrial design training. Chapter 4, "The

Comprehensive Designer: Toward a New Politics of Design," traces the ways in which Papanek's design approach was underpinned by a Cold War research agenda, and the influence of architectural maverick R. Buckminster Fuller on his ideas of comprehensive design. Chapter 5, "From Bionics to Racial Politics: The Making of a Cold War Designer," locates the origins of social design in the military design experiments of leading US design institutions, where interdisciplinary design teams, and research into the potentialities of bionics, formed an integral part of Cold War policy that played out against a backdrop of global decolonization and civil rights movements. Chapter 6, "Prescription for Rebellion? From Styrofoam Domes to Animatronic Women," charts the intersection of radical student activism and design experimentation on campus, and Papanek's attempts to harness these in the creation of a transdisciplinary design course for a new age of industrial design and social action. Chapter 7, "Northern Lights: The Allure of Nordic Design Activism," explores the reliance of Papanek's social design agenda on his interactions with Nordic design activists, and the beginning of his seminal relationship with Finland and a young pan-Scandinavian design group, culminating in the writing of his groundbreaking book *Design for the Real World* in the late 1960s. Chapter 8, "Design for the Real World: A Call to Action," considers the polemical impact of a burgeoning social design agenda, and Papanek's role as dean of design, in the progressive environs of the newly formed design institute at the California Institute of the Arts in the early 1970s. Chapter 9, "Pragmatism before Politics: Social Design Turns Nomadic," traces the beginnings of a backlash against the "real world" polemic, and a critique of its apolitical standpoint. Based in Denmark, as a guest professor, and then in the UK, Papanek produced his coauthored book *Nomadic Furniture 1* as his ideas moved further toward pragmatism and away from activism. The concluding chapter 10, "After *Design for the Real World*," examines the shift of social design from the idealism of 1970s to the terrain of the 1980s "designer decade"—and the shift of Papanek's ideas to a form of ecological spiritualism, published in his final book, *The Green Imperative*, in the last years of his life.

1 Design Clinic: The Early Formation of a Social Designer

In January 1942, *Harper's Bazaar* magazine ran a lush, color-saturated feature choreographed by the leading fashion photographer of the day, Louise Dahl-Wolf. Under the byline "Flight to the Valley of the Sun," the piece centered on the visual drama of Frank Lloyd Wright's iconic Rose Pauson House, its monolithic structure "clinging like an Eagle's nest to a mountainside," with statuesque models drenched in the arid Arizonian desert heat poised precariously upon its parapets.[1] Weeks later, Wright's masterpiece was razed by fire, leaving the fashion magazine's story of jaunty culottes, turban hairbands, and oversized sunglasses as the popular immortalization of the architecture's stunning modernity. The portentous fashion shoot, shortly after which Wright's monument was reduced to a pile of rubble, proved significant in other ways. The make-believe backdrop of jet-set glamour and modish wartime frivolity provided the improbable introduction of a young Viennese émigré into the world of American design.

Victor J. Papanek, a young refugee who had escaped Nazi Austria three years earlier, had traveled from New York City accompanying his girlfriend, a fashion model named Dahira, for the high-end shoot.[2] Later describing this encounter with Wright's work as something akin to an epiphany inspired by the great American master's oeuvre and the transformative power of design, Papanek experienced an awakening as a would-be designer with a determination to integrate the material world into a humanist and social agenda. Remarkably, a few years after this chance desert encounter with America's preeminent architect, Papanek managed to secure himself an apprenticeship with the master himself, taking up residence at the famed Taliesin studios (in Spring Green, Wisconsin, and Taliesin West, in Arizona) with the design mentor and his fellows in the late 1940s. In one fell swoop, then, the young émigré acquired a caliber of design credential and connection that most aspiring architects could only dream of possessing in the course of entire careers. Papanek's extraordinary journey— from a bourgeois life in Vienna to a Lower East Side charitable Jewish hostel in Manhattan and on to apprenticeship with America's revered architect Frank Lloyd Wright—marked the beginning of an exceptional design biography, one shaped and defined by a remarkable network of influences and connections.

While serendipity played a crucial role in bringing Papanek into a firsthand encounter with the work of America's most celebrated architect, the role of the outsider émigré in de-provincializing the American mind, upgrading culture from their marginal position of reluctant exile, is a trope familiar to many twentieth-century émigré biographies, from artists to scholars.[3] The means by which individuals and groups negotiate the vicissitudes of the emigrant experience, mixing strategy, expedience, and improvisation, is particularly apparent in the context of design and architecture: networks of social and intellectual relations become indistinguishable from networks of creative output.[4] In this respect, the early biography of Papanek proves no exception; his influences and connections spanned generations of acclaimed architecture and design "greatness," including Buckminster Fuller, Raymond Loewy, Frank Lloyd Wright, and émigrés Victor Gruen, Frederick Kiesler, George Nelson, and Bernard Rudofsky. There were, however, some startlingly improbable additions to this parade of twentieth-century design icons, among them the maverick American author Henry Miller, psychiatrist and campaigner for legislation of adolescent comic books Fredric Wertham, and, perhaps most unexpectedly, the 1940s bondage-fetish glamour model Bettie Page. Unconventional creative design sources though they were, these latter characters would prove vitally significant in shaping Papanek's creative and intellectual world by mediating his early experiences in an alien culture, and honing his critique around the social imperative of a design culture that looked beyond base commercialism.

Becoming an Émigré: From Bourgeois Vienna to Downtown New York

To describe Papanek's traumatic experience emigrating to the United States as formative would be a stark understatement. After an arduous journey by train and sea, he and his mother arrived on the SS *Pennland* from Vlissingen, Holland, docking in Hoboken, New Jersey, on April 4, 1939. Having been processed at the federal immigration station on Ellis Island, Victor Josef Papanek, a fifteen-year-old youth, accompanied his vulnerable, widowed mother Helene, who prior to their escape from Vienna had been violently stripped of the family business, of property, and of intimate personal belongings.[5] As they stood together in line, anxiously waiting to be processed on Ellis Island along with the thousands of other "Hebrew" refugees with whom they had fled their homeland after the National Socialist *Anschluss* of Austria, Victor's realization of his poignantly shifted status, from affluent bourgeois Viennese to impoverished alien, would have been profoundly visceral.[6] His mother, fragile, non-English-speaking, and entirely dependent on her young son, had left Vienna perilously late after the Nazi Anschluss.

In a later autobiographical account tracing the "Peasant Aristocracy" lineage of his family the "von Papaneks" (written for his two daughters), Papanek obliterated the trauma of this émigré history simply stating that he and his mother "emigrated to the United States for political reasons in 1939, when I was a boy, dropping the 'von' which we considered pretentious in a democratic country."[7]

Born Victor Josef Papanek, November 22, 1923, Victor was the sole offspring of Richard and Helene (née Spitz), the owners of a fine food and wine store on the prestigious Imperial square Am Hof, in the heart of *Innere Stadt* (Inner City) of Vienna.[8] Until the tragic death of his father in 1935, and the horrors of the Anchluss in 1938, his life had been indistinguishable from that of any other well-to-do young *Wiener* (Viennese). A cherished only child, Papanek enjoyed the childhood pleasures of Vienna in the 1920s, its numerous parks, museums, winter ice skating, and seasonal *Christkindlmärkte* (Christmas markets) (see figure 1.1).

Prior to their escape to the United States, Papanek's family home had been a grand apartment on the Ringstrasse, an opulent nineteenth-century development encircling the Innere Stadt, a location popular within the upper echelons of Vienna's Jewish society.[9] Raised in the rarefied upper-bourgeois surroundings of Vienna's premier address, the "grace and favor" apartment granted to his father as a certified "Counsellor of Commerce" (*Kommerzialrat*), occupied a sought-after location a short stroll from the opera house, directly opposite two of the city's most genteel parks, the Burggarten and the Volksgarten. Following his attendance of a local *Volksschule* on nearby Johanngasse in Vienna's first district, Victor gained entry to one of Vienna's most prestigious high schools, the Akademisches Gymnasium, and at the age of ten enrolled in the rigorous Viennese *Gymnasium* school system with its relentless emphasis on rote learning and a classic curriculum of mathematics, modern languages, and the humanities. Renowned for its liberal but rigorous education, the roll call of Akademisches' alumni included Nobel Prize winners, eminent Austro-Hungarian leaders, politicians, scientists, philosophers, economists, composers, and writers. Notably, one of the school's most influential alumni was émigré Paul Lazarsfeld (1901–1976), a member of the Vienna Circle of philosophers (which also included political scientist, sociologist, and information designer Otto Neurath). Lazarsfeld, born to Jewish Viennese parents, left Austria in the 1930s to become the founder of twentieth-century empirical sociology, instigator of mass-media theory based on his pioneering research in 1920s Vienna, where he founded the Wirtschafts- psychologische Forschungsstelle (Business Psychology Research Centre). As a pioneer in the application of psychology to business, he offered Papanek an interesting émigré role model, in that Lazarsfeld's maverick approach, borne of the hotbed of 1920s Viennese intellectualism, applied social analysis to American consumer culture.[10]

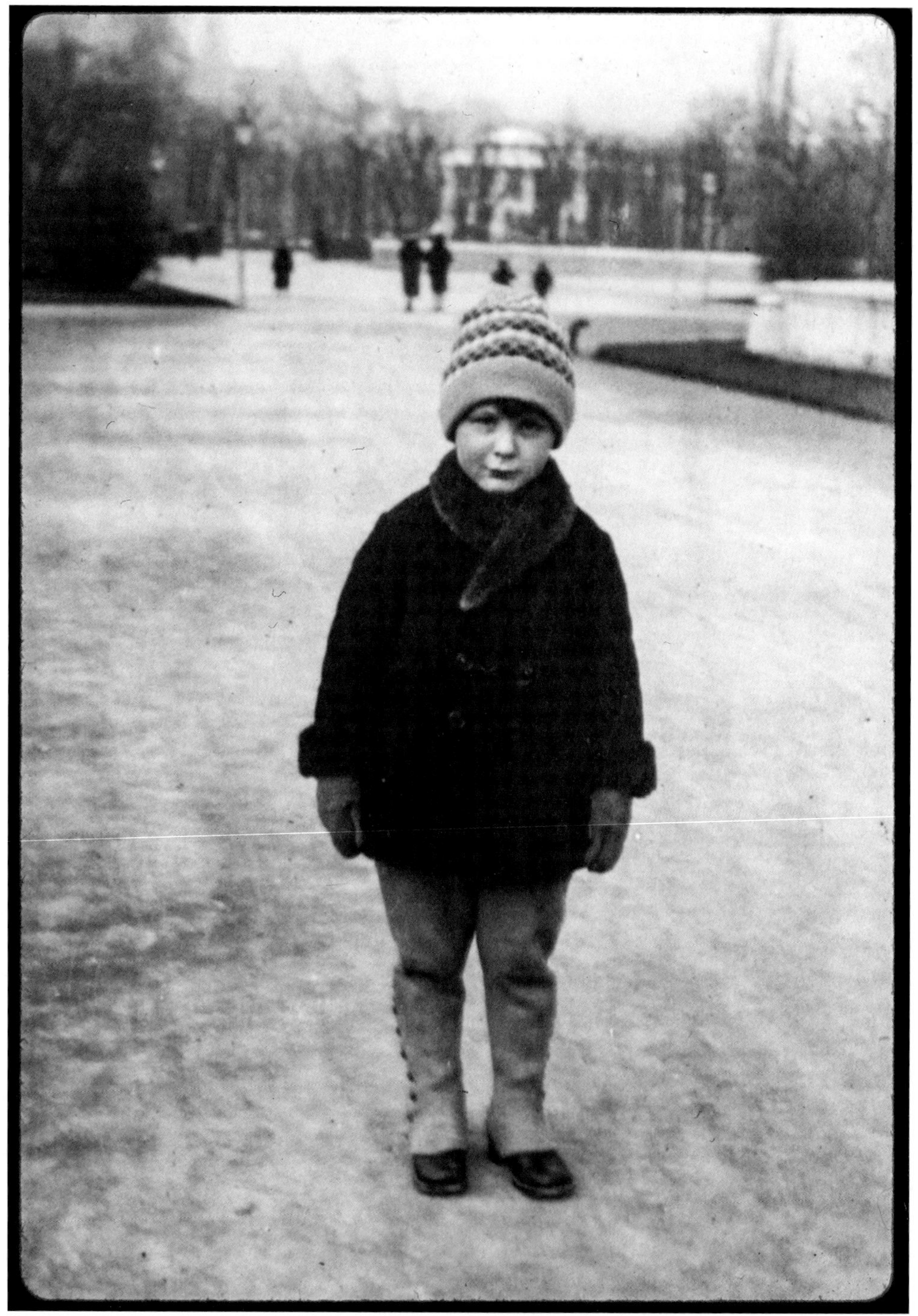

Figure 1.1
Victor Papanek at Volksgarten, Vienna, c. 1927. © Victor J. Papanek Foundation, University of Applied Arts Vienna

Once a fully integrated and progressive high school, by 1938 the Nazi instigation of Jewish segregation led to a 40 percent drop in pupil numbers at the Akademisches. By this point, possibly on account of his foundering grades, and the impact of his father's death (in March 1935) on the family more generally,[11] Papanek had already transferred to a more liberal and less academically demanding high school in the leafy bourgeois suburb of Döbling, allowing more opportunities to study practical subjects such as landscape design.[12] The last record of Papanek's education as a schoolboy in Vienna dates to this school, the Realgymnasium XIX Wien, in February 1937.

In later autobiographical accounts, Papanek referred to his attendance of an elite English boarding school responsible for educating former British prime ministers. No archival evidence has been traced to substantiate this claim, but his faintly British accent (which endured even after decades of life in North America) may support this assertion. Certainly, it was customary for bourgeois Viennese families wishing to ensure a native level of English language to arrange a year-long residency, at minimum, for their offspring in British private schools prior to returning for the completion of the Austrian Matura examinations. Victor's widowed mother, Helene, struggled to run the family business alone and was doubly fraught by raising her young son in an increasingly hostile social and political environment. There is every reason to surmise that sending Papanek to a safe haven in the form of a British boarding school offered an expedient solution to an increasingly urgent situation. Although undoubtedly protected by a certain level of affluence, Victor's early life saw Vienna witness to violent political upheaval: clashes between Socialists and Austrofascists in the revolutionary uprisings of the Austrian Civil War (the so-called February Uprising of 1934) and by 1937 the escalation of National Socialist assaults.

Residing as they did on Vienna's premier imperial thoroughfare, Helene and Victor witnessed firsthand the celebratory Austrian National Socialist Party parades, tumultuous social unrest, demonstrations, and protests. In fact, the family apartment on the Opernring section of the prestigious Ringstrasse sat directly adjacent to Heldenplatz, the notorious site of Adolf Hitler's triumphant speech of March 15, 1938—tens of thousands of National Socialist supporters marching directly before the Papaneks' home. The mounting terrorization, threats, trepidation, and exposure to the horrifying details of newspaper accounts, including the *Kristallnacht* atrocities committed in November that same year, would have proved inescapable. Scenes of routine humiliation, interrogation, arrest, suicides, and escape attempts made by Vienna's Jewish population eradicated any semblance of normal life.[13] The boycotting and defacing of Jewish-owned or -staffed shops had become commonplace across the city, with graffiti "Nicht arisches Geschaeft" (non-Aryan business) and "JUDE" (Jew) daubed across business windows. Stores on the Ringstrasse itself were closed down, their status as "Nichtarisches Geschaeft" (non-Aryan) meant customers were forbidden from using them. As retailers themselves, with their gourmet grocery store situated in the heart of prestigious central Vienna, the Papaneks would have been particularly vulnerable to public persecution.

While the Papanek family was culturally assimilated into mainstream-bourgeois Viennese life, they were clearly integrated into the city's thriving Jewish community. Helene and Richard married at the Schmalzhofgasse Synagogue in Vienna's sixth district, and his birth was registered with the Viennese *Israelitische Kultusgemeinde* (Jewish Community Organization) Seitenstettengasse, in the first district.[14] Papanek viewed himself first and foremost as a Wiener, his Jewish identity an entirely normalized and unexceptional part of his upbringing in Vienna. Despite the trauma of his escape, he held a lifelong love of the city and its culture, having the traditional Viennese cake, *Sachertorte*, delivered to his home in the United States for each of his birthdays in the latter part of his life.[15]

The Specter of Vienna: Loss and the Politics of Material Culture

Pre-1938 Vienna occupies an almost mythological place in memories and historical accounts of émigré experience. The city and its culture, prior to the fleeing of persecuted inhabitants following the Nazi Anschluss, are frequently identified as constituting the locus of intellectual and political inspiration carried forth and translated into new world cultures, to reemerge as a progressive force.

Historian Lisa Silverman has gone as far as to argue that the spaces invoked by émigrés, specifically the Jewish spaces of pre-1938 Vienna, would be better understood as a "formative aspect of material culture." She highlights how Jews "individually and collectively" negotiated their place in the city and later utilized such spaces in the making of their émigré biographies.[16] She has similarly argued the significance of property (material culture, furnishings, and artifacts of mundane and sentimental value alike) confiscated, lost, or destroyed through the "Aryanizations" recounted in émigré narrative histories, as a mistakenly peripheralized aspect in the larger understanding of Jewish identity and memory.[17] American historians, most notably Andrew Heinz in his 1990 *Adapting to Abundance*, have long established the vital role of consumption and material culture in the Jewish diasporic experience of acculturation. But overturning the common-sense, liberal assumption of private property as a naturally divisive entity, Silverman affords a collective imperative to owned goods, arguing that property "reaches far beyond the bounds of the negative, splintering economic effects usually ascribed to it," to the point of actually enabling "the survival of cultural groups."[18] The specificity of the Austrian émigré experience, as outlined through her use of biographical and autobiographical sources, tends in most part toward the representation of the bourgeois émigré experience. Yet the relevance of the material culture of possessions and concepts of taste is, she argues, pertinent to a wider group, given that in the era of the new Austrian republic the majority of Jewish Viennese rented their apartments, meaning "household furnishings became even more important markers of [their] identities as acculturated Austrians."[19] This would certainly have been the case for the Papaneks, wrenched as they were from a genteel Viennese life with only the clothes on their backs to take on board the SS *Pennland*. Revealingly, throughout his career, despite his proselytizing ethos of social equality and inclusion, Victor Papanek associated himself with Austro-Hungarian aristocracy (referring to a noble ancestry through the nomenclature "von Papanek"), and later in life designing his calling cards with the Austrian imperial emblem of an embossed double-headed eagle. His championing of an approach to design focused on the politics of "things" and consumption might arguably be rooted in the trauma of the émigré experience. The sense of loss, and a spatial and material remembrance of possessions, runs through a multitude of émigré biographies. But there is one life story in particular in the area of architecture and design that stands out, specifically in the way it parallels Papanek's own struggle to occupy a professional niche.

Viennese American retail architect and urban planner Victor Gruen (born Viktor David Grünbaum), the famed "inventor" of the American shopping mall, often recounted an apocryphal story of arriving as an architect in New York City with only a T square to his name. Describing Vienna nostalgically as "a haven" he had "never really left," Gruen's life as an émigré was dependent to such an extent on the idealization of his lost city that it evolved to the point of exerting a specific agency on his designs in the States.[20] Historian M. Jeffrey Hardwick argues that it was Gruen's idealized version of Vienna as a diverse but socially intact city that would come to profoundly shape the invention of America's stereotypical form— the shopping mall.[21] Through the process of being stripped of his identity, Gruen constructed an extensive network of fellow émigrés and influential figures on whom he could rely, much as Papanek did in the early stages of his career. As a Jewish architect, he was systematically excluded from the upper echelons of an American architectural practice, which, despite the impact of 1930s progressive modernist émigrés, remained predominantly circumscribed by cultural and social constraints tied to the notion of the "gentleman architect." As Hardwick puts it, "Anti-Semitism drove Gruen from Vienna to New York and being Jewish in America pushed him into the field of retail design."[22] In that sense, outsider status was conferred on Gruen twice over, pushing him to the sphere of commercial practice so derided by more "genteel" architects, a sphere in which he would later innovate a core American trope.

The ambivalence toward commerce, and its association with Jewish identity and the émigré experience, has been dealt with extensively by historians looking at the impact on American culture made by émigrés ranging from marketing psychologist Ernest Dichter (1907–1991) to sociologist Paul Lazarsfeld. Intellectual historian Daniel Horowitz's early work on the émigré as a celebrant of consumer culture emerged as part of a broader shift toward material culture and consumption as key mechanisms of understanding historical agency.[23] Following this earlier work, a more recent generation of scholars has elaborated and complicated the ways in which twentieth-century Jewish immigrants, far from being at odds with American consumerism, actively embraced and shaped it as psychologists, intellectuals, sociologists, educators, architects, planners, retailers, advertisers, graphic and furnishing designers, and so forth.[24] Yet the fervent critique of modern consumer culture in the Frankfurt School mold is still commonly viewed as the prevailing condition of the Jewish European émigré and exile experience. Regarding the prevalence of a reductionist understanding of the émigré relation to New World, historian Jan Logeman argues that "Adorno, Marcuse, and his colleagues present us with an image of the European émigré at odds with mid-century consumerism and its manipulative capabilities in the United States."[25]

Victor Papanek's work as a designer and design theorist, his critique of popular culture, and his best-known work *Design for the Real World: Human Ecology and Social Change* (1971) all lead to the drawing of neat parallels between his discourse and the idea that the European intellectual sensibility has been "at odds" with US mass culture—an idea in keeping with the Frankfurt School émigrés (Theodor Adorno, Max Horkheimer, et al.). Papanek's relentless critique of American corporate and consumer culture has also tended to frame him in the tradition of postwar US anticorporate lobbyists such as Vance Packard and Ralph Nader, whose best-selling exposés *The Waste Makers* (1960) and *Unsafe at Any Speed* (1965) revealed the machinations of built-in product obsolescence in the consumer goods industry and negligent design within car manufacture, respectively. Yet examination of his early life opens up a far more complex understanding of the ways in which his polemic was shaped by émigré experience, from personal trauma to disjuncture in his material life and exposure to an extraordinary array of discordant sources and influences through which he carved his identity and livelihood. There are clearly trajectories that Papanek held in common with fellow émigré designers and architects, but the specificity of his experience surely centers on the shift in social class and ethnic identity, and the violent unanchoring of his material, sensory, and aesthetic world.

Total Immersion: From the Ringstrasse to the Lower East Side

Initially lodging with his mother at the apartment of his maternal great uncle and aunt Edward and Clara Strauss in Harlem, upper Manhattan, within a few months of his arrival in the United States Papanek found himself immersed in the throes of a harsh metropolis at a time of profound economic depression. The reasons behind his move to the Hannah Lavanburg Jewish charitable hostel remain unclear: it may have been a necessary step toward independence, citizenship, social support, and employment. What is evident is that the transition from the opulence of the Viennese Ringstrasse, to a Jewish hostel bordering the Lower East Side of Manhattan, where he lived side by side with impoverished Jewish émigrés working in low-salary positions as apprentices, clerks, and errand boys, must have proved a shockingly eye-opening experience for a sheltered European upper middle-class youth.[26] Bewildering in its vastness, and sensorially overwhelming, New York City offered a harsh transition to émigré urban life. The ethnically vivid Lower East Side that Papanek rubbed up against shortly after his arrival in the States could not have been further removed from his former life in the genteel first district of Vienna, with its decadent opera house, imperial architecture, and fine cafés.

As a young Viennese bourgeois, if Papanek had any familiarity with ethnic Jewish culture at all it would have been confined to a distant, almost mythical knowledge of the Leopold-stadt, well known as Vienna's working-class Jewish district, situated on the opposite side of the Danube Canal from his home in the affluent Innere Stadt. Had Papanek ever glimpsed the pre-1938 atmosphere of Leopoldstadt, as described by historian Silverman, it would have been faintly recognizable in the ethnically Jewish neighborhood of Manhattan's Lower East Side: "the influx of Jews from the East, many of whom were religious and more traditionally-garbed, ensured that [Leopoldstadt], with its high concentration of Jewish residents, signs with Hebrew letters advertising kosher shops, and Yiddish newspapers, theaters, cafes, and synagogues, continued to thrive and evoke shtetl life."[27] Although by 1939 a large proportion of New York's Jewish population had moved out of Manhattan into areas such as Brooklyn and the Bronx, according to historian Peter Buhle, the Lower East Side became the symbolic home of Jewishness, defined in the popular imagination by "crowded streets, small peddlers in their stalls, factories, penny candy stores . . . scenes reimagined by the emerging [1940s] Jewish artists of Hollywood, therefore almost as real as if they had taken place."[28]

These were formative years for Papanek, immersed in the throng and tumultuousness of the metropolis, no longer constrained by the trappings and strictly delineated behaviors of his conventional bourgeois Viennese life. A contemporary marketing analysis of Manhattan and its various neighborhoods vividly captures the cultural diversity and vibrancy of the area in the early 1940s:

> Visitors to New York City find the Lower East Side an amazing show. There is nothing comparable in America. It is the most populace, most crowded, most Old-World district in New York City. Its more than 100,000 foreign-born population gives the Lower East Side a tinge that is essentially alien. The pushcart markets, Chinatown, the Bowery, barber colleges, tattoo shops, second-hand clothes exchanges provide color and atmosphere seldom encountered in the American scene.[29]

Although anxiety-provoking, Papanek's new-found neighborhood must also have proved an exhilarating and transgressive experience. As a refugee outsider in the alien spaces of a vast city, caught up in a maelstrom of high and low culture, Papanek's firsthand experiences of a popular culture and material life, so glaringly at odds with the stasis of a Viennese bourgeois upbringing, shaped a mode of design criticism that would come to transform the notion of "social" within design.

Dawn of the New Day: The Promises and Contradictions of Design

Victor Papanek's arrival in the United States auspiciously coincided with the opening of the spectacular 1939 New York World's Fair: a showcase of the power of industrial design in shaping social futures.[30] The widespread promotion of the spectacle's "futurama" iconography, and the presence of its leading designers on the pages of newspapers and magazines, would have made this an inescapable spectacle and talking point. Under the title "Dawn of the New Day," the World's Fair theme of progressive democracy and the envisaging of futuristic living stood in stark contrast to the oppressive visions of the fascistic regime in war-torn Europe; this vision would have been particularly potent to a newly arrived immigrant.[31] The gleaming architectonic structures of the Trylon and Perisphere, the symbols of the 1939 World's Fair, popularized design as an agent of modernity for the masses. These two symbols, the cone and the sphere, graced the covers of the national, popular, and design press from the *New York Times* to *Popular Mechanics* and *Vogue as* icons of a spectacle whose theme "The World of Tomorrow" would attract forty-four million visitors.[32] Just as significantly, the event propelled industrial design as a profession, and the figures of Raymond Loewy and his contemporaries—Norman Bel Geddes, Henry Dreyfuss, and Walter Dorwin Teague, specifically—to even greater heights of popular fame as the choreographers of the future, and the harbingers of a new post-Depression world.

Although it is unclear whether Papanek personally visited the 1939 World's Fair, it is certain that its famous choreographer Raymond Loewy (himself a émigré) was present throughout Papanek's career, taking on the mantel of a bête noire of design. Papanek claimed to have worked with America's premier design-branding entrepreneur in his New York–based practice Loewy Inc, identifying this experience as the impetus for his rejection of commercial product design, and his turn toward socially inclusive design. An anecdote often repeated in press articles relayed how Papanek quit the Loewy office following its reluctance to take seriously his proposal to gear design to those outside the norm—those, for example, with shorter stature, like Victor's own petite mother, who encountered practical difficulties with basic tasks.[33] Inspired by Japanese *geta* (a type of elevated wooden sandal), Papanek offered Loewy a series of renderings depicting how a traditional cultural trope, the geta, might be appropriated for contemporary use in solving social and practical issues of short stature (see figure 1.2). This alternative pseudo-anthropological solution (uncannily redolent of Bernard

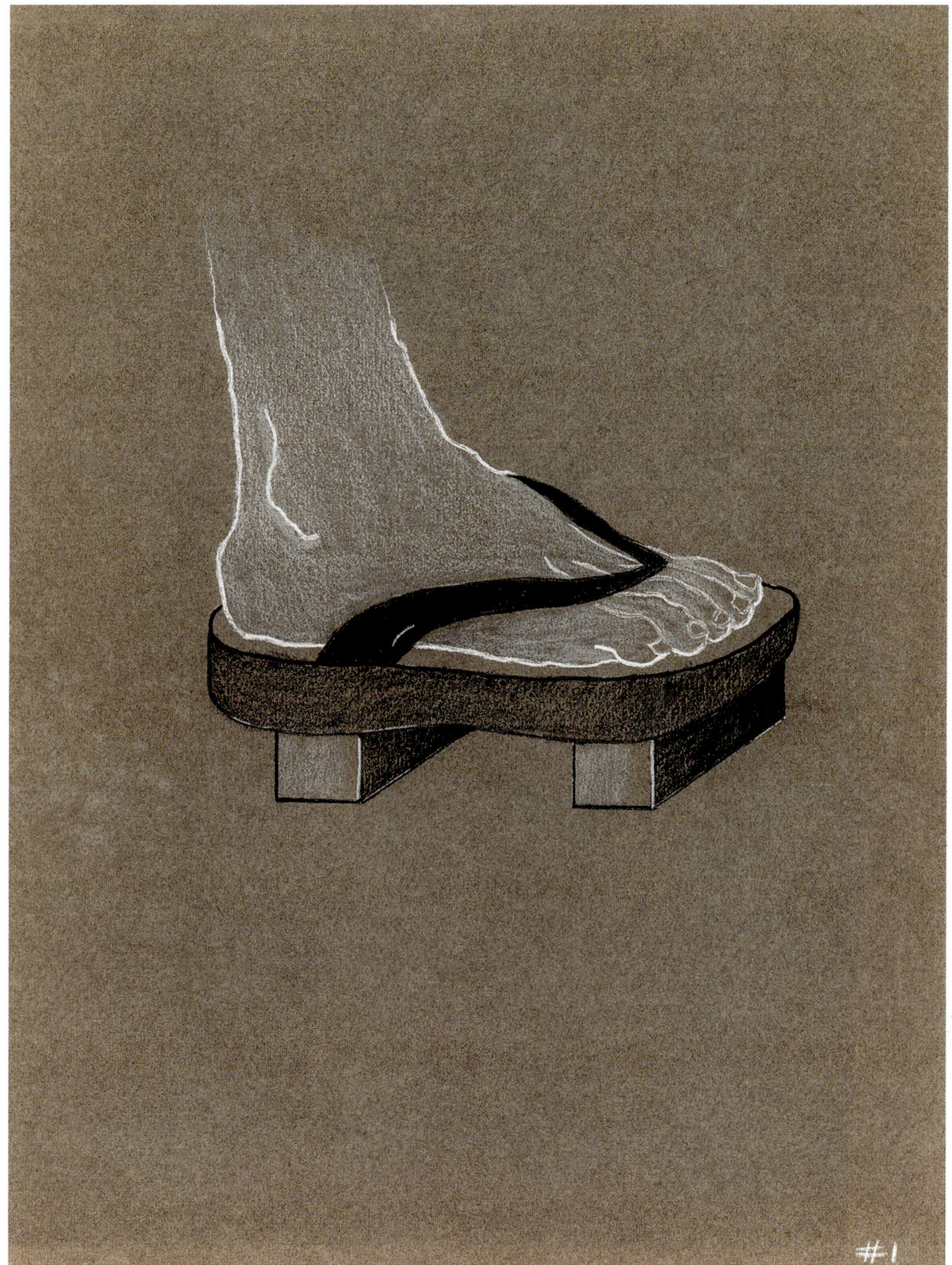

Figure 1.2
Platform sandals designed by Victor Papanek for people of short stature, c. 1946. © Victor J. Papanek Foundation, University of Applied Arts Vienna

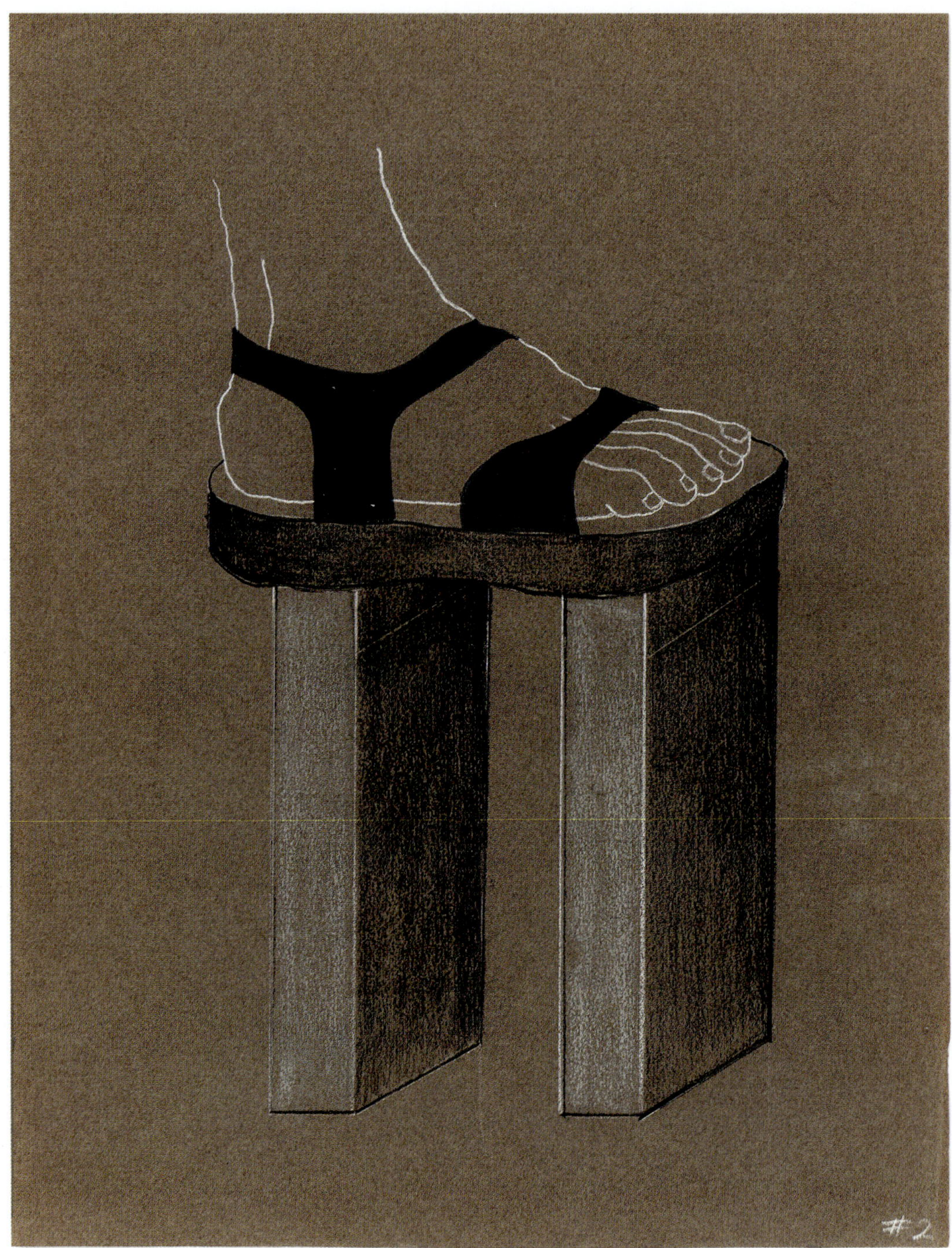

Figure 1.2
Continued

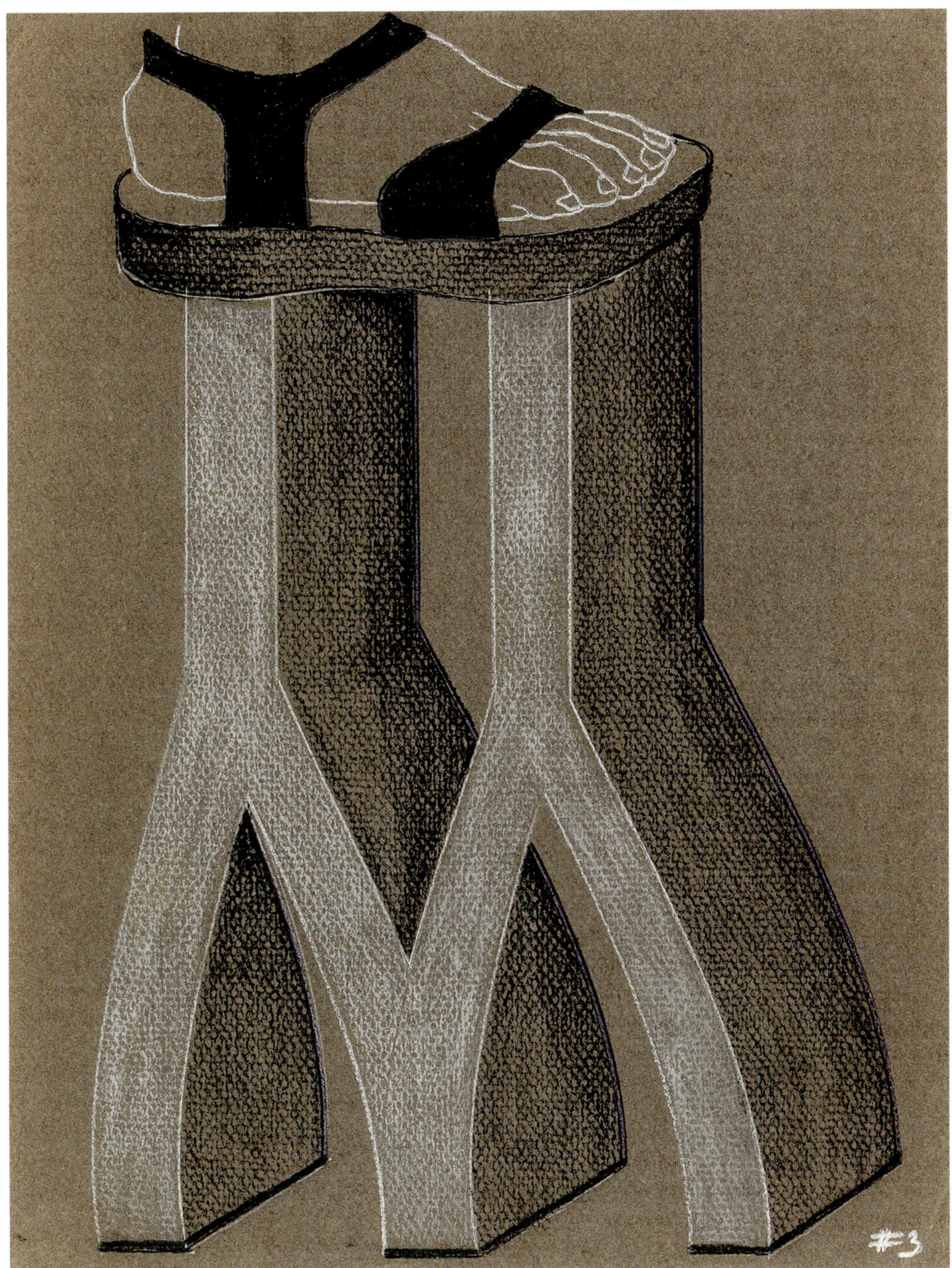

Figure 1.2
Continued

Rudofsky's famous 1940s antifashion sandal design, the Bernardo), purportedly proved laughable to Loewy.

This oft-told vignette of Loewy's disparaging rejection of Papanek's empathetic design innovation was retold to emphasize the gulf between "profit-driven" and "socially driven" design. This leitmotif of Papanek's autobiographical accounts served a dual purpose: it lent Papanek insider professional status by association with America's great industrial design maestro and, inversely, the credentials of a rebellious outsider whose superior ethical stance resulted in the ultimate sacrifice of a prestigious job in the United States' premier design consultancy. As an unapologetic technocrat and corporate stylist set on boosting the sales of consumer commodities, Loewy became Papanek's proverbial straw man. Yet reading this oppositional relation between Papanek and Loewy at face value (set up as a convenient short-hand to promote Papanek's social design reform agenda) actually overlooks the startling similarities in the two émigrés' design agendas.

As a French-Jewish émigré, Raymond Loewy used his outsider status to bring to account a modern industrialized nation that failed to provide an enhanced material standard of living to the mass of its population. As a designer tasked with bringing a substandard material life to account, his moral crusade focused on increasing consumption rather than curtailing it. This is most evident in his 1951 autobiography *Never Leave Well Enough Alone*. Widely credited as the founder of the twentieth-century US industrial design profession and zealot of US consumer expansionism, Loewy sketches out a vision of 1940s New York as a metropolis in radical transition, the mundaneness of everyday city life transformed by a sudden dislocation from tradition to populist modernity.[34] A thinly veiled critique of figures such as fellow émigré designer Paul T. Frankl, whose success depended on the patronage of a wave of nouveau riche Hollywood clientele hungry for "jazz moderne" furnishings, Loewy drew on the outsider status of his upper-middle-class French émigré background to position himself and his entourage of "apostles" at the center of a counter movement of "simplicity and restraint."

Taming the excesses of a vulgar New World culture intoxicated by an unstoppable groundswell of modernity, Loewy used his mid-century autobiography to reframe himself as an opponent of the "modernistic cataract" that had beset the less principled members of the design profession, whose base commercial intent and lowbrow taste Loewy strove to distance himself from.[35] The publication of *Never Leave Well Enough Alone*, part autobiography, part design treatise, formed an integral part of a broader public relations coup that led to Loewy's appearance on the front cover of *Time* magazine in 1949, an accolade serving to cement his reputation as the post-Depression era's outspoken proponent of consumer abundance and leading advocate of branded corporate design.

Dubbed the "Father of Streamlining," his totalizing product- and brand-restyling approach boosted sales on a mass scale. *Time* portrayed Loewy as a dapper and besuited figure set against a floating backdrop of automobiles, airplanes, consumer durables, and packaged everyday American products (including his celebrated packaging designs for Lucky Strike cigarettes and Coca-Cola).[36] As early as 1946, in an interview with the British newspaper the *Times*, Loewy embraced an entrepreneurial, anti-intellectual, self-parodying approach dismissing European modernist aesthetic debate as anathema to an American industrial designer for whom "a conception of aesthetic consists of a beautiful sales curve, shooting upwards."[37] Opposing the trope of the European high-modernist designer, he borrowed from the caricature of the self-made assimilated émigré, proud of his modest beginnings as a window dresser for the department stores he would go on to design in their entirety.[38] The self-conscious construction of Loewy's prominent public relations profile crucially relied on the expertise and innovation of the publicist, Elizabeth Reese, who fastidiously curated the heroic "streamlined" designer's image through her gendered understanding of popular consumption. As well as securing the renowned *Time* magazine portrait, Reese conceived and staged the extraordinarily ambitious product photo shoot (shipping a Coldspot refrigerator and Studebaker automobile to Pennsylvania) that saw Loewy flanked by his streamlined design trophies, including gargantuan streamlined locomotives, effectively instilling in the popular imagination a "design empire, in stark black and white."[39] This carefully engineered profile not only benefited Loewy as an individual designer but also lent the profession of "the designer" an enhanced cultural status. Papanek would also nurture a self-fashioned European, cultured émigré outsider stance toward US design culture, and in keeping with Loewy (and Rudofksy, the architectural figure who would also come to have so much influence on Papanek), he rejected the rationalism of high-grounded modernism.

Unlike other prominent entrepreneurial figures, such as Gruen, who famously instigated the rise of the American shopping mall and revolutionized the application of design to retail space, minimal attention has been paid in contemporary scholarship to the relevance of Loewy's émigré background.[40] The French-born son of a Jewish Viennese father, Loewy's posturing as a purveyor of superior European discernment was integral to his role as America's master of modernity.[41] Critiquing, for example, the phenomenon of ready-made convenience foods as an abomination, Loewy drolly observed that "'the American' was fond of eating certain types of building materials."[42] Yet, in pronounced contrast to many of his contemporaries, and the artistic and modernist elite he enjoyed parodying, Loewy openly celebrated aspects of US popular consumer culture and the unprecedented scale of its influence. He praised the movie industry for encouraging good table manners in "the smallest farming belt or coal town," allowing average Americans to share in the nation's aspirations while happily boosting the sale of "millions of home and labor saving devices."[43] Crucially,

Loewy sought to differentiate himself from "stage-set" designers like Frankl, who were busy decorating the opulent interiors of Hollywood starlets' homes rather than immersing themselves in the crusade to reform design for the masses through the mechanisms of free-market, consumer-capitalist choice.

Two decades after the publication of Loewy's best-selling autobiography, Papanek would single him out in his writings as a wanton "peddler" of false desire whose only "crusade" had been "a crusade to get clients."[44] He cast Loewy as the ringleader of an era of outmoded designers who had willingly collaborated with the large-scale corporations, enhancing their profit margins on an unprecedented scale and at the expense of the consumer masses they purported to serve. Outlining an ersatz manifesto for a radical social-design movement, Papanek, in a style redolent of Ayn Rand's bombastic, ardent individualist architect (based on the figure of Frank Lloyd Wright) Howard Roark in the 1943 novel *The Fountainhead,* condemned designers for selling their souls to commerce: "Industrial design differs from its sister arts of architecture and engineering in one basic way: it is the only profession that has moved from discovery to degeneracy in one generation . . . members of the profession have lost integrity and responsibility and become purveyors of trivia, the tawdry and the shoddy, the inventors of toys for adults."[45]

Supplanting the pivotal position of the "Father of Streamlining" as the exulted design pioneer of the early twentieth century, Papanek proffered himself as the alternative "social" design guru replete with newly converted disciples whom he instructed to "knock on doors that had never been opened before." Papanek wrote, "Just as Raymond Loewy and others had visited potential clients in the 1920s and 1930s to show them what industrial designers could do," the new apostles of design would oppose profit-driven design culture in favor of spreading the word of socially responsible design. Papanek's apostles would be "knocking on the doors [of] developing countries, clinics, hospitals," rather than the corporations.[46]

In keeping with other US émigré designers, Papanek's status as an outsider lent him a fresh perspective on a nation at odds with Old World European culture, sharing with Loewy a sensibility of the amateur anthropologist: making acute social observations of a host nation rife with cultural contradictions (an approach also recognizable in the writing and designs of émigré architect and design critic Rudofsky). Loewy, representing as he did the apotheosis of commercial design that Papanek would later pit himself against, nevertheless shared with him the mantle of crusader against New World vulgarity. Tellingly, Papanek held a first-edition copy of *Never Leave Well Enough Alone* in his personal library, and flicking through the pages he would have read Loewy's treatise on the nonprogressive horrors of "nouveau riche" consumption: "nightmares of vulgarity were sweeping the country, what were the odds against us, the pure boys, apostles of simplicity and restraint? Our only hope was that some men of taste would revolt against their gilt outrages and turn to us for help."[47]

As a figure who by the late 1960s would condemn design as "the single most danger-ous profession in the world," Papanek's formative émigré experience stood in pronounced contrast to that of his imagined rival, Loewy, who, hailing from an affluent Jewish-Parisian background, had settled in New York in the early twentieth century already equipped with an extensive network and upper-echelon social connections. Yet the two designers shared a strikingly similar tone—that of the zealous reformer, stirred up by the contradictory experi-ence of inhabiting a new culture at once alienating and awe-inspiring. Loewy's mission was to transform "bad" consumer culture (in the form of middlebrow modernistic design) into "good" streamlined consumer culture for the masses. "We [serious designers] had to contend with a group of twenty or so crackpot commercial artists, decorators, etc., without expe-rience, taste, talent, or integrity, who called themselves industrial designers," complained Loewy.[48] Papanek's mission was to "heal the world" through design, a notion he had in mind when he launched his first design practice, *Design Clinic*, in Manhattan in 1946.

In the course of a career spanning five decades, Papanek's background and formative expe-riences as an Austrian émigré remained entirely unreferenced: distancing himself from any semblance of identity as an émigré, he aligned himself instead with a hybridized identity forged from liberal counterculture and Old World European intellectualism. Following the success of *Design for the Real World* in the 1970s, press interviews and biographical intro-ductions accompanying his work invariably referenced the two prominent Amerian design figures Raymond Loewy, constructed as his archrival, and Frank Lloyd Wright, depicted as his supporter and abiding influence. Yet Papanek's entrance into design was far from con-ventional, and a great deal less glamorous than the enhanced autobiographical accounts he offered in the design media.

The Air-Conditioned Nightmare: The Critics and Mavericks from Within

Critics and mavericks who at once celebrated and critically exposed the ironies of American culture fascinated the young émigré Papanek. One such eminent figure was US author Henry Miller. In his postexile travelogue of 1945, *The Air-Conditioned Nightmare*, Miller famously appropriated the stance of the émigré "outsider" collating observations and reflections in the course of a year's journey across his estranged homeland. Miller had repatriated from Paris to America shortly before the outbreak of World War II, following a decade-long self-imposed exile in response to a ban: his work was censored for obscenity in his home country. Often cited as the precursor to Jack Kerouac's *On the Road* (1957), *The Air-Conditioned Nightmare* read as subjectivist social critique exposing the alienating effects of his nation's conformist, consumer-driven culture. "Walt Disney . . . is the master of the nightmare," scoffed Miller, before going on to condemn American mass media, and the spiritual bankruptcy of a nation driven by utilitarian, economic concern.[49] As a critic from within, Miller's self-declaredly antibourgeois, artistic hedonism represented a model of the "reverse-émigré" as an empowered commentator liberated from social convention, pitted against the effects of the ceaselessly expanding and seemingly inauthentic consumer culture of his native homeland.

The tirades against the vacuity of design and consumer culture that came to define Papanek's polemical style in the late 1960s and early 1970s are a testament to Miller's formative influence on his early life as an émigré. So keen was Papanek to shape himself in the mold of the American bohemian literary figure that, in a similar way to his retelling of Loewy vignettes, he cited Miller as a personal, not just a cultural, connection. Papanek even suggested that Henry Miller had used his Manhattan apartment as a "bolt-hole" shortly after his return from European exile in 1940.[50] One of Miller's later publications, a booklet exploring the work of his friend the outlandish avant-garde collage artist Jean Varda, was indeed signed with a dedication to Papanek reading: "for 'Vic,' Henry." The veracity of Papanek's claim of friendship with Miller aside, the writer occupied the unique role of being a disaffected American antihero whose "outsider" status was voluntarily endowed and self-defined rather than externally imposed like that of an émigré refugee. In this respect, his status as a real or an imagined part of Papanek's network did not substantially affect his efficacy as the model "insider-outsider" critic.

Bohemianism vs. Ethnic Integration: Surviving the Émigré Experience

Despite his bohemian creative aspirations, by 1943, just weeks before celebrating his twentieth birthday, Papanek enrolled in the US Army (see figure 1.3). Enlisting in the military was a tried and tested means for recently arrived immigrants to acquire citizenship, social acceptance, and related social benefits. Beyond basic facts, the content of Papanek's army documents provide an intriguing insight into a the newly burgeoning identity he was carving for himself. The only son in a conservative Viennese gourmet merchant's family (which prior to the Nazi Anschluss had traded under the coveted imperial Austro-Hungarian warrant), Papanek had, within a few years of arriving at Ellis Island and with limited financial means and meagre social connections at his disposal, confidently identified his profession as "artist." Intriguingly, his military documentation also indicates Papanek was "married"; the first of four marriages, this "hidden" marriage remained entirely unacknowledged throughout Papanek's life.[51]

Personal details such as those regarding Papanek's marriages and relationships, routinely omitted from the standard hagiographies of iconic designers, offer vital insights into the ways in which professional biographies are negotiated and engineered by their owners. These historical details are particularly pertinent in the context of émigré lives in which entire

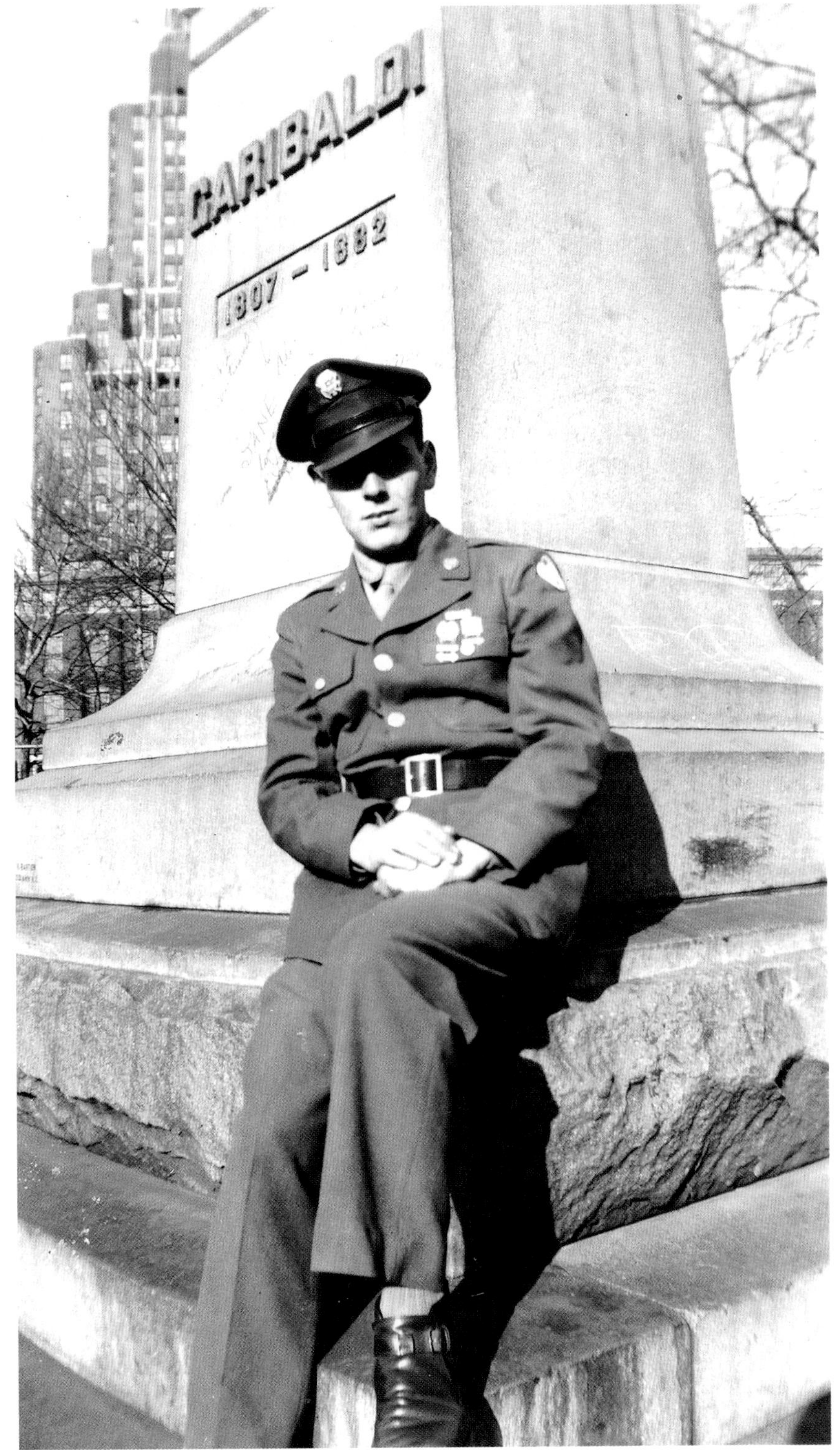

Figure 1.3
Victor Papanek in army uniform, Washington Square Park, New York City. Photograph c. 1943–1944.
© Victor J. Papanek Foundation, University of Applied Arts Vienna

biographical identities and social and cultural networks are eradicated and denormalized, obliging individuals and communities to remake or improvise them. As the earliest, foundational studies of the American-Jewish émigré experience in the United States recognized, the internalized machinations involved in this process became a defining feature of the émigré experience: "The true drama of the immigrant's life was not reflected in his *external* success or failure. . . . What made his story memorable was the complex *internal* process, his *seelische Einordnung* that went on beneath the socio-economic adjustment, yet conditioned it all the same. Veterans of this experience share memories that set them apart as a special fraternity among the victims of the twentieth century."[52]

In Papanek's case, remaking his biography involved marriage into a social world that, although vaguely culturally and religiously recognizable, was to all intents and purposes more alien to him than the maelstrom of American popular culture of New York City that might have overwhelmed him. As a fluent English speaker who had enjoyed the exotic novelties of American consumer culture from afar through magazines and Hollywood films, and even close up in the installation of an ice-cream soda fountain in his family's high-end Viennese food emporium in the 1930s, the home of a Russian-Jewish working-class family in Brooklyn constituted a far more alien experience.

But marriage offered a vital stepping-stone into a newly acculturated life as an Austrian American that would benefit young Papanek's mother as much as himself, through the nurturing of an invaluable support network fostered by a thriving American-Jewish community. In the summer of 1944, a Brooklyn rabbi of Hungarian descent presided over the marriage ceremony of Victor Joseph Papanek to a young Jewish bookkeeper of Russian heritage named Anna Lipschitz. This alliance formally integrated Papanek into New York's Jewish community.[53] The reasons behind the nondisclosure of this union remain unclear; it may have been a marriage of convenience. Or it could be that Papanek viewed this marriage into a traditional working-class ethnic-Jewish Brooklyn family as being sorely at odds with his burgeoning self-identity as an artist and bourgeois European intellectual.[54] Certainly, Papanek's reluctance to expose aspects of his émigré life echo the biographies of émigrés such as Gruen, who "rarely spoke about his refugee experience" or publicly about his immigration to the United States.[55] Comparable artists and designers avoided mention of their early émigré lives and censored their own biographies through a desire to distance themselves from the associated and ongoing trauma or, in the case of figures Saul Steinberg and Xanti Schawinsky, due to their politically contradictory lives in the early 1940s.[56]

Papanek's relationship to this tight-knit traditional Jewish community, the duration of the marriage, and factors contributing to its dissolution (divorce or death) remain unclear.[57] But marriage to a US citizen, and a year-long tenure with the US Army (serving as a German-speaking interpreter with the 417th Engineers Assault Unit, on the Alaska and Aleutian

Islands) made him considerably more eligible for naturalization. On June 25, 1945, Papanek was successfully granted full US citizenship. As well as aiding his bid for citizenship, his experience in the remote Alaska and Aleutian islands proved formative in inspiring his interest in alternative forms of "design." In this harsh volcanic archipelago Papanek first encountered the Inuit culture. The ethnographic artifacts he collected here, including snow goggles, a child's toy, and a mask that later appeared in his publications, formed the basis of a holistic design approach incorporating the anthropology of "things": the cosmological meanings of design taking precedence over rational, functional, modernist problem-solving.

Design Clinic: The Birth of a Social Design Studio

Newly conferred citizenship and a growing interest in the social context of design led to Papanek's first foray into formal design when upon his return from military service in 1946, he established his own design studio, with the provocative appellation *Design Clinic*. Using the address of his newly rented efficiency apartment in Lower Manhattan, and equipped with little more than a self-styled business card (the words "Modern, Interior, Furniture, Design, Display, Consultations" spelled out in uppercase), Design Clinic was a bold first attempt at carving out a career as a modern New York designer, the studio's name hinting at design's ability to "heal" or resolve social (as much as aesthetic) problems (see figure 1.4).[58]

By this point, Papanek had already generated a number of furniture designs, charting their creation in a self-made design scrapbook. A prototype for a "Pick-Up" portable occasional table captioned with "*Interiors* December 1944," suggested the item had been featured in the pages of a prominent design magazine. Yet no such design appears to have ever actually been published there.[59] The adjacent page of the scrapbook features a "four-way dining table and chairs" with the credit "for Paul Bry." Puzzlingly, the image (originating from the designer's commercial showroom in Manhattan) clearly shows furniture designed "by" rather than "for" the German émigré designer Paul Bry. It is unclear whether Papanek intentionally sought to attribute Bry's designs to his own hand, using this scrapbook as a type of portfolio to show potential clients in his pursuit of freelance work, or whether he merely considered figures such Bry as a further extension of his network of design connections, and inspirational sources. Either way, the designer Paul Bry made an interesting addition to Papanek's network, not least because of his association with the design émigré figures Victor Gruen and George Nelson (1908–1986).

In 1940, Gruen had secured a prestigious commission for the interior design of the Park Avenue apartment belonging to celebrity industrial designer George Nelson and his wife, Ilse.[60] While Gruen planned the dining room, master bedroom, and living room, Paul Bry worked on furnishing the nursery, the highlight of which was an avant-garde hybrid (part

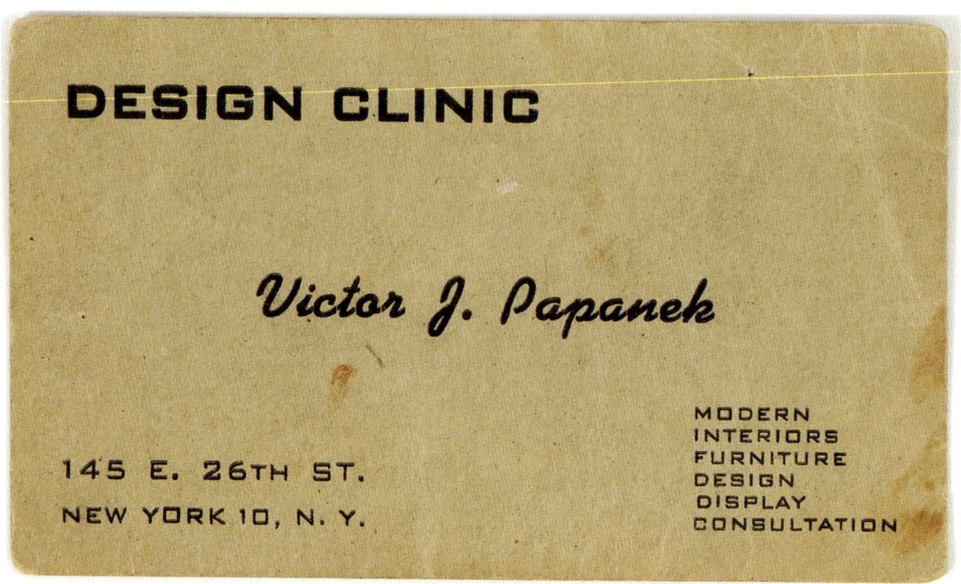

Figure 1.4
Design Clinic, business card, c. 1946. © Victor J. Papanek Foundation, University of Applied Arts Vienna

Bauhaus, part Streamline Moderne) sky-blue leatherette-lined infant bassinet combining the conventional and contemporary materials of bleached walnut, woven Tenite (an early transparent plastic), and acetate. A sensation in the popular press, the "washable" and "hygienic" streamlined design was featured in *Newsweek* and *Modern Plastics* as an example of plastic's transformative potential in the domestic living space.[61] Like Loewy and his ilk, Bry had been involved in designing for the New York World's Fair, moving to the United States from Europe in 1938 specifically to take up the commission to design interiors for the "small House of Brick" featured in the World's Fair's "Town of Tomorrow."[62] As a fellow émigré, he offered a perfect role model for a newly arrived refugee starting a design business from scratch in a notoriously competitive city, during an economic depression.

Metadesign and Expansionist Design: 1940s New York as a Creative Crucible

The appearance in the popular and design press of an ultramodern, celebrity Park Avenue design project commissioned by a high-profile second-generation émigré designer (Nelson), and carried out by a first-generation Jewish Viennese refugee (Gruen), would surely have attracted Papanek's attention as a new émigré arrival to the city. While the furniture of Paul Bry represented a quintessentially New York trope of modernity, it was George Nelson's humanistic design thinking and his expanded career, not just as a designer but also as a design critic, television producer, and artist that proved most influential—this and Nelson's undisguised ambivalence toward design. Best known as a champion of the mid-century modern aesthetic, in fact Nelson displayed a consistent ambivalence toward design practice throughout his career. In the words of his biographer, Nelson's "is the career of an architect who advocated the end of architecture, of a furniture designer who imagined rooms without furniture, an urban designer who contemplated the hidden city, an industrial designer who questioned the future of the object and hated products."[63]

It has been argued that Nelson the design critic and prolific writer is better understood as a "metadesigner." As opposed to a conventional, corporate industrial designer like Loewy, Nelson's "object of study and practice [was] neither the building nor the product but rather the process of design itself."[64] Nelson's *Problems of Design* (1957), which included chapter titles such as "Obsolescence" and "The Designer in the Modern World," clearly preempts Papanek's *Design for the Real World* (1971); both design polemics feature analysis of the societal role of design and the wastefulness of commercial design with its unerring emphasis on novelty. It is also surely no coincidence that the provocative title of Victor Papanek's earliest essay criticizing design, "Do It Yourself Murder" (1968), echoes that of Nelson's *How to Kill People*, a design show aired by CBS in 1960.

Largely self-educated prior to establishing his Design Clinic, the backdrop of early 1940s New York offered Papanek unmitigated access to a thriving critical design culture. During this period Papanek claimed to have assisted contemporary architect and designer Frederick Kiesler (1890–1965) building structures for his Art of This Century gallery, opened by Peggy Guggenheim in October 1942 to wide critical acclaim. As a fellow Viennese émigré, Kiesler had a reputation as "Design's Bad Boy" that preceded him, as did the nonconformist impact of his design theorizing. "Kiesler's U.S. history is that of the avant-garde European gone astray in the American commercial woods," commented one contemporary architectural journalist, who went on to describe how the experimental thinking of European modernists stood in stark contrast to the profit-motivated machinations of their Depression-era American contemporaries.[65] Significantly, émigrés like Kiesler were viewed as direct beneficiaries of a brand of Viennese "hot-house" creativity, exemplified by a lineage of social engagement and design theorizing that followed Adolf Loos (Kiesler himself had trained in Loos's office), through to mass social housing projects and design manifestos. The US perception of this "cultural capital" allowed figures such as Kiesler to maintain intellectual standing while taking on commissions from department stores and consumer catalogues. If Nelson's design approach could be termed "metadesign," Kiesler's might be described as "expansionist": his writings littered with "abstruse terms and coined words" such as "co-realism; biotechnique; time-space-continuity," he defied definition by mainstream US architectural standards, and (rather like his contemporary Buckminster Fuller) inhabited a space between genius and screwball.[66]

Against Modernism: Humane and Anthropological Design

While Frederick Kiesler occupied the position of "bad boy" of design in 1940s New York, the controversial Viennese émigré Bernard Rudofsky (1905–1988), was counterpoised in the position of "humane designer."[67] Rudofksy's design approach, as reflective of an architect, curator, design theorist, writer, and designer, challenged stylistic and technological formalism, embracing instead a humanistic interpretation of the material world (underpinned by his classic European scholastic training), which emphasized culture, history, and environmental context. Graduating with a degree in architecture and engineering from the prestigious Technische Hochschule in Vienna (1929), he completed a doctorate on the early concrete structures of Greece's Cyclades Islands (1931); his Viennese architectural education placed him at considerable advantage over his US contemporaries.[68] This rigorous training, coupled with extensive travel, lent him an anthropological perspective that put local customs, rituals, and behaviors, and the Indigenous and vernacular, at the forefront of his design thinking. After an initial nine-month stay in New York in 1935, and subsequently working in Italy, Argentina, and Brazil, Rudofsky was drawn back to the city as the Latin American category winner of the 1940 inter-American Museum of Modern Art "Organic Design" competition. The launch of the follow-up *Organic Design in Home Furnishings* exhibition of 1941, arranged by Eliot Noyes, MoMA's first curator of Industrial Design, and conceived by Edgar Kaufmann Jr. (who had suggested the idea to MoMA Director Alfred Barr) aimed at bringing designers into direct contact with department stores and commercial manufacturers through the contracting of the winning products.[69] *Organic Design* propelled his work into the city's thriving design scene, Rudofsky finding in New York a rich cultural life in the 1940s unparalleled anywhere in the world.[70]

In 1944, Rudofsky himself curated a MoMA exhibition: the format-busting *"Are Clothes Modern?"* scrutinized the phenomenon of fashion, opening up to critical review the underlying hierarchies of design, the formalist paradigms that engendered high modernism in a Western capitalist economy, and the ironies of a feckless fashion system driven by the fickle concerns of style consciousness. Ultimately intended as an antidote to arch-Modernism's rationalism, the self-declared intent was "to take the blinders of tradition off modern eyes so that they can see that certain conventions . . . are in fact useless, impractical, irrational, harmful and unbeautiful."[71] Using anthropological paradigms that foregrounded the local and the vernacular, the exhibition's intention was to expose the disjuncture between conventional "modern" fashion and the historical culture and traditions of the Mediterranean and Europe.[72] The overarching aim of this comparative rhetorical device was to cast light on the absurdity of modern Western culture's unique claim on "progress" through the urbane observations of everyday cultural life and its artifacts, with the backdrop of wartime material and cultural insecurity as a potential driver for change. As the bastion of modernist values and, since Henry-Russell Hitchcock and Philip Johnson's 1932 *Modern Architecture: International Exhibition,* the flag bearer of the International Style, Rudofsky's *"Are Clothes Modern?"* constituted a provocative addition to MoMA's repertoire. Like the maverick author Henry Miller, Rudofsky acted as the "critic from within," bringing to account the effects of the American cultural mainstream and establishment alike.

Despite his biographical parallels to well-connected émigré figures such as Rudofsky, the self-taught designer Papanek struggled to position himself in the burgeoning field of industrial design. Rudofsky had in a short period of time made his name as a curator and design editor by "quickly befriending European émigré artists and architects," among them Walter Gropius, Richard Neutra, Gio Ponti, Xanti Schawinsky, and Saul Steinberg. Papanek, on the other hand, pieced together his design profile through a relatively ad hoc series of encounters with New York's design scene using the media of showrooms, exhibitions, and publications.[73] Yet, it is arguable that in creating Design Clinic Victor Papanek shared with Rudofsky and Kiesler the inconvertible influence of Vienna, drawing on "public discourses central to the formation of Viennese modernism, such as the reclamation of the body, the relationship between fashion and architecture, and the identification of bad design with social decline."[74]

Function Unit X-3: The Social Agenda of Affordable Furniture Design

Along with the writings and works of leading 1940s designers, Papanek drew on his informal research "window-shopping" Manhattan design stores from the Paul Bry showroom, to Victor Gruen's extravagant brand design for Barton Bonbonniere candy stores commissioned by a fellow Viennese émigré, and the more austere Finnish design showroom that showcased Artek furniture.[75] A blockbuster MoMA exhibition of Finnish architect and designer Alvar Aalto had taken place the year prior to Papanek's arrival in New York, and Aalto's design for the Finnish Pavilion at the New York World's Fair attracted widespread critical acclaim.[76] Aalto, and a Finnish design model underpinned by sensitivity to vernacular cultures and demotic ideals, remained an abiding influence on Papanek's notion of the social in design that began with Aalto and Artek in the 1940s and culminated in the 1960s with his engagement with pan-Scandinavian student design activists.

The first solid impetus behind Papanek's "design for the real world" polemic may actually have been ignited by a small-scale but highly impactful MoMA postwar design initiative dealing with design for the masses, rather than the social idealism of Finnish design, or Kiesler and Rudofsky's grand societal design gestures. Inaugurated in 1947, the *International Competition for Low-Cost Furniture Design* exhibition and its chief protagonist Edgar Kaufmann Jr. had a notable impact on Papanek's understanding of design's social potential, crucially shaping his ideas around the politics of everyday design aimed at the non-elite sections of society.

In the introduction to the exhibition's catalogue *Prize Designs for Modern Furniture* (1950), MoMA's recently appointed director of industrial design (son of renowned philanthropist, department store magnate, and modern architecture patron Edgar J. Kaufmann Sr.) claimed an integral role in improving the public's welfare and the standard of living on a mass scale: "Low cost housing and home furnishings are among the most important factor in the national economy and the welfare of the peoples of all countries."[77] Linking up commercial furniture retailers with forward-thinking designers, the affordable furniture project prompted in the area of home furnishings a response proportionate to that of the postwar housing shortage within architecture. The MoMA press release outlined that a selection of winning designs would be shown both in the gallery setting of the museum, and would also be commercially available to purchase in home furnishing showrooms nationwide as a means of influencing public taste: "the museum hopes to encourage the wide-spread use of well-designed furniture in the home through wide-spread availability."[78]

While building on MoMA's didactic agenda to educate the public and influence retail taste, the competition responded to straitened living conditions by calling for entries under three functional categories: designs for seating, designs for storage pieces, and designs for upholstered dual-purpose living-bedroom units. Despite the austerity theme of the contest, in 1949 the ten winners from Europe and the United States (all male, and including Robin Day and Charles Eames) were stylishly photographed by the celebrity and society darling Cecil Beaton, and featured in the up-market Condé Nast publication *House and Garden*. There was, however, one notably innovative aspect of the competition: its inclusion of "Design-Research Team Entries" that comprised interdisciplinary groups, with the designer being one expert of several rather than a singular driving force. As a zealous advocate of design research and transdisciplinarity, which was the cornerstone of his approach as a practitioner from the mid-1950s onward, Papanek used early experiments in furniture design (inspired by the MoMA competition) as his earliest creative practice ground.

The *Prize Designs for Modern Furniture* catalogue that accompanied the *Low-Cost Furniture Design* exhibition, along with the publication *Introduction to Modern Design: What Is Design?*

(1950) and a series of other volumes authored by the younger Kaufmann, feature in Papanek's personal library as early influential reference sources. Papanek determinedly looked beyond the modernist machine aesthetic promulgated by MoMA's "Good Design" movement in the postwar period in favor of a more accessible version of modernity, his earliest design portfolio revealing experiments with themes overtly tied to events such as the low-cost furniture initiative.[79]

An abiding influence, not least due to his endeavor to reject cultural elitism in favor of the promotion of a democratic utilitarianism of everyday industrial design objects aimed at the greater good, Kaufmann Jr. (third-generation Jewish-German émigré) shared other aspects of his career with Papanek. Kaufmann apprenticed with Wright at the Taliesin studios from 1933 to 1934, his stint there giving rise to an enduring relationship he and his father forged with Frank Lloyd Wright and famously leading to the commissioning of Wright's iconic Fallingwater (or Kaufmann Residence) in Pennsylvania, 1935. Kauffmann Jr. also studied in Papanek's home city, Vienna, at the Hochschule für angewandte Kunst, during his travels in the 1920s; his father maintained numerous business contacts and artistic connections in the city.[80]

But Papanek was almost certainly working on affordable furniture prior to Kaufmann's exhibition. One of his earliest designs, which he first mentions as the "Pick-Up Table" dated 1944 in his design scrapbook, was a self-assembly item titled "E-1: Carry About Table" (see figure 1.5), which had all the hallmarks of a low-cost furnishing solution: it was movable, adaptable, and easy to manufacture.

Using the industrial aesthetic of stenciled typography and prototype numbering to present his furnishing designs, the young designer built up series of objects intended to work together as a furnishing system for the contemporary home. These included "Function Unit X-3," a rendering for an abstract-shaped chair for "reading, sleeping, resting, lounging," signed "Victor Papanek + Design Clinic" (see figure 1.6). The drawing included a version of the distinctive three-legged Hindu antaḥkaraṇa symbol in one corner, representing the inner psyche functions or senses, which Papanek appropriated intermittently as a type of logo for Design Clinic during this period. The "ABX Webbed Unit Day-Bed" (see figure 1.7), a type of day bed with a woven textile and wooden frame, available in interchangeable sizes, formed part of the series along with the "X-& Chairside Table," a magazine rack–occasional table hydrid illustrated in situ with a glass and an ashtray containing a burning cigarette. Additional adaptive and portable furniture in his late 1940s portfolio were "E-2 Dual Height Table," which could be utilized at dining table or lounging table height, and the "37-A Fire Squat," the rendering for which showed a Sears and Roebuck–type collage of a young girl relaxing in pajamas, with a description of its assembly parts for the chair: "Molded ply-wood, rubber foam, metal tubing, 'B-X' cable, rubber shock mount" (see figure 1.8).

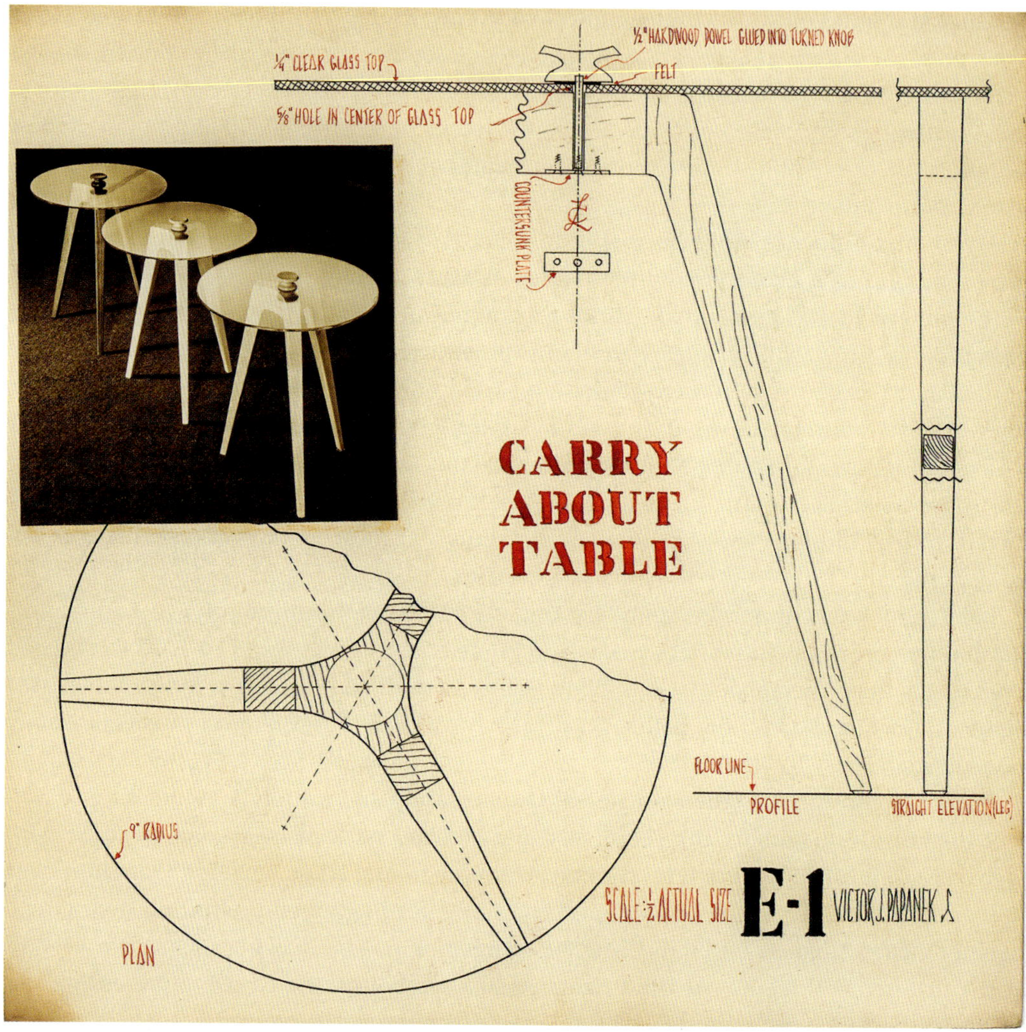

Figure 1.5

"E-1: Carry About Table," Victor Papanek, c. 1944. © Victor J. Papanek Foundation, University of Applied Arts Vienna

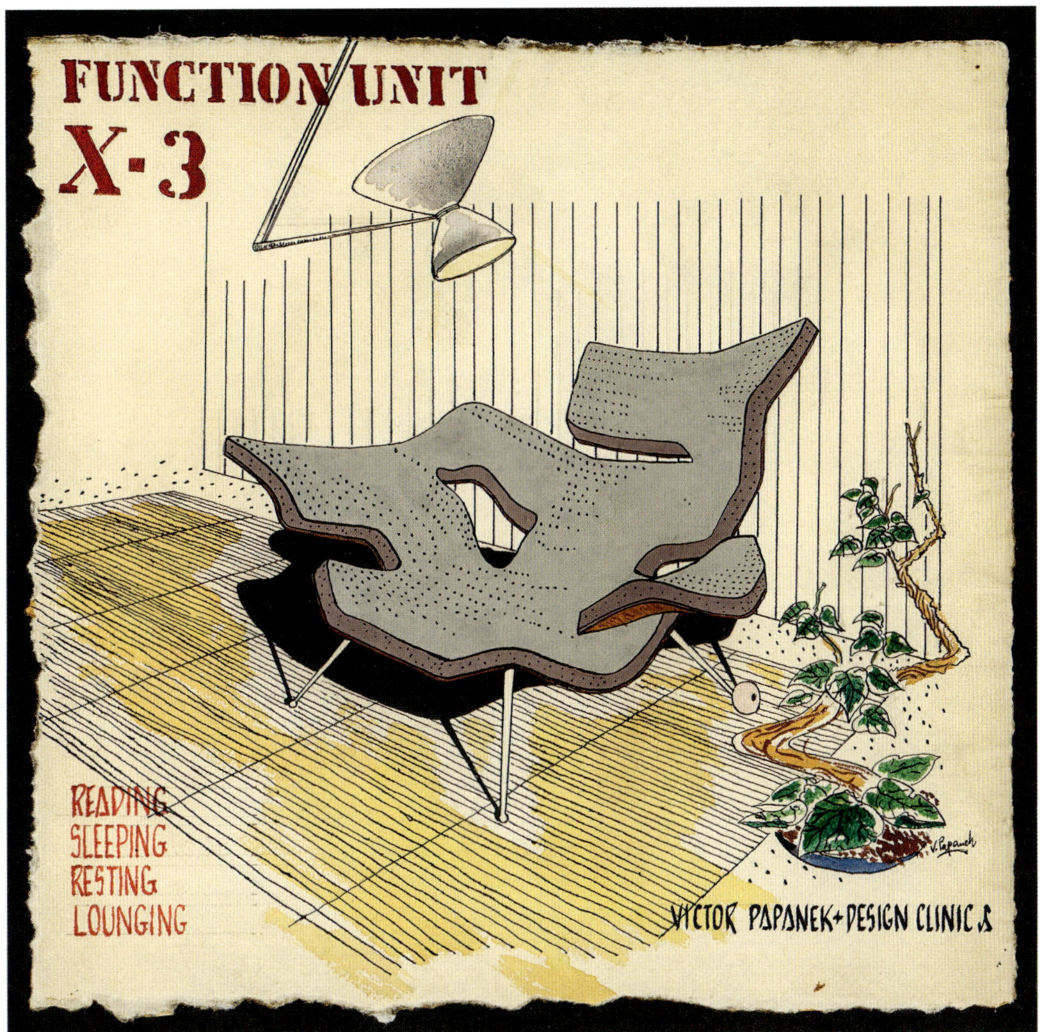

Figure 1.6

"Function Unit X-3, for Reading, Sleeping, Resting, Lounging," Victor Papanek, Design Clinic, c. 1944–1948. © Victor J. Papanek Foundation, University of Applied Arts Vienna

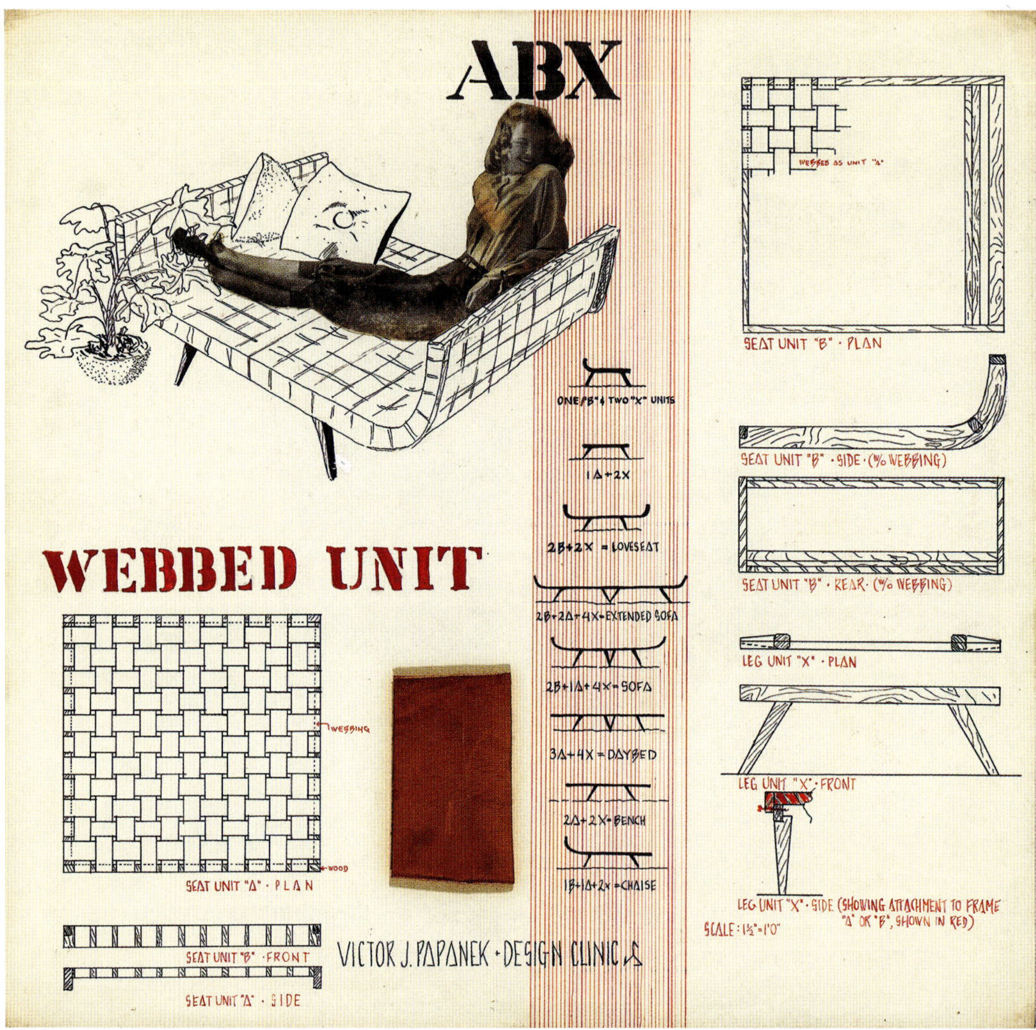

Figure 1.7
"ABX Webbed Unit Day-Bed," Victor Papanek, Design Clinic, c. 1944–1948. © Victor J. Papanek Foundation, University of Applied Arts Vienna

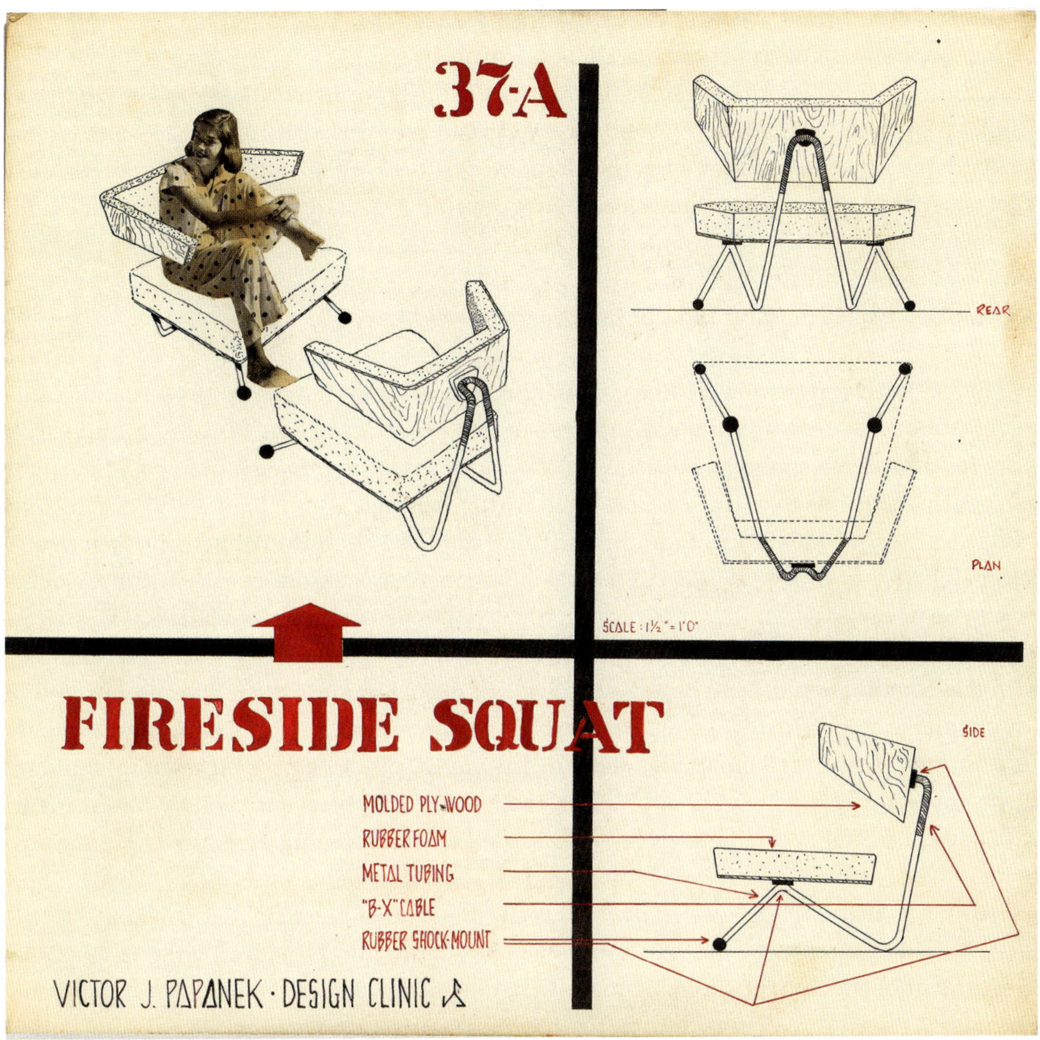

Figure 1.8
"37-A Fire Squat," Victor Papanek, Design Clinic, c. 1944–1946. © Victor J. Papanek Foundation, University of Applied Arts Vienna

Design Clinic's principle aim, at least in its early inception, was to resolve the problem of contemporary living with low-cost, adaptable, easily assembled series of potentially customizable furnishing types. Although the first IKEA furniture appeared in 1948, it was traditional and local in style, and certainly leaned less toward all-out modernity or the sleekly contemporary as depicted in Papanek's Design Clinic works: his early furniture system preempted a massive shift in design consumption toward a more transient and moveable idea of the home interior.

A Foundation in Humanism: The Cooper Union School of Art

In tandem with a relatively ad hoc, self-taught approach to design, in 1946 Papanek formally enrolled in evening classes at the Cooper Union School of Art, a twenty-minute walk from his downtown studio and apartment. The Cooper Union offered daytime and evening courses to applicants over the age of sixteen with good high school grades and evidence of "diligence." Exceptional in its founding principle to make education "open and free to all" and its provision of outstanding-caliber teaching aimed at supporting talented individuals otherwise precluded from higher education, the Cooper Union had a reputation as a progressive institution focused on a generalist critical and creative approach to the fine arts, offering a unique combination of both professional and vocational training. Producing an impressive list of notable alumni across the fields of graphic design, sculpture, filmmaking, and architecture, it offered an extraordinarily varied curriculum. Despite lacking a high school diploma (his completion of the Austrian equivalent, the Matura, had been curtailed by his escape to the States), Papanek secured himself a place in the four-year evening program in architecture in 1946.[81]

In the late 1940s, the School of Art faculty was made up of practicing designers and architects, as well as theorists, with the strong liberal arts provision provided by the Humanities Department of the Cooper Union School of Engineering. The architecture evening program of courses attended by Papanek did not confer professional architectural degrees but rather, in the words of the prospectus, prepared "the student for the architectural field" by covering subjects included in the New York State Licensing Examination for Architects.[82] Despite (or possibly due to) its generalist approach, the program offered a highly distinctive gateway into design and architectural discourse, which began with a one-year foundation course loosely based on the Bauhaus principle of total immersion across the art and design forms. By the second year, the architecture program had a strong historical component led by the Berlin Jewish émigré professor Paul Zucker who, having originally been invited to join New York's New School for Social Research (a renowned institutional haven for progressive-thinking émigrés in the late 1930s), eventually escaped National Socialist Germany in 1938 to join the faculty of the Cooper Union School of Art. There he instigated comparative art and literary history, stridently advocating the subjects as an integral part of a broader humanist pedagogic agenda: "Without the understanding of history and human motivation in historical perspective conveyed to us only through art and literature," he later wrote, "our practical and mechanical progress leads to complete loss of humaneness."[83]

While fellow émigré Zucker provided a strong humanist-based liberal arts education, Michael Radoslovich (an instructor teaching in the curriculum section called Architectural Design) introduced Papanek to the social and political dimensions of practice. Radoslovich was a key contributor to the New York World's Fair Community Interest Zone, putting his humanistic design theories firmly into action. As well as designing the modernistic stainless steel spiral fountain outside the Halls of Science, Education, Medicine and Public Health he was responsible for crucial aspects of the World's Fair's infrastructure, including the pedestrian overpass from the Administration building to the World's Fair Amusement Grounds and the Long Island Railroad World's Fair station, "enabling the railroad to handle 18,000 passengers an hour."[84] In one press interview praising the "harmoniously arranged homes" of the "Democracity" town of tomorrow (and integral part of the Community Interest Zone), Radoslovich pointedly criticized profit-oriented New York developers whose substandard housing design generated "the foundation for slums to come." Instead (preempting the reform polemic of Papanek's *Design for the Real World*), he urged the public to be "educated not to buy homes because of the inlaid linoleum on the floor of the kitchen, or by the color of the bathroom fixtures."[85] Most significantly in terms of his influence on Papanek's early forays into design, Radoslovich exerted an overtly politicized view of design, which he put directly into practice when he later became the chief architect of the New York City Board of Education (1952–1967), steering progressive ideas around pedagogic architecture.

"Formal Academic Degrees: None." The Biography of an Outsider

Despite Papanek's enrollment at the Cooper Union, there is no archival trace of Papanek having formally graduated from the School of Art as either architect or designer. Indeed, a number of his self-penned résumés from the 1950s and 1960s open with the deliberately provocative statement "Formal Academic Degrees: None." Papanek wore his lack of formal qualification as a badge of honor in his early career, taking every opportunity in media interviews to flag this as a means of further conferring his rebellion within an otherwise cliquish architectural and design fraternity. Yet his earliest résumé does list a diploma of industrial design awarded by the Cooper Union School of Art in 1948, and intriguingly the completion of a course in general semantics with the feted advocate of the field, Polish American Alfred Korzybski, at the Institute of General Semantics, Chicago, in 1944. Korzybski, whose most famous dictum was "the map is not the territory," advanced the ideas of semantics beyond the academic or specialist spheres of the period. Semantics as a topic accrued growing popularity in the 1940s, with articles in *Time* magazine and "book of the month" clubs featuring series such as Korzybski's student S. I. Hayakawa's *Language in Action* (1941) reaching a mass general readership. The controversial American writer William Burroughs famously attended Korzybski's courses in the late 1930s, and Buckminster Fuller, who shared with Korzybski a number of similarities in structural approach, delivered the Alfred Korzybski Memorial Lecture 1955.[86] Considering Fuller's later close connections with Papanek and his work, Korzybski indisputably acted as conceptual bridge between the two designers.

Far more significant, then, than any formal academic degree in design or architecture, Papanek's immersion in New York City with its groundswell and cultural clash of ideas and visual cultures—popular and high-end, middlebrow and intellectual—untethered him from the strictures of a prior identity as a bourgeois young Wiener; his new émigré life, anxiety filled as it was, offered him nonprescriptive opportunities to sample the intellectually and sensorially engrossing environment in which he had been reluctantly immersed. Papanek used his immersion in the full spectrum of popular and avant-garde contemporary culture, ranging from Peggy Guggenheim's cutting-edge Art of This Century gallery to glamour pin-up stores, to push far beyond the conventional and culturally sanctioned spaces of the Cooper Union, MoMA, and the like. Instead, as a young émigré forcibly and traumatically unrooted from any semblance of a secure biographical trajectory, Papanek inverted the precariousness of his refugee status and, in the process, self-consciously reimpowered himself as a renegade "insider-outsider."

2 Seduction of the Innocent: From Bondage Fetishism to Frank Lloyd Wright

In the course of his career, Victor Papanek accrued a reputation as a vehement critic and castigator of vulgar commercial design and popular culture. By the 1960s, he initiated and presented public radio and television shows condemning multiple facets of popular culture, from new-fangled gadgets and plastic gimmicks to lurid girlie magazines. All the more incongruous, then, that a dimly lit, dingy basement store on the edge of the Lower East Side dealing in comic books and movie-star and bondage-fetishist photographic stills provided his first paid design commission. This foray into the underworld of American culture lent Papanek a unique insight that would later help launch his career as a cultural critic of mass consumerism and its discontents.

Bettie Page and Tempest Storm: Burlesque, Bondage, and the Stirrings of a Pop Culture Critic

The proprietor of this store was the infamous bondage-fetish impresario Irving Klaw, a Jewish Brooklynite dubbed the "Pin-Up King," who oversaw a soft-porn "glamour" business that had expanded exponentially from a regular film-still and comic book outlet. Along with his sister, Paula, Klaw observed that movie stills fetched higher prices when they featured "a twist" (for example, a woman wearing high-heeled shoes to model a swimsuit), and so they launched their own business imitating this genre of pin-up image, ultimately spawning a booming mail-order enterprise purveying bondage-fetish soft pornography.[1] Loosely based on a burlesque aesthetic, mixing satin, leather, and rubber costume details with fastidiously choreographed bondage and flagellation imagery, the extraordinarily risqué business nevertheless managed to evade strict censorship laws throughout the 1940s.

Attracted by the New World novelty of the Klaws' emporium of erotic ephemera just a block away from the efficiency apartment that acted as Irving Klaw's home and design studio, Papanek became a regular visitor. As well as instructing him in the finer points of

American popular culture, here was the location where the budding social designer met one of America's most infamous and admired glamour pin-up models, Bettie Page. Both the phenomenon of popular culture and Page herself (whom he later described as a "humane, human, pleasant and genuine . . . unspoiled pleasant young woman from a Tennessee farm" unfairly exploited by Klaw) made abiding impressions on Papanek.[2]

Bettie Page had emerged from an early career as a "saucy beach-blanket" model, arriving in New York in 1947, where she quickly evolved as the Klaws' principle fetish and sado-masochist starlet. She accrued extraordinary underground cult status, her thick, long dark hair and blunt bangs lending her a type of risqué modernity, veering away from the conventional glamour model look and making her instantly identifiable as the notorious Bettie Page. Contemporary cultural critic Margaret Talbot attributes Page's enduring appeal, well into the twenty-first century (with over a hundred fan clubs dedicated to her worldwide) to the broader phenomenon of "ironic preservationism" that offers a younger generation a version of postwar kitsch wrapped in a regime value wherein "the more crudely legible an artifact, the better."[3] Talbot argues further that the depiction of Page with women rather than men (censorship law forbade women's interaction with males in accessible erotic and pornographic materials), along with her unique ability to combine ingénue wholesomeness with subversive eroticism, reinvented her as a proto-feminist and a pro-sex figure who defied conventional pornographic repertoire. "What lifts [Bettie Page's] photos above the blankness and the deadness of so much porn and pseudo-porn," writes Talbot, "is the stubborn, exuberant persistence of a self, the irreducible core of a personality shining through. Bettie is never inert; she is always Bettie; the girl can't help it."[4] Page, then, possessed an unparalleled allure, and belonged to an evolving genre of outsider subcultures the Klaws were quick to commoditize.

Ever-enterprising, they transformed a series of "cat-fighting" women films into comic book format. The "Battling Women" graphic novels (derived from erotic films of the same theme) used burlesque models including Bettie Page and Tempest Storm as their source, requiring an illustrator to render their transition from real women to comic characters. Papanek was commissioned to translate the filmic stills into these comic book fetish characters, as he himself later recounted: "Irving Klaw would pay me $175.00 for a 15 × 20 inch black and white page for a comic book which took me less than a day to do. At that time the average pay for a normal illustrator was only $20.00 a week! Still later I developed two reasonably ravishing looking women heroines called "Gail" and "Stormy" for him as well. One of these women was modelled after Page" (see figure 2.1).[5]

In addition to indirectly providing Papanek with his first design commission, Bettie Page remained a lifelong influence, as would the psychosexual imagery of sadomasochistic fetishism, which cropped up time and time again in his lecture slides, and critiques of popular

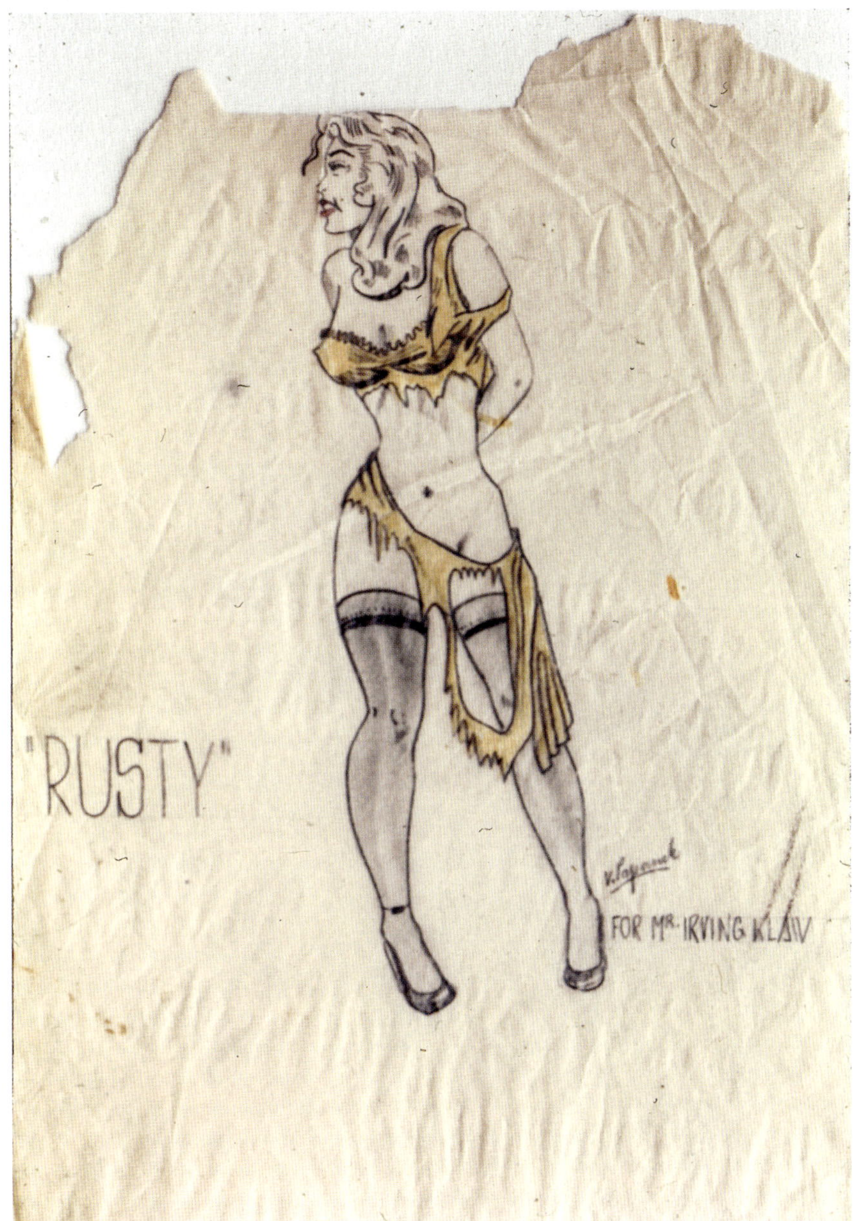

Figure 2.1
"Battling Women," illustration by Victor Papanek, c. 1941–1944. © Victor J. Papanek Foundation, University of Applied Arts Vienna

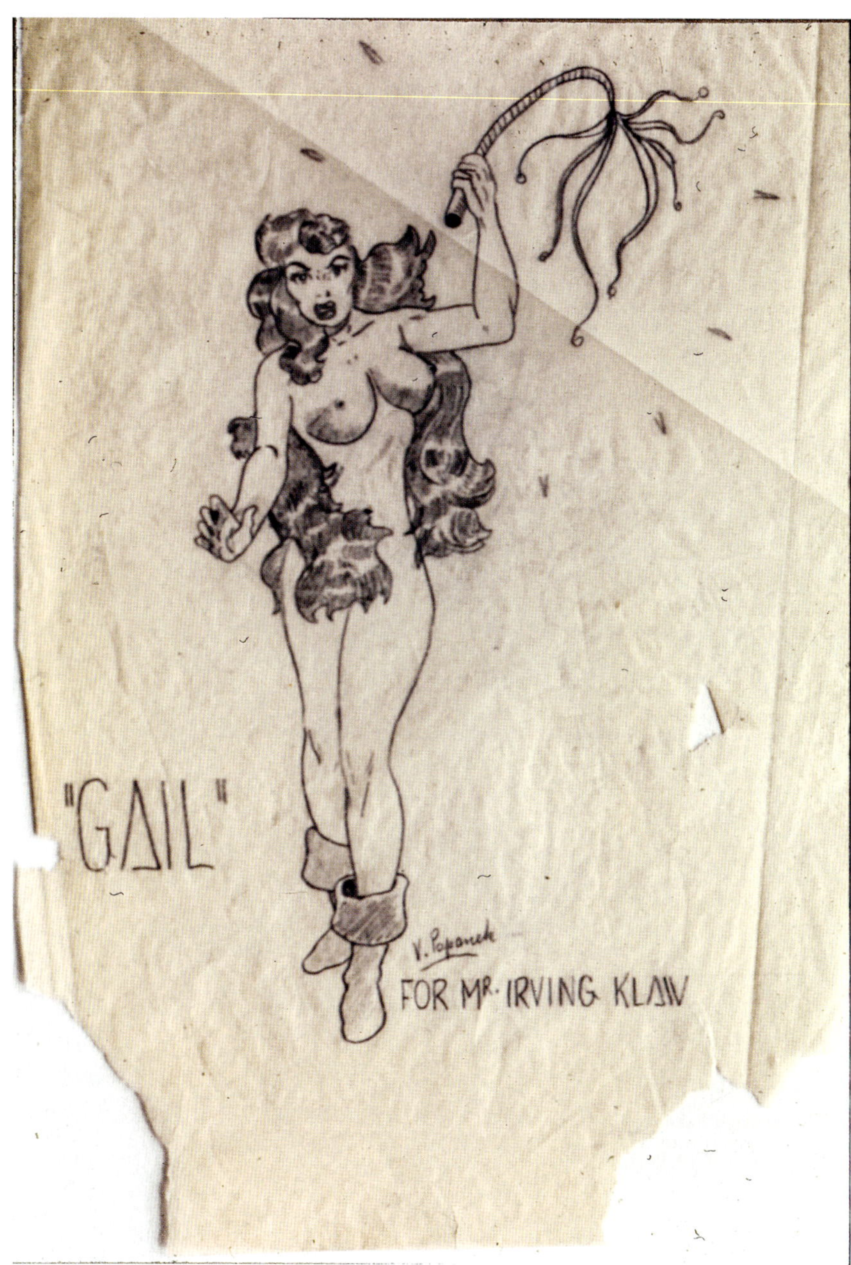

Figure 2.1
Continued

culture and the representation of women.[6] This penchant for all things "Bettie Page" seemed to extend to his personal relationships.

During his enrollment at the Cooper Union he met and dated (concurrent with his marriage to Anna Lipschitz) a fellow student, twenty-two-year-old Jeanne C. Hertle, who bore an uncanny resemblance to the pin-up. With her bright blue eyes, jet-black hair, and blunt-cut bangs she became Papanek's design muse.[7] Signed "For J.C.H," he dedicated a particularly detailed interior design to her, dated between 1946–1948 and titled "Sunken Hearth in Bedroom," that featured a "TV–music console" inset with the muse's photographic portrait for the screen, embedded in a sunken hearth with steps leading up to a spot-lit, wall-mounted, cantilevered, neon-yellow, rhombus-shaped bed hovering above a Mondrian-style area rug (see figure 2.2). Despite the superficial similarity between their looks, "Jeannie" Hertle ("an advertising artist by day, and a fine artist by night" as the 1948 Cooper Union year book described) appears to have been a reserved character, keen on classical music such as Bach, and sharing Papanek's interest in non-Western culture with her enthusiasim for "India, its philosophy and its culture."[8] The resemblance of his 1940s student girlfriend to Page could be passed off as an uncanny coincidence were it not for the fact that one of his future wives, Diana Jean Kleefeld (whom he married in the 1960s) also bore a startlingly close resemblance to the model; and that in his later life, Papanek subscribed to the Bettie Page appreciation fanzine. Beyond this personal infatuation established in his youth, Bettie Page's national notoriety and the broader debates around the socially detrimental effects of popular culture projected onto her had the greatest impact on Papanek.

Fredric Wertham, the Comic Book Critique, and the Moral Dilemma of Mass Culture

Coinciding with Page's heightened profile as Playboy Playmate of the Year in 1955, the Klaws released the film *Teaserama*, featuring their premier star alongside the burlesque starlet Tempest Storm ("Stormy"), as well as a drag performance by Vicki Lynn. Although fastidiously avoiding nudity, the risqué low-budget movie openly flaunted fetish and dominatrix genres, showing Page dressed in high-heeled shoes and eroticized maid's outfit attending in the "boudoir" to her costar "Stormy." With the shift from mail-order basement business to movie distribution, the Klaw siblings gained nationwide infamy.

Unfortunately for the Klaws, *Teaserama*'s release coincided with the so-called Kefauver Hearings of the Senate Subcommittee on Juvenile Delinquency, which extended investigation (first initiated in April 1954) from comic books and television to the link between youth criminality and pornography.[9] As part of the legal process, a number of semiunderground figures, including Irving Klaw, were subpoenaed to elaborate on their dealing in and production of pornography and lubricious materials—all of whom pleaded the Fifth Amendment.[10]

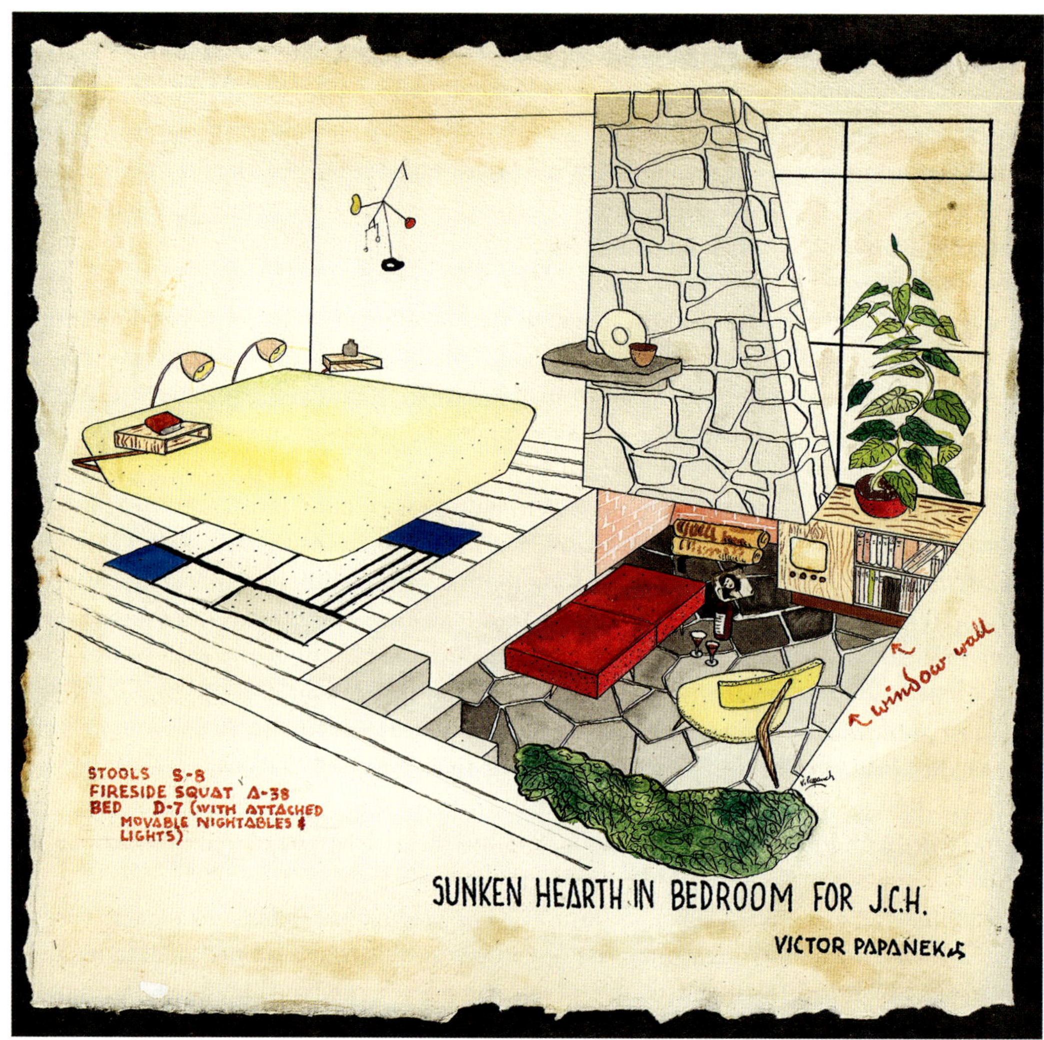

Figure 2.2
"Sunken Hearth in Bedroom, for J.C.H," design by Victor Papanek, c. 1946–1948. © Victor J. Papanek Foundation, University of Applied Arts Vienna

The resultant high-profile court case curtailed the Klaw enterprise and Bettie Page's career as an underground fetish model. But her influence, and that of Klaw, extended well beyond Papanek's first design commission as illustrator for titillating comic books.

The encounters with Klaw and Page brought Papanek into contact with the works of the leading American contemporary critic of comic books and their impact on juveniles, German-born psychiatrist Fredric Wertham, author of the seminal and controversial study published as a best-selling book *Seduction of the Innocent* in 1954. In a further astonishingly ironic twist, responding to a newspaper advertisement in the late 1940s, Papanek applied the expertise gleaned from his work in interpreting erotic stills for Klaw's comic book cartoons to assisting their greatest public detractor, Fredric Wertham. Together they constructed an exhibition exploring the corrosive effects of juvenile comic books through the depiction of violence toward women and children (the major theme of Wertham's *Seduction of the Innocent*).

For an émigré, this intense roller-coaster ride amid the cultural politics of the mid-century United States offered an extraordinary hands-on experience of the complexities, contradictions, and moral quandaries of fully fledged consumer-driven society. If one were to attribute Papanek's anticonsumption discourse and socially inclusive rhetoric to the ideas of a single theorist, it would surely be those of Wertham. A pioneering German American psychiatrist, psychoanalyst, and media theorist, Wertham's highly influential work, arguing the correlation between juvenile delinquency and insidious and socially detrimental racial stereotyping, would ultimately fall out of fashion as an overly reactionary response to the rise of popular culture.

As is the case within the field of architecture, the impact of émigrés and their European intellectual traditions of critique is a familiar trope of historical analysis regarding the rise of psychoanalysis in 1930s and 1940s United States. Prominent figures such as Austrian émigré Ernest Dichter, dubbed the "Freud of Madison Avenue" for his pioneering application of psychoanalysis to advertising, substantiate conjecture that the "two biggest forces brought from Europe" in the second quarter of the twentieth century were "atomic science and psychoanalysis."[11] Since he was a German immigrant, Wertham's research and role as a public intellectual advocating strongly for the societal benefit of psychoanalysis and psychiatry also fit neatly into to this summation. Historian Bart Beaty further argues that the rapid influx of German and Austrian psychiatrists arriving in the United States between 1932 and 1941 "undeniably shifted the center of global psychoanalysis from Europe to America," and with it a newly honed critique of popular culture as a phenomenon worthy of sociopolitical analysis.[12]

Wertham's groundbreaking inquiry into the psychological effects of printed media's unchecked violent and sexualized imagery, freely available to children and youths from the shelves of drugstores and newsstands, was part of a broader application of psychiatry to the endeavor of progressive social reform. His testimony at the Senate Subcommittee on Juvenile Delinquency in April 1954 was integral to the proceedings, and his work—*Seduction of the Innocent* was fortuitously released to coincide with the proceedings—formed the bedrock of the case for closer scrutiny and control of the comic book industry. Despite being attacked by "New York Intellectuals" for his overly censorious and commonsensical interpretative methods and research findings, Wertham in fact steadfastly resisted any simplistic attribution of blame to the comic book genre: "nobody would claim comic books alone are the cause of juvenile delinquency," he declared in his formal testimony as an expert in the Kefauver Hearings.[13] Yet his insistence on reiterating the seductive power of the rhetoric and imagery of comic books as part of a media-effects paradigm, particularly in relation to the negative influence on the psychologically and socially vulnerable, made him unpopular with the avant-garde of literary and cultural theory. Ultimately, his liberal-progressive reform agenda led to his alienation among the more conservative censors. Wertham's principle objection centered on a culture that prioritized profit, commerce, and the market above the collective benefit of social good, and in this sense his writing formed a critique of capitalism's wanton profiteering as much as morally degenerate media.

Cultural critique of popular culture and mass culture, and an understanding of taste as an expression of social class discernment, first became mainstream in the 1930s and 1940s. In the sphere of design, the "Good Design" movement promulgated by institutions such as MoMA was an integral part of the discourse. By the 1950s, when figures such as media and social theorist Paul Lazarsfeld (fellow émigré and alumnus of Papanek's first Viennese high school) were brought to testify at the Kefauver Hearings on matters of methodology in mass-communication research in relations, the media's relationship to "taste" had taken on a distinctly political edge.[14] By mid-century, debates around mass culture spanned the gamut of the political spectrum; "from ardent conservatives to liberal reformers and radical Marxists."[15] Refusing to compromise, Wertham's ideas resided between, and on the periphery of, delineated political positioning rather than being redolent of any specific ideological stance. His empirical experience as a practicing psychiatrist, rather than the purist critiques of the Frankfurt School émigrés, shaped his media theory. Despite writing extensively for the liberal progressive magazine the *New Republic* and the fact that he "shared many of the same traits and aesthetic predispositions as the New York Intellectuals," Wertham was never actively integrated into their circle but was instead frequently the object of their criticism.[16] Keen to apply his findings and create societal change as a practitioner rather than part of literary or public intellectual elite, Wertham remained an "insider-outsider" within US culture; a public position Papanek, his accidental protégé, would occupy as a designer-critic later in the twentieth century. As an advocate of a user-based, empirical research approach to social psychiatry, Wertham indisputably pioneered urban mental-health accessibility as a vital facet of broader civil rights. He acted as a significant role model for Papanek in his capacity as a principled, vociferous public intellectual with an unerring dedication to the critique of mass culture's effect on socialization.

The Harlem Project: Social Inclusion, Antiracist Psychiatry, and the Papanek-Wertham Connection

Papanek's formative connection with progressive social psychiatrist Fredric Wertham stretched way beyond a distant engagement with his theories and public declarations; for it was Wertham who gave the young Viennese émigré firsthand insight into the social activism of New York's African American avant-garde as well as broader theories around the intersection of race, class, and societal alienation. In addition to assisting the progressive psychiatrist in the preparation of an educational exhibition on the damaging effects of mass media, remarkably, Papanek also personally volunteered at celebrated social psychiatry initiative, the Lafargue Clinic, in the late 1940s.

Having occupied senior positions in New York as an academic professor and then as director of Bellevue's Mental Hygiene Clinic, in March 1946, Wertham cofounded (with German American psychiatrist Hilde L. Mosse) the Lafargue Mental Hygiene Clinic in the basement of the parish house belonging to the influential Harlem church of St. Philips Episcopal; considering the sheer magnitude of Harlem's social deprivation, this progressive and "improbable little clinic," in the words of one historian, defied the odds by surviving until forced closure in 1958.[17] The major premise of Wertham's argument was that poor, urban Black children and youth, already systematically socially excluded in their own country of birth, were further dehumanized by their depiction within, and through the consumption of, popular media and comic books. The Lafargue Clinic was set up as an intervention directly in the heart of urban poverty—Harlem. Offering a volunteer-run psychiatric service set-up specifically to support African Americans, it operated outside of governmental or philanthropic funding or governance, relying instead on the public support of influential figures including actor Paul Robeson, and writer and political activist Ralph Ellison. The Lafargue Clinic, often engaged in the psychiatric treatment of young Black Americans defined by the white establishment as delinquent and violent, evolved as a focus of civil rights debates, and the progressive Harlem institution was discussed widely in the national press at a crucial point in US postwar race relations. In his groundbreaking essay "Harlem Is Nowhere" describing the neighborhood's decaying, impoverished urban landscape—with its "crumbling buildings with littered areaways, ill-smelling halls and vermin-invaded rooms"—Ellison identifies the clinic as an intervention whose "importance transcends even its value as a center for psychotherapy," seeing it instead as "an underground extension of democracy."[18] Inspiration for the clinic came from the shared interests of Wertham and author Richard Wright, whose powerful 1940 book *Native Son* dealt with the damage of racism and capitalism to the psyche. Wertham had also engaged in research for a book on this theme, exploring the violent, unhappy life of a young migrant boy from Italy. *Dark Legend: A Study in Murder* was published in 1941. According to historian Dennis A. Doyle, "representations of the black male psyche became a matter of real debate within postwar Manhattan psychiatry."[19] The parallels in Wertham's and Wright's objects of study, in particular a shared interest in the relationships between racial stereotyping and violence and the potentials of psychiatry, led the two men to work together to create a low-cost community clinic for individuals disempowered through destitution and racism—those who had the greatest need, yet conversely, the least access to the benefits of psychotherapy.[20]

Turning his back on the dominant contemporary intellectual discourse surrounding psychiatry, through the Lafargue Clinic, Wertham offered a model of how practice (not mere ideology or discourse) could affect change. Buoyed by the initiatives and support of the African American literary and activist figures, he drew attention to the paucity of research on racism within psychiatric practice and advocated social psychiatry as a means of intervening in broader socioeconomic indices that underpinned racism. Following an invitation from the National Association for Advancement of Colored People to research the psychological effects of school segregation on children in 1954, Wertham testified as an expert at a US Supreme Court hearing for *Brown v. Board of Education*, which resulted in the landmark ruling that found racial segregation in public schools unconstitutional.

Father Figure, Therapist, and Inspiration: An Outsider Adrift

A German American whose caustic critique and evocative book titles (*The Show of Violence, The Circle of Guilt, Seduction of the Innocent*) caught the attention of a popular audience, who set up a pioneering community intervention in the form of an urban drop-in mental health clinic for a section of the population otherwise precluded from the elite practice of psychoanalysis—could it be that Fredric Wertham was Papanek's principal role model? Was it Wertham who inspired the young émigré Papanek to imagine a world in which community-based design healed social ills and inequalities? It is surely no coincidence that in 1946, the same year that Wertham launched the celebrated Lafargue Clinic, Papanek launched his first design practice, the Design Clinic, aimed at the social inclusion through access to affordable furnishings?

Wertham's clinic offered Papanek a unique insight into an exceptional, progressive model of socially inclusivity through empirical practice such as playroom therapies for younger children, a model Papanek himself would later adapt in his transdisciplinary designs for children with disabilities. But the two men also appear to have established a significant personal relationship, with Wertham ultimately offering professional help with Papanek's own beleaguered mental health as an émigré and social outsider. Victor's only close blood relatives in the United States, the Strauss family, lived in Harlem, Upper Manhattan. It was the Strausses' address and family connection that had "allowed" Victor and his mother to escape Nazi Vienna in 1939. Situated just four blocks from the Lafargue Clinic, the Strauss household acted as Papanek's ersatz family home during the first months of his arrival in the United States, and he remained a regular visitor to his family there. As a native German speaker, Papanek almost certainly attended lectures or events that Wertham organized within the German-speaking community a short walk from the Strausses' home.[21] Other personal ties to Wertham include the psychiatrist's familiarity with Vienna and his personal visits and correspondence with Sigmund Freud; according to the art college prospectus, Wertham was also a regular guest lecturer at the Cooper Union, Papanek's alma mater.

But it is Papanek's description of conducting Rorschach inkblot tests as a volunteer at the Lafargue Clinic and, most intriguing of all, having undergone psychoanalysis with Wertham himself that unequivocally attest to the influence of progressive ideas of social psychiatry on Papanek's budding concept of a socially responsible design practice that placed the socially excluded at its heart.[22] Papanek was also keen on promoting this exceptional association that few, if any, designers of the postwar period could have matched. On his résumé, submitted in a successful bid to secure a senior design faculty position at the newly founded California Institute of Arts (CalArts), Los Angeles (opened in 1970), he states the following: "Worked with Prof Frederick [sic] Wertham (former Chief psychiatrist of New York, Bellevue mental clinic and Wertham-Lafarge [sic] clinic) on new method of psychological testing; helped as lay analyst at Lafarge Clinic."[23]

Further, biographical details contained in Papanek's unfinished autobiography link Papanek and Wertham. These include the hazy circumstances surrounding his hasty departure from New York in 1950 following the purported death of his girlfriend there after a routine hospital operation on Christmas Eve 1949. Thrown into inconsolable grief, Papanek retold the trauma of checking himself into the Bellevue Hospital following an extended drinking session and complete mental breakdown.[24]

The psychiatric ward of the public Bellevue Hospital (where Wertham had directed a unit until 1940) was a notorious cultural icon of New York, infamous as the main treatment center for alcoholism and related addictions, and for treating the socially excluded and destitute. A well-publicized refuge for artists and creatives who had gone off the rails, its reputation was cemented by Viennese-émigré Hollywood director Billy Wilder's uncompromising Oscar-winning 1945 film noir *The Lost Weekend*, which depicted an alcoholic writer embarking on a four-day drinking binge that culminated in a stint at the Bellevue psychiatric unit. In fact, Wilder's movie strikes an uncannily similar chord to the narrative recounted in Papanek's own tragic story of grief and mental illness. If he had been as personally close to Wertham as the evidence suggests, as a former senior manager at the hospital, Wertham would certainly have been in a position to facilitate Papanek's mental health care there in an emergency, which adds veracity to Papanek's version of events regarding his retreat from New York.

Furthermore, twenty-eight years his senior, and a native German speaker, there is every possibility that Wertham acted as a father figure to Papanek. A bourgeois young man and an only child in his early twenties and a refugee in a strange metropolis, Papanek was in dire need of a guiding hand to navigate the tumult of working-class life in 1940s New York. Shortly before his traumatic escape from Nazi Vienna, Papanek's father had died, leaving him fatherless at the tender age of twelve years, in a period of extraordinary political upheaval and violence.

Another compelling reason to believe Papanek was influenced in his ideas by firsthand insight into the discourse of antiracist, social psychiatry in 1940s Manhattan is a connection with one of its leading advocates—the ardent socialist and reformer Ernst Papanek (1900–1973). Although the Strausses were the only close relatives Victor and his mother Helene relied upon in New York, a distant elder relative was Ernst Papanek, who also lived and worked in New York in the late 1940s. In fact, Ernst and Victor's listings sat side by side in the 1948 New York City residential directory. Ernst Papanek and Fredric Wertham were also connected as proponents of antiracist psychiatry and as directors of leading progressive New York institutions enacting the principles of this newly honed and hotly debated approach.

Arriving in the United States from France in 1940, Viennese émigré Ernst Papanek was a committed socialist treating refugee children during and immediately after the Second World War, with particular focus on the horrific impacts of the Holocaust on infants and youths.[25] As a prominent figure in the field of psychology and pedagogy, Ernst Papanek took up the position of executive director at Wiltwyck School for Boys in Esopus, New York, from 1949 to 1958.[26] An interracial reform school for underprivileged boys with emotional and psychological difficulties, the Wiltwyck School, founded by the Episcopal City Mission Society and later supported by the Eleanor Roosevelt Foundation, became a cause célèbre attracting support from famous figures such as the Hollywood performer Harry Belafonte, who was an early supporter of the Black civil rights movement.[27]

Ernst Papanek had joined the Wiltwyck School shortly after the release of *The Quiet One* (1948), an Academy Award-winning documentary exploring the life story of a ten-year-old pupil at the reform school, and the impact of poverty and social neglect on the psyche of children. These issues, then, were not confined to a handful of specialists and activists but were firmly in the public realm. Both Wertham and Ernst Papanek belonged to a burgeoning antiracist psychiatry movement in postwar Manhattan, exploring the damage of racial stereotypes and the impact of racism on African American children and youths, which had the potential for social impact on the "real world"; Wertham, though not belonging to the Communist Party, embraced Marxist ideas within his theoretical approach to mental illness, while Ernst Papanek was a committed socialist who had been highly active in Viennese socialist politics before his exile from Austria.

Despite the commonality of interests and politics, the two figures came to represent differing approaches and emphases within the field, with members of the Lafargue Clinic describing the team at Wiltwyck pejoratively as "La Guardia Liberals" who were too little engaged in frontline politics. Unlike the Wiltwyck School, Lafargue Clinic was volunteer-run and ultimately placed less emphasis on children, being predominantly concerned with Black adults, "the poor, the middle class, whites, Latinos, war veterans, the elderly and children."[28] A rift existed between the two institutions, partly based on Wertham's mistrust of the bureaucratic state dependency of the Wiltwyck School (prior to Ernst Papanek's directorship of the institution) and the approach to racial inequality it depicted in the *The Quiet One* documentary. Wertham and the Lafargue Clinic foregrounded race and place (specifically ghettoized areas such as Harlem) as factors in the making of delinquency, whereas the Wiltwyck School approach, as showcased in the 1948 documentary and the ensuing publicity, consciously sidestepped this. Wertham, according to Doyle, was not arguing for racial determinism, but rather was one of the few within the "wartime medical community to publicly declare that biological race was not 'real,' it was a social construct."[29] *The Quiet One* presented to a predominantly white audience the idea that it was family neglect, rather than a broader structural neglect tied to racial construct on a mass scale, that had made its "quiet," disturbed protagonist Donald delinquent. Although the rift between the Lafargue and Wiltwyck institutions began prior to Ernst Papanek's tenure at the latter, clearly the figures knew each other's work and profiles at the highest level and were rivals in their shared field of practice. Significantly, the debate around race and social inclusion moved beyond psychiatry toward the newly emerging cultural politics of media, and its role in promulgating social exclusion.[30]

Victor Papanek, connected to Wertham as a public intellectual in the field of media and consumer critique, as well as the everyday running of the Lafargue Clinic in Harlem, must have also been aware of his distant relative Ernst Papanek's work and ideas in late 1940s New York. Understanding these nuanced details of Victor Papanek's biography casts an entirely fresh light on the origins of his ideas of social inclusivity: cutting-edge ideas founded in the specificity of Harlem's 1940s racial politics and the growing critique of popular media buoyed his vision for a "Design Clinic" for healing social ills, and the thesis of social inclusion that resulted in his definitive work *Design for the Real World: Human Ecology and Social Change.*

It is hard to imagine a more extreme immersion in postwar cultural politics in the United States than that experienced by the young émigré Victor Papanek. Between 1946 and 1949, he was concurrently a friend and admirer of the nation's premier pin-up fetish porn and the most notorious, high-profile pornographers of the land, a volunteer in a psychiatric clinic at the forefront of the nation's Black-activist politics, and at the polar end of the spectrum, an apprentice to America's leading figure of architecture: all facilitated by the vicissitudes of his émigré identity.

"The Pencil and the Plow": An Apprenticeship with Frank Lloyd Wright

The perversely contradictory experience of designing for pornography impresario Irving Klaw's erotic comic book series while assisting on Wertham's polemical *Seduction of the Innocent* exhibition (which vilified the self-same subject) placed Papanek at the center of debates around American popular culture and its side effects. Quite astonishingly, it was around this frenetic time in Papanek's life that he also secured an apprenticeship with the nation's most renowned architect, Frank Lloyd Wright.

Wright would remain Papanek's most revered role model. He amassed an enormous collection of the architect's publications, including drawings, an autobiography, and related monographs, and mentioned his affiliation with the much-admired "genius" as the cornerstone of his career in design. Following Wright's death in 1959, Papanek had accrued a strong enough reputation as an affiliate to the creative genius as to be invited to pen obituaries of his former master in the design press. A number of Papanek's musings relied on his memories of life as a fellow at Wright's famed Taliesin Studios, and a tempestuous relationship with a figure Papanek himself dubbed the "great meister." The means by which Papanek became a fellow with Wright, and the exact dates of his apprenticeship with the architect, remain, however, decidedly opaque. Though it might be surmised that he was present there for some unspecified duration in 1948, despite his lifelong reference to the life-changing experience of Taliesin, his accounts of this period remain vague.

In 1992, for example, the Austrian architectural magazine *Architektur und Bauforum* ran a special edition dedicated to Wright, in which Papanek invoked his memories of his formative Taliesin days under the title "Mit Bleistift und Pflugschar: Erinnerungen und Anekdotisches zu FLW" ("The Pencil and the Plow: Personal Memories of Frank Lloyd Wright"), providing one of only a handful of detailed written testaments of his stay at Taliesin.[31] The article also served as the basis for a lecture Papanek gave regarding Wright's influence on his life and works, in which the social designer notably avoided mention of specific dates pertaining to the length or year of his apprenticeship. In a preface to the talk, Papanek explained away this lack of clarity as being symptomatic of a Viennese style of narration, simultaneously positioning himself in a lineage of Vienna's leading auteurs and letting himself off the hook should he be pressed for further details: "After all, Viennese have been known for over a century now as enriching their writings with little 'stories' and anecdotes, from Peter Altenberg to Adolf Loos. This is the reason why these memories of Frank Lloyd Wright and life at Taliesin and Taliesin West are kaleidoscopic fragments, not necessarily in chronological order."[32]

Pieced together, Papanek's "kaleidoscopic fragments" told an idealized story of himself as an enthusiastic acolyte, an émigré designer serendipitously stumbling into the creative realm of his hero. Inspired by his first material encounter, in 1942, with Wright's Pauson House in Phoenix, Papanek recounted how he then "spent many days and weeks bicycling or hitchhiking along the East Coast of the United States, as well as in the Midwest taking photographs of some of his houses."[33] Upon arriving at Taliesin in Spring Green, Wisconsin, he began to photograph Wright's private residence and studio, whereupon he was apprehended by one of the members of the Taliesin Fellowship: "William Wesley Peters, Mr. Wright's right-hand man . . . grabbed me by the back of my shirt and brought me into Wright's studio," explained Papanek. After convincing Wright and Peters of the innocence of his motives ("I was not after some media scoop"), the dramatic interlude "ended up with Mr. Wright accepting me as an apprentice to work and study with him," concluded Papanek neatly. The entire retelling of the encounter was framed by an acute awareness (echoed in his lecture's allusion to the narrative styles of Altenberg and Loos) of the cultural pedigree invoked by the Viennese émigré architects in whose steps he trod. "Considering the age difference of nearly sixty years, a reasonably warm relationship developed," continued Papanek in his account, "since Wright enjoyed Viennese people. Both Richard Neutra and R. M. Schindler are merely the two best-known Austrians to have been associated with him."[34]

Although Papanek's retelling, verbal and written, of his time with Wright took on a romanticized and mythical tone, the most convincing evidence of his presence at Taliesin presents itself in his acute descriptions of everyday life there. By all accounts, the Taliesin Fellowship resembled rather more a cultish experiment in collective living than an architectural school: "To be an apprentice at Taliesin and Taliesin West," Papanek observed, "did not primarily mean that one studied Architecture. . . . Instead we were called upon to help give birth to baby piglets, milk the cows, build additions to the barns, sow and harvest the crops, cook and work in the kitchen, serve meals and set the table, burn quicklime and lay brick walls, and so on."[35]

Induction into the "inner circle" of fellows involved manual labor ranging from ploughing fields and other heavy farm work to the tedium of domestic duties; each student was allocated a part of the residence or studio, and assigned housekeeping duties pertaining to that specific space. "The few hours one spent in the studio," reminisced Papanek wistfully, "were taken up largely by lettering and drawing borders around other people's drawings."[36]

For all but a few longer-term fellows, "working" with Frank Lloyd Wright rarely led to direct contributions to the design of architectural masterpieces. But the pursuit of creativity at Taliesin manifested itself in other ways, as Papanek beautifully illuminates in the following extract:

> One of the duties expected of the apprentices assigned to dining table preparation, was to go out into the fields and gather and collate natural materials, designing "au naturel" center pieces for each table. These arrangements of rocks, snail shells, driftwood, possibly the bleached white skull of a fox or squirrel, twigs, flowers, stag antlers, minerals and other materials were related in spirit to Japanese Ikebana flower arrangements, and Suiseki, the art of stone appreciation. In the course of the meal, they were discussed and critiqued as stringently as any apprentice's architectural presentation.[37]

But, rather than providing a master class in visionary American architecture, of far greater significance to Papanek's future as a designer was Taliesin's unique model of alternative design culture and thinking. From their inception in the early 1930s, the studios relied on an ostensibly "applied arts" approach more in keeping with the Arts and Crafts tradition from which Wright's first work emerged, than a contemporary modernist architectural workshop. The antiestablishment embrace of communal living, nature, shared creative enterprise, and cross-cultural holistic approaches offered young aspiring architects a life philosophy, as much as practical training. One of the few comparable pedagogic initiatives of the period was architect Eliel Saarinen's private training center in Michigan, which offered on-site postgraduate workshops. Unlike Taliesin, though, in the words of one former fellow, "in no sense was it a commune."[38] Far removed from Wright's gritty "back to the earth" immersive strategy, Saarinen's school best resembled an exclusive country club, replete with luxury amenities including swimming pool, tennis courts, manicured lawns, and servants.

Given the influence of Olgivanna Wright's (Wright's third wife) taste for mysticism, Taliesin garnered a reputation for its bohemianism, yet it was still an unabashedly commercial enterprise charging fees to apprentices and advertising the positions publicly.[39] In his biographical accounts of his time spent in Wright's inner circle, Papanek omits the fact that entrance to the Taliesin Fellowship relied on a student's ability to pay the ascribed fees, rather than competitive selection alone. According to an early 1930s brochure for the fellowship scheme, the application procedure alone commanded the sum of $200, and the overall tuition fee came to $1,100; students could be rejected at any time during their apprenticeship, and were required to supply their own clothing, bed linens, towels, blankets, and tools, and to labor for a minimum four hours each day "at whatever work may be assigned to them."[40] Despite its lofty ideals, the fellowship scheme principally sought to remedy the financial devastation wrought by the diminution of Wright's architectural commissions during the depth of the Depression years. It also provided much-needed labor for Wright's Wisconsin farmstead. While many fellows arrived at the Taliesin gates through word of mouth or personal recommendation, in the earliest days of the school, newspaper features and promotional brochures drummed up initial interest in the venture.

As a wealth of former Taliesin fellows attest, its commercial intent did not render it any less radical as a pedagogic proposition. It nurtured an international intake of fellows, including many émigrés (notably the Austrian architects Richard Neutra and Rudolph M. Schindler), and was markedly progressive in its acceptance of women as architectural students (20–25 percent of the fellow quota) at a time when most architectural schools actively prohibited them; Wright famously coining the phrase at Taliesin "a girl is a fellow here."[41]

The range of honorary visitors to the studios, from authors to musicians, offered apprentices an unrivaled exposure to contemporary creative culture. Papanek himself singled out the innovative American weaver Dorothy Liebes, responsible for many of Wright's upholstery designs, and the Indigenous American pottery artist Maria Martinez, as visitors whom he particularly admired during his fellowship there. Inspired by the rituals, customs, and seasonal celebrations of Taliesin, these lent him a sense of community that stood in stark relief to his Lower East Side life in the city. Papanek described working life being split between the two holidays of Easter and Christmas, which Olgivanna (a Russian-educated Montenegrin) and the architect presided over, respectively. They were located at Spring Green until Thanksgiving, and immediately after that holiday a "motorcade of cars, vans and trucks would begin the long trek to where Taliesin West lay nestled in the desert near Phoenix. Having avoided the rigors of the northern winter, work then continued until Easter Monday, when the entire group would start on the long ride back to the north for the late spring, summer and autumn."[42]

Papanek's account of both Easter and Christmas, and his claim to have resided at both the summer studio Taliesin in Wisconsin, and the winter residence Taliesin West in Arizona, would suggest that he must have spent at least nine months as a fellow before (purportedly) quitting due to the strictly regulated and insular lifestyle. Jazz and smoking, both of which Papanek partook in enthusiastically, were prohibited. A strict "lights out" policy at 9:30 p.m. meant the young men and women were ready for a dawn start for their farming and domestic duties. Papanek admired Wright's eccentricity and charisma, commenting on his "meticulous attention to his somewhat flamboyant clothing style—flowing cravats, silk suits, hand-tailored shirts, the flowing cape made according to his own design, his 'prairie' hat, a cane nourished with bravura, and everything neatly pressed."[43] By his own account, ultimately, he was forced to rebel against the "autocrat," asking himself, "Did I want to become a third-rate Frank Lloyd Wright or be a first-rate Papanek!"[44] He left Taliesin in haste with Wright accusing him, with feigned theatricality, of being an untalented, ungrateful traitor and further asking that his bags be searched to ensure the boy "he'd picked out of the gutter" had not made off with any of the precious artifacts that decorated the shelves and walls of Taliesin. "It doesn't need the good Dr. Freud's psychoanalytical skills, nor Carl Jung's Unconscious," commented Papanek in his reminiscing, "to deduce that what was really happening in this theatrical episode was a re-acting of the 'primal scene,' the ultimate rejection of the Father figure, a rebellion against the guru."[45]

Whatever had actually taken place between the two men, Papanek had crucially failed to secure a much-coveted letter of reference from his influential master. Although details regarding his Taliesin fellowship remain hazy and uncorroborated by contemporary records, Papanek keenly understood the significance of an affiliation with America's greatest architect. Wright routinely provided references for his apprentices and assistants, often imbuing them with a slightly ironic tone of address—for example, "To Anyone, Anywhere." Papanek needed these validating credentials in order to secure his future prospects as a designer, and to prove his fellowship at Taliesin had taken place.

The Frank Lloyd Wright Letter: Seeking Legitimacy

In early 1949, Papanek received his ultimate trophy, an addition to his growing portfolio of design and architectural connections: a markedly personal and supportive letter from Frank Lloyd Wright. "Dear Victor," it read, "At your age, had Lieber Meister brought me to Taliesin, I too would have fought, rebelled as you did. Mrs. Wright and I will remember you as part of the stones you helped to lay here and at Taliesin West, for you have always tried to be the 'Real Thing'" (see figure 2.3).[46] Completing the extraordinary network Papanek had generated as a young would-be designer, Wright's words instantly conferred on him the Howard Roark–style rebellious edge he would foster in his future polemicizing and self-promotion as an acerbic design critic. An "inside-outsider" in US culture, Papanek relished having been given the nickname "Howard" during his studies at the Cooper Union School of Art in the late 1940s—and now he was the rebel the great American master architect himself had come to admire as the "Real Thing."

This letter remains, however, the only formal evidence of Papanek's relationship with Wright. And, intriguingly, a question mark hangs over the authenticity of this prized document, an artifact that hung framed above the social designer's desk, like a talisman, for the entirety of his career. No carbon copy of it can be traced in the otherwise meticulously kept records of the architect's Taliesin correspondence, and according to Frank Lloyd Wright experts the nonspecific date ("Easter 1949") is notably atypical of the architect and his assistant Gene Masselink's communications. Further, the familiar tone of the missive is entirely out of keeping with Wright's formal writing style, and at least one leading expert questions

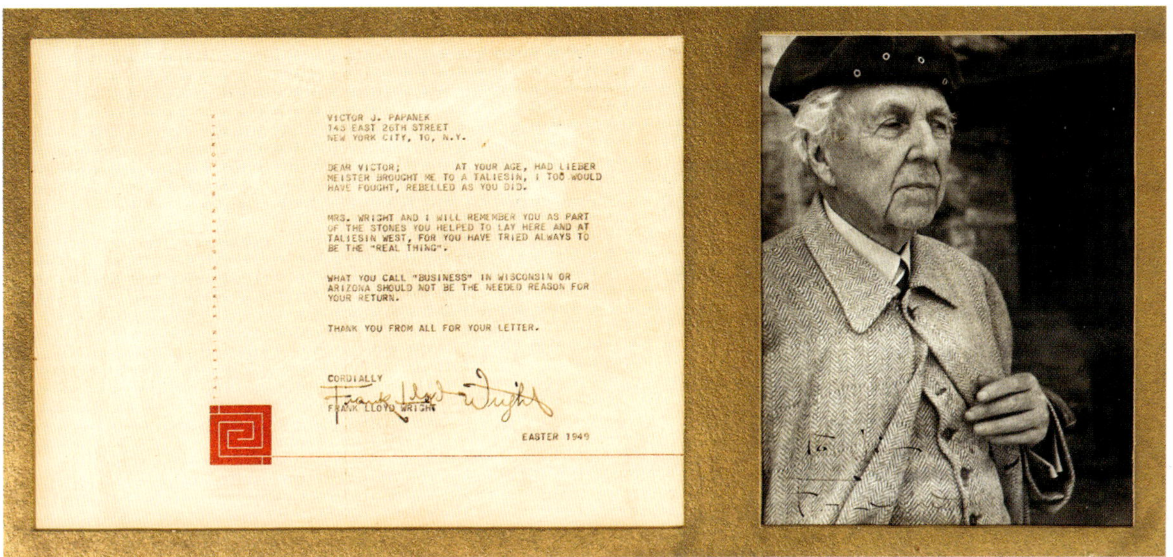

Figure 2.3
Framed letter (with autographed photograph) to Victor Papanek, Veterans Art Circle, 145 East 26th Street, New York, dated Easter 1949, from Frank Lloyd Wright. © Victor J. Papanek Foundation, University of Applied Arts Vienna

the authenticity of the document's signature. With Wright signing off with the words "thank you from all for your letter," the correspondence suggests that it is a response to an earlier communication sent by Papanek to Taliesin. But, again, there is no archival trace of any personal letter sent from Papanek to Wright that might have prompted this warm-hearted response.

One fact stands out, left untold in any of Papanek's accounts of his encounters with Wright. In April 1948, under the guise of the "Veteran's [*sic*] Art Circle" (an organization Papanek founded using the address of his New York City efficiency apartment) he penned a letter to Frank Lloyd Wright requesting literature of the "great maestro" for the "non-profit, co-operative group of young people here in this city, who get together for the study of art and architecture on evenings."[47] The Veteran's Art Circle (VCA) may have been inspired by the postwar initiatives of major cultural institutions, such as MoMA's Veteran Art Program and War Veterans' Art Center, focused on reintegrating returning conscripts through exposure to art. Papanek had similarly reached out to Henry Miller using VCA letterhead to contact the author's office, successfully acquiring for his library a personally dedicated copy (with the inscription "To Vic, from Henry") of *Varda: The Master Builder*, helping substantiate his claim to have personally known America's leading avant-garde author.

While there is solid archival evidence showing that Masselink received the request from Papanek's veteran enterprise, and generously responded with an extensive bundle of Taleisin materials, there is no trace of the VCA ever having existed as a formally registered or bona-fide organization or charity. As an isolated individual desperately searching for professional connections and networks, the invention of this entity appears to have been a savvy under-taking on Papanek's part, enabling him to reach out to leading creative figures under the auspices of a worthy cause, from the address of his studio apartment on the Lower East Side.

Is it conceivable that Papanek's ambitions as a budding designer-architect spurred him on to audaciously repurpose the letterhead received as part of Masselink's correspondence with the faux veterans' organization, using it to type a falsified letter of endorsement from America's premier architectural figure? Might this explain why, despite Papanek putting his apprenticeship and work with Frank Lloyd Wright center-stage in his career biography, he remained studiously vague regarding the dates of his involvement with the architect—always refusing to pin it down?

In a further intriguing twist to the tale of Papanek and Wright's real or imagined relation-ship, two notable pieces of evidence seem to substantiate that Papanek's apprenticeship with Wright did actually take place (though neither unequivocally supports it either). From the early 1940s onward, Papanek used Kodachrome color slide film to document his designs and aspects of his working life. Few slides remain from this period in his personal archive, but one in particular stands out—an intimate image of a lavishly decked breakfast table with Wright and Olgivanna seated among an entourage of a dozen guests, some of whom were possibly apprentices invited as a privilege. One place is left empty, its napkin strewn across the seat, suggesting the photographer vacated the seat in haste to take the impromptu image as a souvenir of the special moment.[48] Could Papanek himself have been the photographer? Cer-tainly he used this same image to illustrate his essay on Wright in *Architektur und Bauforum*, and though he refrains from formally crediting the picture there, the strong implication is that it is Papanek's personal record of his time at Taliesin.[49]

The second detail of evidence, though much less compelling, that lends weight to Papa-nek's claim of having been a fellow with Wright is a fundraising letter he received in 1985, from The Frank Lloyd Wright Foundation, Taliesin West, addressed to "all Taliesin appren-tices," containing an invitation to celebrate the "beginning of the second half-century of Taliesin West."[50] Papanek might possibly have registered himself in later life as a former Taliesin apprentice; certainly there remains an incongruity between his receipt of this let-ter and the entire absence of records supporting his being an apprentice there. As a young, fatherless refugee supporting his elderly mother and making his way in an alien environ-ment, Papanek's relationship with Wright was undoubtedly the most creatively formative one of his early career—whether or not he ever actually met or worked with him.

An Émigré Rebel Retreats: *The Fountainhead* and the Harried Escape

Wright's highly valued letter of endorsement—which Papanek kept as a treasured possession, framed, and hung on his study wall throughout his life—arrived (purportedly) just a couple of months prior to the 1949 commercial release of Hollywood director King Vidor's *The Fountainhead*. This controversial interpretation of Ayn Rand's best-selling novel depicted a maverick architect as the personification of the Objectivism philosophy she championed. Rand—Russian émigré, rabid anticommunist, and advocate of libertarianism—openly modeled the character of Howard Roark on Frank Lloyd Wright, whom she had met through mutual Hollywood connections. Wright's granddaughter, Anne Baxter (daughter of Wright's son Lloyd), had become a movie star with Fox Studios by the mid-1940s, leading to his sporadic attendance of Los Angeles parties and social events where he eventually crossed paths with Rand.

As her success as a scriptwriter in the mid-1940s led to greater prominence and financial security, Rand self-consciously configured herself as an advocate of modernist architecture. Promotional images related to *The Fountainhead*'s release featured her (captured by the celebrated photographer of modernist architecture Julius Shulman) seated by the picture window of the recently acquired Sternberg house: an ambitious, ultramodern residential design (1934–1935) by the former Taliesin fellow and Viennese émigré Richard Neutra, acquired by Rand from Viennese émigré Hollywood director Josef von Sternberg in 1944.[51] Despite several meetings with Rand, and correspondence, Wright refused to allow his architecture or plans to be featured in *The Fountainhead* movie production, and he along with other members of the architectural profession denounced Rand's portrayal of contemporary architectural practice as an expression of libertarianism and unchecked individualism. In commercial terms, *The Fountainhead*, starring Gary Cooper and Patricia Neal, also failed to impress. It contained a notoriously ambiguous treatment of a rape scene, bombed at the box office, and was broadly panned by critics. One typical review, in the *New York Times*, described the movie as "wordy, involved and pretentious" going on to summarize the plot in one acerbic paragraph: "A long-winded, complicated preachment on the rights of the individual in society—and also upon the privilege of a lady to change her mind— . . . put on the screen with more fervor than compelling conviction, it seems, in the Warner Brothers' film version of Ayn Rand's 'The Fountainhead.'"[52]

Nevertheless, for a young Viennese émigré who had failed to complete his Cooper Union School of Art diploma and was ejected from Taliesin studios for rebelling against the "Meister," *The Fountainhead*'s steely character of Howard Roark (the fictional character himself expelled from architectural school) bore some uncanny similarities to Papanek in his early career trajectory.

Despite the proven endurance of his admiration for Wright, at the height of Papanek's success as a socially responsible designer in the early 1970s, he began to recast the then thoroughly establishment figure of Frank Lloyd Wright (rather as he had with Raymond Loewy) as an autocratic counterinfluence. In 1970, a newly formed "interprofessional" environmental design magazine, *Design and Environment*, ran a profile piece typifying this revisionist approach: "As a negative influence the egocentric and domineering Wright catalyzed Papanek's social and design views . . . some who studied with him crumbled into imitative slaves. The courageous, like Papanek, rebelled."[53]

Beyond the "kaleidoscopic fragments" of his own hazy retrospective accounts, exactly how did Victor Papanek come to achieve familiarity with Lloyd Wright? Was he really accosted by Wright's right-hand man, Wesley Peters, taking pictures at Taliesin, on his bicycle tour of the architect's masterpieces, leading to his serendipitous enrollment as a fellow and a lifelong association with America's great creative genius? Was it the MoMA exhibition *Frank Lloyd Wright: Taliesin and Taliesin West* (April 15–June 15, 1947), featuring architectural photographer Ezra Stoller's images of the architect's winter and summer homes, which piqued Papanek's imagination, and determination to become an apprentice (real or imagined)?

How to Get Ahead in Design: Professional Networks Real and Imagined

Victor Papanek's network of carefully chosen (as well as serendipitous) figures of influence, from Frederick Kiesler and Henry Miller through to Frank Lloyd Wright, helped forge his design career. Papanek's claim of a personal and enduring association with Wright extended to his retelling that his first experience as a designer involved model making for Kiesler and then during his apprenticeship at Taliesin, for Wright's newly commissioned Guggenheim Museum in the late 1940s: All a far cry from sketching cartoons for Irving Klaw's pornographic comic books and hanging out with the pin-up Bettie Page. As Papanek put it in his own words:

> Some of the most interesting work was done on the models for the Solomon Guggenheim Museum of Non-Objective Art that would eventually be built on Upper Fifth Avenue. I found it to be a remarkable coincidence that as a fourteen-year-old boy I had worked for Friedrich Kiesler ("The Viennese Dwarf"), building his oddly shaped "multi-functional furniture" (really exhibition structures), for Guggenheim's niece Peggy, and her "Art of this Century" gallery of Surrealist Art in New York.[54]

The truth regarding the origins and nature of his first design commission proved far from prosaic, but worlds away from the upper echelons of cultural production suggested by his association with the studios of Frederick Kiesler and the designs of Frank Lloyd Wright's iconic monolith, the Guggenheim Museum. Nevertheless, by 1949 Papanek had firmly embedded himself amid an outstanding network of real and imagined design connections; a web of associations that set the young émigré up for a new career when he transplanted himself, in 1950, to the other side of the United States—close to his hero Henry Miller's famous Californian haunt, Big Sur.

From Artist to Architectural Designer: Quitting the Convolutions of New York City

Following his breakdown and hospitalization at Bellevue over Christmas 1949, Papanek would refrain from revisiting New York City—the creative crucible of his early life—until a decade and a half later when he delivered a keynote lecture on the occasion of the Industrial Designers Society of America's annual conference at Parsons School of Design in 1966. Despite never having formally graduated as an architect or designer, he did visit the city again in the late 1980s, when the Cooper Union School of Art granted him an award for "outstanding professional achievement in industrial design and education."

Earlier in the year of Papanek's breakdown, a thirty-one-year-old Jewish woman from Brooklyn, Ada M. Epstein, became his second wife. Married in Manhattan on February 25, 1949, the "Affidavit for License to Marry" indicates Papanek's previous marriage to Anna Lipschitz (in 1944) was "annulled" on his initiative; the reason cited is "fraud."[55] It might be surmised, then, that mention of the tragic death of a girlfriend on Christmas Eve 1949 in his autobiographical accounts pertains to Ada M. Epstein (his recent wife), or possibly the young woman he dated during his time at Cooper Union, Jeanne Hertle. However, Ada remarried less than a year after her union with Papanek, and records reveal Jeanne boarded a ship to Bermuda in Spring 1950.[56] A further twist to the background of Papanek's second marriage is that the Epstein and Lipschitz families were not only connected by religious and ethnic origin (Russian-Jewish descent), but also, far more significantly, as revealed by the 1925 New York census, as neighbors living side-by-side in Brooklyn tenements.[57] As such they may have had longstanding relationships, with Papanek's marriage first to Anna, then subsequently Ada, serving a purpose to both families and helping a newly arrived Jewish émigré in New York—or it could simply be that Papanek met his second wife Ada through her family's connection with the Epsteins in the 1940s. What is certain is that his life in New York at the end of the 1940s was becoming increasingly complex both personally and financially. In 1948, Papanek had momentary success securing his first formal teaching position as a part-time art tutor at the College of Forestry, State University of New York (SUNY) in Syracuse; the yearbook shows a serious and dapper-looking figure, waist-coated and bespectacled, cigarette nonchalantly hanging from his hand, exhibiting a "trade-mark" surly, smooth, intellectual look that would endure well in the 1960s.[58] But, nevertheless, Papanek's Design Clinic struggled to find clients, and failed to operate as a commercially successful venture. Despite his promising position at SUNY, having accrued and then lost two wives and having endured a break-up with a significant girlfriend, by the end of the decade the possibility of a fresh start for the émigré and his mother proved a tempting proposition.[59] For, throughout this tumultuous period in New York, his mother Helene stood by him, living alone in Harlem, sharing the traumatic memory of their escape from Nazi persecution, but also offering him a lifeline in the form of the memories of his comfortable Viennese childhood. Papanek's relationships help frame the deep sense of precarity that underpinned his survival as an émigré in the New World, and the crucial importance of the network of women that enabled his professional ambitions.

By 1949, Papanek dropped the profession of "artist" from his documentation, replacing it with that of "Architectural Designer" written with the flourish of a confident draughtsman, invoking distinctive architectonic handwriting that would evolve as another stylistic hallmark of his lifelong career.[60] In a bid to follow in the footsteps of the multitude of famed Austrian émigré architects and designers, from Richard Neutra to Victor Gruen, who had excelled in their professions following their move to California, Papanek left New York City with his mother in January 1950, and within weeks was turning his hand to speculative interior design for wealthy West Coast clients.

3   Studio 44: Mid-Century California Calling

Leaving behind the throng of America's premier metropolis, and following in the footsteps of a multitude of émigrés before him, Victor Papanek made the move from New York City to California, settling in San Francisco in 1950. As he adjusted to a design culture dominated by lifestyle, the conceptual and theoretical wrangling over issues from the politics of low-cost furnishings to the contradictions of design's claim to modernity faded into the background. Instead, the casual aesthetic of the "California Look" came hastily into view.

Examples of the gilded career route taken from New York to California abounded in the popular design press. The glamorous 1947 Palm Springs home of Raymond Loewy (a man who stood as both Papanek's erstwhile role model and, later in the social designer's career, as his nemesis), designed by Swiss-émigré architect Albert Frey, featured in glossy magazines of the period as the ultimate in high-end Californian mid-century living. Another success story featuring one of Papanek's early role models garnered much press coverage: Viennese émigré Victor Gruen relocated from New York to Los Angeles in 1941, his career taking off there with multiple commissions for his ever-expanding regional retail business. A well-traveled émigré route, framed by inspirational tales of commercial triumph, taking the metaphorical "bridge" between the East and West Coast garnered a mythical status in the making of design and architectural careers.

Design Clinic, Papanek's first studio, had emerged in response to the vicissitudes of late-1940s New York City life, and the quandaries around design's social and aesthetic role. San Francisco, dubbed "Baghdad by the Bay" in light of its multicultural, "exoticized" reputation, offered up an entirely different set of cultural indices. So it was, then, that Design Clinic continued to operate, running alongside Papanek's sleeker-sounding Californian initiative—*Studio 44*. With a striking contemporary-style, electric-blue, blocked-color design identity, Studio 44 complemented the Design Clinic practice by providing a new gallery space in downtown San Francisco (see figure 3.1). Crucially, the metamorphosis of Papanek's design practice was intrinsically intertwined with his own self-rebranding project. Following his

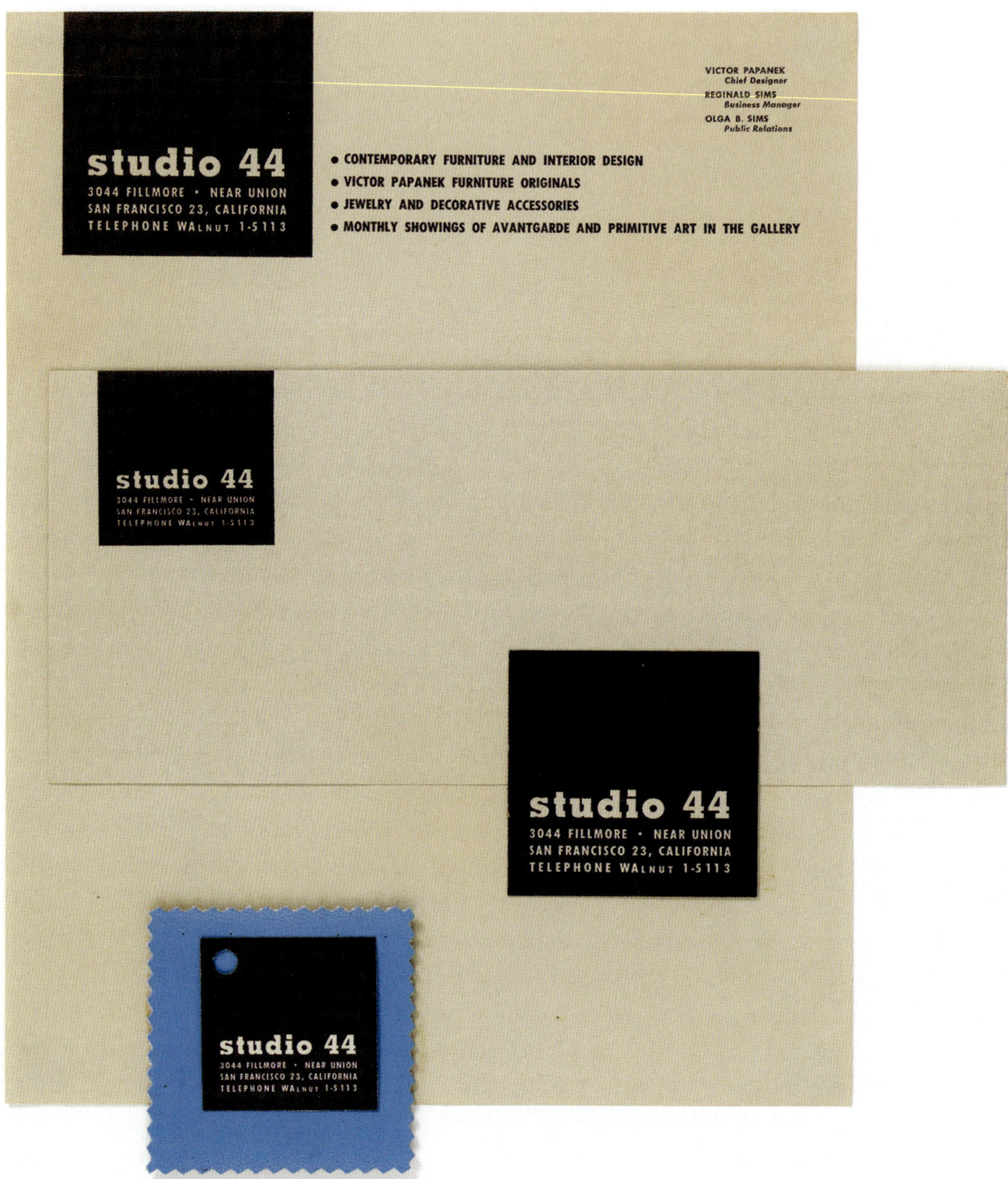

Figure 3.1
Studio 44 brand identity, c. 1950–1952. © Victor J. Papanek Foundation, University of Applied Arts Vienna

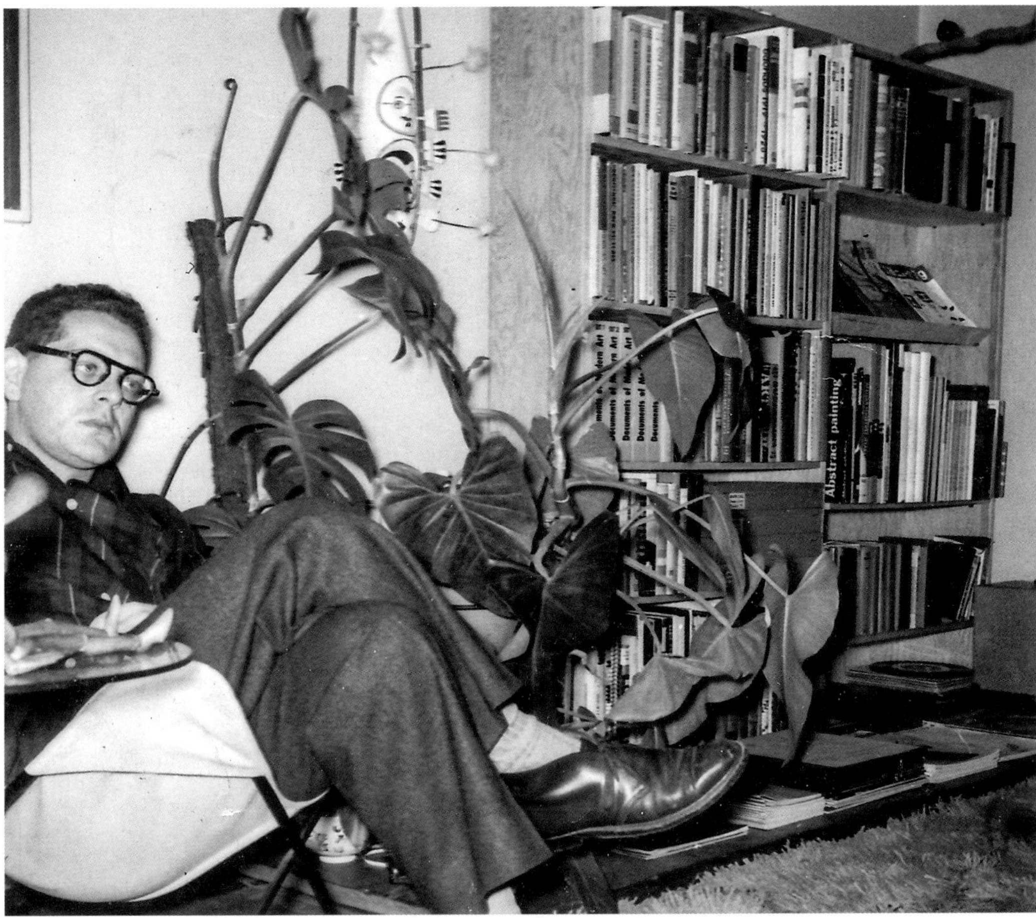

Figure 3.2
Victor Papanek, San Francisco, photograph c. 1952. © Victor J. Papanek Foundation, University of Applied Arts Vienna

harried departure from New York at the beginning of the decade, his emergence into the world of Californian design required urgent personal, as well as professional, consideration (see figure 3.2).

But Papanek did not embark on his westward venture alone. As well as being accompanied by his mother, he brought with him a demure young divorcée, hailing from a traditional Protestant New England background, named Winifred Nelson Higginbotham.[1] Papanek's life and circumstances had changed indelibly. Following their arrival in California in the early 1950s, Victor and Winifred married in June 1951, and their daughter Nicolette was born in San Francisco later that fall.[2]

Throughout his life, Papanek referred to Winifred as his first wife. Firmly consigned to his past, the previous marriages to Russian-Jewish Brooklynites Anna Lipschitz and Ada M. Epstein had brought Papanek into the fold of a quintessentially ethnic, émigré New York life. But their existence in Papanek's life remained undisclosed—unknown to anyone besides the brides themselves, their immediate families, and, almost certainly, Victor's greatest confidante—his devoted mother, Helene. Whatever the motive for embarking on those first two marriages, by consciously editing them from his life story and moving across the expanse of the United States, he was liberated from the early and traumatic machinations of his early émigré life. Starting fresh, Papanek and his young family set up home in an apartment on Union Street a couple of blocks from their Studio 44 gallery on Fillmore Street (as a young ceramacist Winifred shared the gallery enterprise, using it to showcase her works), his mother living nearby to support them. With a hard-won network of connections and a degree of experience behind him, the young designer could finally make an impact on American design.

Studio 44 Meets "Californian Chinoiserie"

California's architectural and design world had undoubtedly become familiar to Papanek during his time studying at the Cooper Union School of Art. But a particularly intriguing link to San Francisco emerges by way of Frank Lloyd's Wright's innovative 1948 design for the V. C. Morris Gift Shop on Maiden Lane, a few blocks away from Papanek's gallery-cum-retail space Studio 44. Wright's design incorporated an internal circular ramp, widely regarded by architectural critics as the precursor to his radical spiral concept for the Solomon R. Guggenheim Museum in New York completed four years after the Maiden Lane space. In autobiographical accounts, Papanek claimed to have worked on the revolutionary Guggenheim Museum in the course of his fellowship with Wright at Taliesin. Dated July 23, 1951, a letter to Papanek sent by the upmarket design store's proprietor, V. C. Morris, regarding the opening of a charge account not only proves Papanek's firsthand knowledge of Wright's design and the architect's continued influence, but also shows the extent to which the young ambitious designer was involved, at least aspirationally, in the city's flourishing contemporary design scene.[3]

There may, however, have been another bluntly instrumental reason behind his frequenting the iconic V. C. Morris store: his visits provided direct research for his own entrepreneurial venture. Studio 44 positioned itself both as a design studio through which commissions were made and as an outlet for select designer-craft objects ("jewelry and decorative accessories") and ready-made design pieces ("Victor Papanek furniture originals"); the space on Fillmore Street also doubled as a gallery, featuring "monthly showings of avant-garde and primitive art."[4] Historical mapping of the area using San Francisco business directories between 1951 and 1953 reveals Fillmore Street as an up-and-coming hub for craft and design: Studio 44 was flanked by boutique retailers including the Schumacher Pottery and Sculpture Studio, Natalie Grunewald Furniture Store, the Ashley C. Drusill Antique Shop, and Jon's Furniture shop.[5] The Papanek family's apartment on Union Street was located in the same neighborhood, Cow Hollow, known today as the hub of the city's contemporary "technogentrification" phenomenon. Papanek definitely had his finger on the pulse of an expanding design culture, and on the city as a leader in contemporary lifestyle trends.

In the early 1950s San Francisco had a vibrant creative community, which spawned a variety of high-end, high-profile design and crafts emporia and specialist design stores more generally. These included the Pacific Shop and Amberg-Hirth, both of which showcased local designer-craftspeople, as well as the work of seminal design figures such as Alvar Aalto (who, along with Frank Lloyd Wright, was a lifelong influence on Papanek). Though Papanek was fully versed in the work of Aalto from the extensive press coverage of the Finnish Pavilion at the New York World's Fair of 1939, and the Aalto exhibition at MoMA the year prior to his departure from New York, prevalence of Scandinavian design on the West Coast may have furthered what would become his lifelong affinity with Finnish design.[6] Papanek also befriended a sculptor of Finnish descent, Walter Muhonen, during his time in California, who provided him with contacts in the local art and craft scene, possibly even supplying items for Studio 44. As well as acting as a showcase for the work of "many outstanding craftsmen on the West Coast" (modern jewelry, artisanal, weavings, wood-turned and ceramic objects)—just as Papanek ambitiously envisaged with Studio 44—Amberg-Hirth operated as a tastemaker: "It is the enthusiasm of Amberg-Hirth plus their appreciation and understanding of modern design, and fine craftsmanship that has led them to build up this remarkable organization, . . . helping to raise the standards for public taste," observed *Design*, the retailer's magazine.[7]

High-end outlets such as V. C. Morris Gift Shop, the Pacific Shop, and Amberg-Hirth collectively raised the profile of a Californian style and bolstered San Francisco's reputation as the epicenter of the West Coast lifestyle. But it was Gump's renowned department store on Post Street, purveyor of Eastern objects, art, and exotica, which exerted an equal, if not greater, influence on the trajectory of Californian design, an influence that certainly found its way into Papanek's early 1950s interior design drawings (see figure 3.3). In 1952, Papanek hitherto unknown as a painter, exhibited a series of abstract painted panels there in a group show.[8]

Gump's fame as a national treasure was such that in 1949 *Time* magazine ran an article regarding its controversial refurbishment from an emporium of Eastern objets d'art to a modern design location, offering unique insight into its reputation as a purveyor of Eastern and

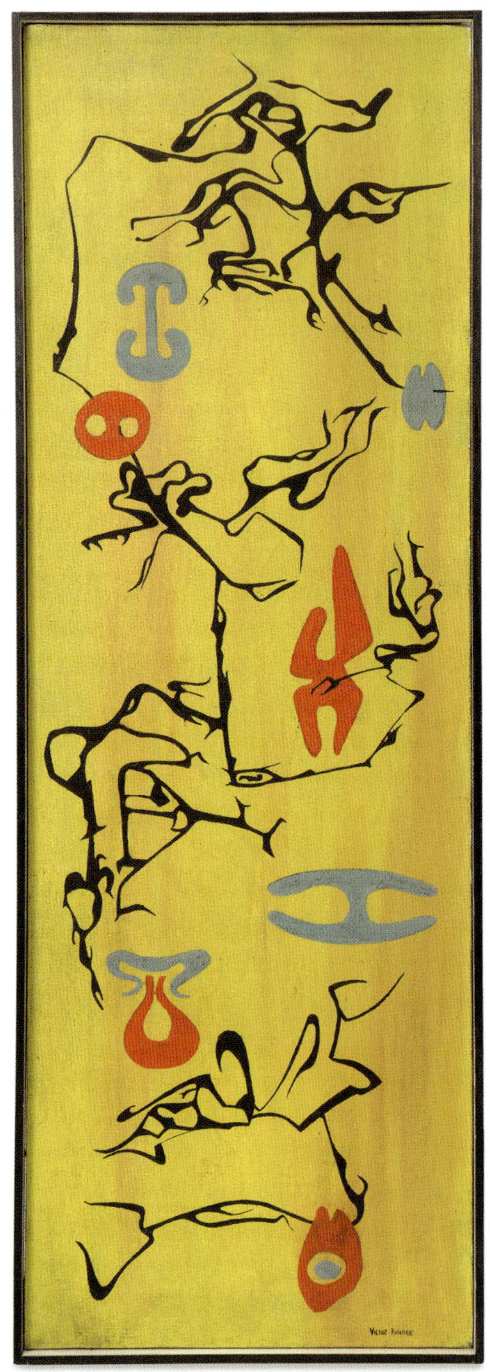

Figure 3.3
Abstract painting, Victor Papanek, c. 1951–1952. © Victor J. Papanek Foundation, University of Applied Arts Vienna

regional design: "Gump's will give shelf room to a $12.50 piece of California pottery—if it is esthetically good—alongside a $1,800 piece of Ming Dynasty porcelain, and encourages its customers to do the same."[9]

Perfectly encapsulating *Time* magazine's observation, a type of mid-century "Californian Chinoiserie" pervaded Papanek's designs during the years he spent on the West Coast, between 1951 and 1954. This eclectic middlebrow aesthetic is perfectly illustrative of the broader phenomenon of "Cold War Orientalism" historian Christina Klein identifies as emerging in the postwar period to stymie hostility toward the "East" by reframing Asia, through global imaginary of integration, as a potential US comrade as opposed to an alien force to be feared.[10]

By the late 1950s, popular reference to San Francisco as "Baghdad of the Bay" (or "Baghdad by the Bay") invoked the sense of a culturally and ethnically mixed city with an intellectual scene bubbling just under the surface.[11] Although the city boasted design outlets that surpassed the contemporary sophistication of any New York equivalent, its famed institution, Gump's department store, filled with Asian masks, trinkets, vases, textiles, and ethnographic objects collected on the owner's travels, fostered a mass of loyal consumers.

Richard Gump, the descendent of the nineteenth-century store's original founder, purportedly "projected through his books, store, and catalogues an image of Baghdad San Francisco as a city of taste, catering to a worldwide clientele."[12] As a lifelong collector of Asian antiques and artifacts, and avid reader on topics of Asian anthropology, Papanek's penchant for collecting may have begun with his first visit to Gump's "Exotic Treasury." Evidently, Gump's dissemination of a distinctive upscale interior-design style, the blending of Asian and European influence for which San Francisco was renowned, impacted Papanek's early interior and furnishing designs, the verve of Baghdad by the Bay seeping into a rapidly evolving genre of design.

Papanek regularly designed interiors for his own use, indulging his most extravagant design fantasies. One such design, "Low Dining Group for Self," features flat, tasseled floor cushions around a slightly elevated square table upon which are placed a pale green placemat, rice bowls, and chopsticks. A Turkish-style tea urn sits upon a small modern chest with bright-red slatted drawers; a hanging drapery of a rampant dragon adorns the far wall; a non-Western musical instrument leans in the corner; a floating, suspended lamp formed of driftwood attached to a large white metal disk redolent of a Japanese parasol hovers over the table; and the ensemble is completed by an animal rug and wrought-iron candlestick. Details from this dining ensemble (Asian ornamental hangings, rice bowl, ikebana style driftwood) reemerge in "Lighting Unit XM-7," "Designed for Hansen of 978 1st Ave, NYC, N.Y.," and signed "Victor Papanek." The leaning wrought-iron lighting unit, featuring a "24 inch aluminum reflector disk, baked white enamel finish," was clearly imitative of a parasol, with a "Harp 'baby spot' bulb," and it incorporated the stand of a non-Western musical instrument that featured on various other drawings. George W. Hansen (an American of Danish origin) famously designed the swing-arm lamp mechanism, and his eponymous lighting company had a showroom in New York City. Like many of Papanek's works, "Lighting Unit XM-7" was undoubtedly a speculative rather than a commissioned and realized design (see figure 3.4).

This succession of ostensibly middlebrow designs, first originating in his observations and immersion in a New York popular-design culture and then blossoming into a West Coast style, is testament to Papanek's unerring ambition to carve out a commercially success profile as a designer keyed into the nuances of modernity and contemporary living. Nowhere is this more apparent than in the evocative lighting design that Papanek incorporated into one of his contemporary interior designs. Simply titled "Living Room Corner," it depicts a Jean Arp-style painting suspended from metal ceiling rods, in the foreground an undulating blood-red fabric weave and wooden double day bed, a leaning suspended lamp with organic-shaped appendage, and an angular "radio-tele table" console with black-and-white images of his wife, Winifred, collaged to its screen (echoing a similar audiovisual furniture design he had made for his then-girlfriend Jeanne Hertle in New York). The embedding of a combination of radio and television technologies into interior design sets was notably innovative, with sound systems more conventionally encased in furnishing forms such as side cabinets.[13] This quirky *Gesamtkunstwerk* design approach evolved as Papanek's signature style at this period.

Sunken Dens and Japanese Flourishes: Frank Lloyd Wright Goes West Coast

Victor Papanek's penchant for populist "Orientalist" style pastiche, sunken dens, floating neon beds, and integrated audiovisual technology appeared as far removed as possible from any aesthetic principle or vernacular sensitivity imparted by Frank Lloyd Wright to his Taliesin fellows in the late 1940s, and the vernacular design pedigree Papanek outspokenly embraced. Likewise, his studied attempts at mastering a popular rendition of contemporary lifestyle during this period failed to impart even a glimmer of the socially responsible design approach he would later champion with unmitigated zeal. Rather, his designs relish the foibles of modernity and consumer culture lifestyle—the very same frivolous consumer lifestyle Papanek would later lead a relentless vitriolic campaign against in his polemical writing, media broadcasts, and pedagogy.

Through closer scrutiny of his work, and specifically the repeated reference to the hearth and the den, one catches glimpses of Wright's enduring influence, albeit heavily disguised in the Californian casual look. The fashion for open-plan spaces of the early 1950s meant

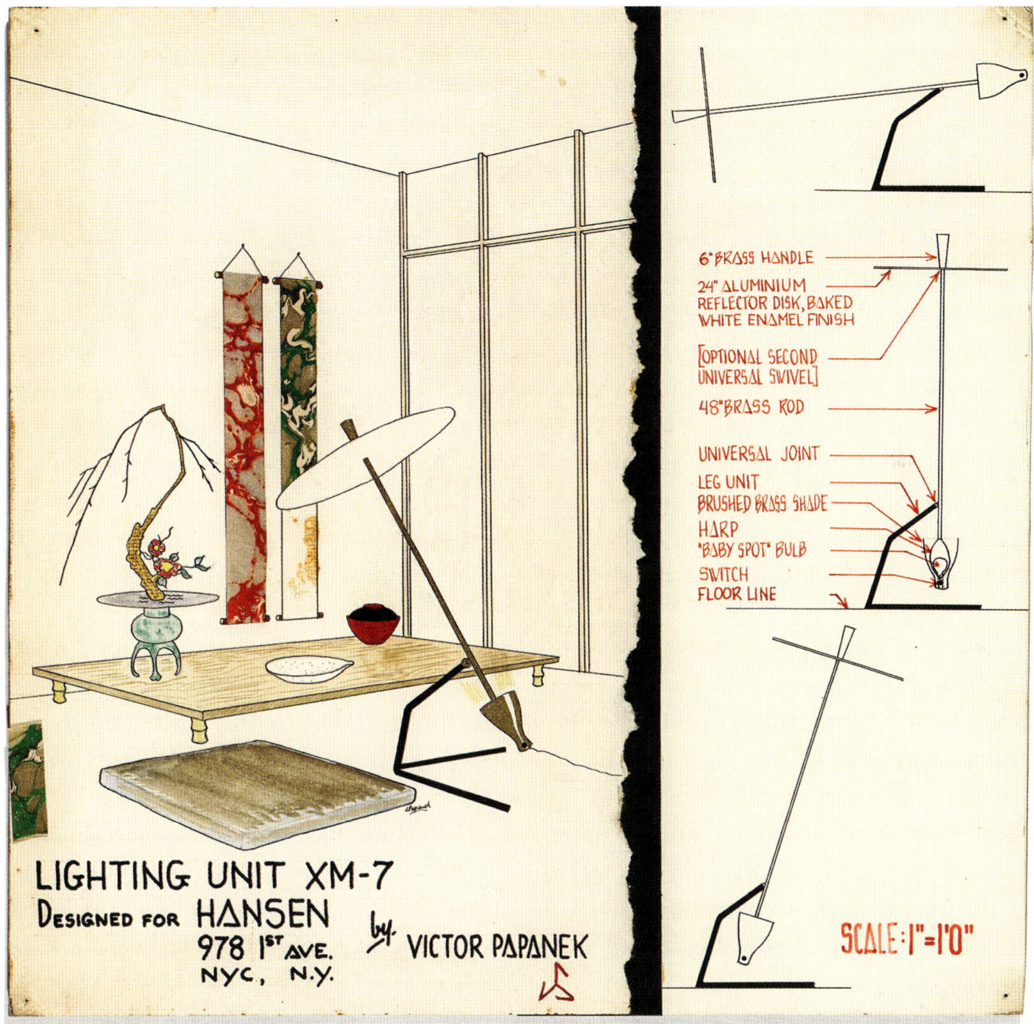

Figure 3.4
"Lighting Unit XM-7 designed for Hansen, 978 1st Ave, NYC, N.Y. by Victor Papanek," c. 1946–1948.
© Victor J. Papanek Foundation, University of Applied Arts Vienna

low-set interior zones were incorporated in order to partition and demark spaces. So sunken spaces became a familiar aesthetic and structural device in the Californian high-end interiors that filled the pages of design and lifestyle publications. According to historian Pat Kirkham, despite or because of Californian outdoor living, the idea of enclosure prevailed, and was best encapsulated in the low-set "cave-like" interiors of John Lautner (1911–1994), whose mentor Frank Lloyd Wright "grasped better than most the importance of psychological 'markers' of the home."[14] Of Austrian and Irish descent, Lautner had in the 1930s apprenticed with Wright for a sustained period, and continued to work with him on a range of projects. A prominent Californian architect, he courted a high media profile, with his work for domestic residences featured widely in periodicals from the *Los Angeles Times* and *House Beautiful* to *Architectural Forum* and even *A Guide to Contemporary Architecture in Southern California* (1951), with which Papanek would certainly have been familiar. Lautner also pushed the edges of establishment taste in his involvement with so-called Googie architecture of postwar popular commercial culture, following his design for the eponymous architect of the Lautner strip-mall coffee shop in 1949 Los Angeles: an architectural genre celebrating the dominance of West Coast car culture and consumerism and a far cry from the Frank Lloyd Wright architectural dictates with which he had begun his architectural career.

As well as summoning up the architectural emphasis on den and hearth, Design Clinic's most pronounced nod to Wright's oeuvre was perhaps in its persistent reference to all things Japanese: from the samisen (or shaminsen) three-stringed musical instrument—the inspiration for a series of dining chairs—to the nomenclature of its designs. The "*Shibumi* Series #2 and #3" of wooden chairs, for example, borrowed from the Japanese noun *Shibumi* indicating a thing of subtle, simple, understated beauty (see figure 3.5). Later referring to it as the Samisen chair in the journal *Interiors* in 1956, Papanek drew attention to qualities his chair shared with an organic chair design of Carlo Mollino showcased in the same magazine under the title "Mollino's Flexible Fantasies" in 1953—possibly to preempt any potential accusations of plagiarism.[15]

For another seating line, designed for the Dependable Furniture Company of San Francisco, Papanek used the Japanese term *kakujitsu*, translatable as "reliability," "robustness," and "certainty." Wright's enthusiasm for Japanese culture, art, and aesthetics infused his life's work, and in the first half of the twentieth century helped popularize Japan more generally, following his best-known commission there in 1922: Tokyo's highly intricate and iconic Imperial Hotel. Eschewing the fashionable architectural scene of Western Europe, Japan remained the only country, outside the United States, in which Wright ever worked and lived. He famously attributed his "all-American" aesthetic to the influence of the Asian nation's art and design, its stunning woodblock prints, and its unique approach to nature. While *Japonisme*, a taste of the exoticism of Japan, had fueled a craze among the European avant-garde artists of the

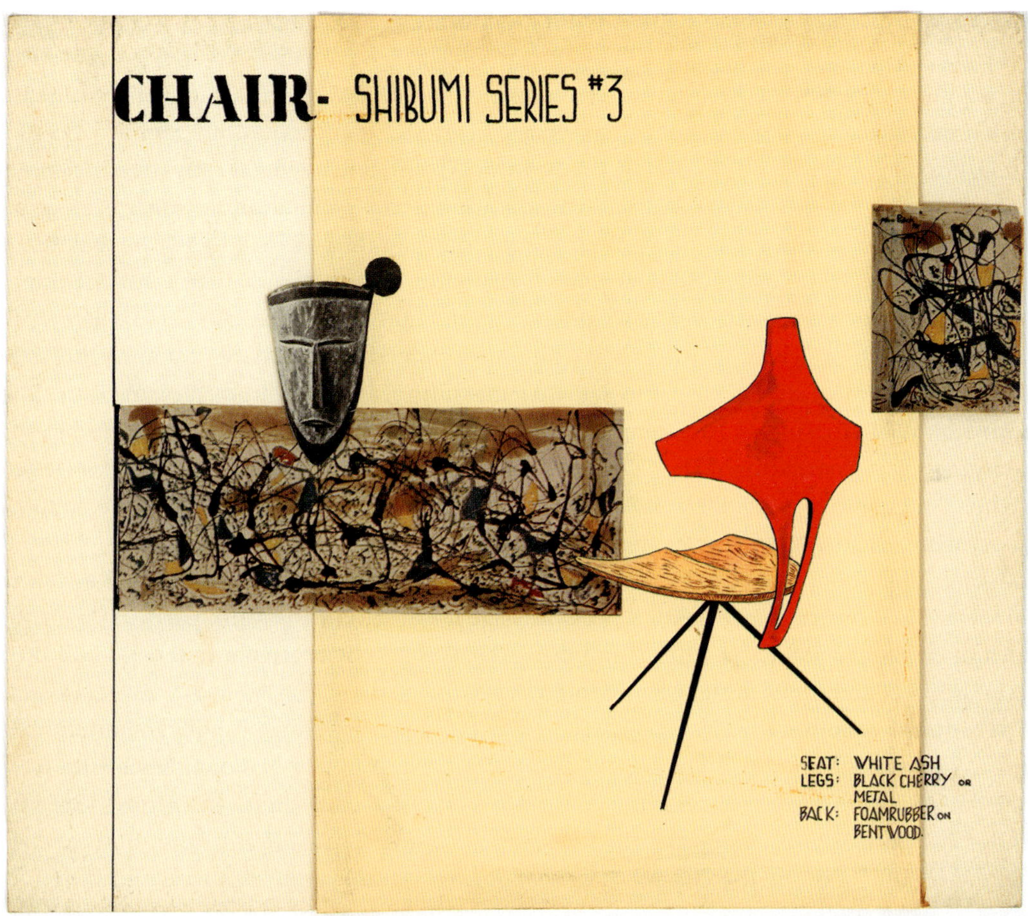

Figure 3.5

Chair, *Shibumi* Series # 3, designed by Victor Papanek, c. 1948–1952. © Victor J. Papanek Foundation, University of Applied Arts Vienna

late nineteenth century, by the 1940s, when Papanek first encountered Wright, the wartime conflict with Japan had dulled its cultural profile in the States. Nevertheless, Papanek maintained a lifelong interest in Japan's design, culture, and art, beginning in the 1940s with one of his earliest designs that was based on Japanese *geta* footwear and through to his designs in San Francisco; he was undoubtedly inspired by the large population of Japanese immigrants and their historical relationship to the city and the burgeoning "Cold War Orientalism" that Klein argues dominated the middlebrow US cultural imagination of this period.[16]

Émigré Allure and Alienation in Mid-Century California

Although lacking New York's intellectual credentials, by 1950 California had garnered a reputation as a place of "optimism and democracy, fearless experimentation" and a concerted "rejection of conventions." Stoked by the postwar G.I. Bill and a federal campaign of "boosterism," the region underwent widespread expansion and regeneration, luring a generation of designers and architects to its coast, with its myriad possibilities making it a "repository of the most intense longings for reinvention."[17] Historian Wendy Kaplan highlights the intensity of the region's magnetic pull through the figure of prominent designer Henry Dreyfuss, who along with designers such as Loewy famously shaped the futuristic vision of the 1939 New York World's Fair. Dreyfuss crossed that metaphorical bridge from New York to California in 1944, arriving in Pasadena, inspired by the idea of the state's counterpoise to the renowned East Coast uptightness. Under the evocative title "California: World's New Design Center," a couple of years after making the move, he observed in an article for a regional advertising journal: "On the Pacific Coast there are fewer shackles on tradition. There is unslackening development of new thought. There is a decided willingness to take a chance on new ideas."[18] As a newly defined postwar stage for design innovation, particularly in the area of consumer goods, and home interiors, the West Coast exerted a magnetic pull on a generation of ambitious designers and architects.

California occupies a unique position in twentieth-century North American cultural history. A refuge for European émigrés escaping the persecution of National Socialism, from the late 1930s onward it was home to a swathe of musicians, writers, and Hollywood film industry figures, all of whom breathed cultural and intellectual life into the sun-drenched tourist region.[19] Despite the Golden State's status as a fecund creative hotbed and haven, in the 1930s and 1940s the freshly arrived inhabitants often encountered California as a barren and alienating environment at odds with their cultural values, rather than a stepping-stone on the route to newfound freedom and success. Transplanted into a brash, consumer-driven, climatically arid landscape, some newcomers were subject to an extreme form of culture shock that compounded their broader sense of loss and uprootedness.

Bertolt Brecht's 1941 poetic treatment of the city of Los Angeles, provocatively titled "On Thinking about Hell," perfectly encapsulates the émigré's jarring transition to the scale and brashness of American life. For Brecht, himself an exile, the cultural dissonance was coupled with an impinging sense of European cultural superiority.[20] Nevertheless, "On Thinking about Hell" is an evocative lament, which describes an illusionary, urban sprawl punctuated by artificially opulent gardens perishing instantly without their constant supply of expensively administrated water, the incessant automobile traffic transporting "rosy people, coming from nowhere, going nowhere" like fleeting "foolish thoughts."[21] Exiled with Brecht on the West Coast in what Thomas Mann described as German California, the philosopher Theodor W. Adorno (1903–1969) epitomized the alienation and disjuncture of the émigré experience. This condition manifested in Adorno's highly influential writings, including *Dialektic der Aufklärung* (*Dialectic of Enlightenment*, 1944) coauthored with fellow Frankfurt School social theorist and émigré Max Horkheimer.[22] In a seminal essay of twentieth-century thinking, Adorno and Horkhemier brought to bear a pioneering mode of interwar European critical theory, which, stoked by an émigré's observation of the contradictions and vicissitudes of their host culture, resulted in the full-scale condemnation of popular culture (from jazz to advertising) and the sharpening of a theoretical treatise on the politically degenerative effects of consumer culture. Brecht, Adorno, Horkheimer, and their ilk offered an intellectualized response to a shared sense of dislocation through their depiction of a society defined by anomie, eroding social relations, and an overreliance on vacuous commercialism. As such, they captured the internalized sense of estrangement many émigrés experienced upon arrival in the jarringly alien Southern Californian desert, offering a counterbalance to the plethora of success stories regarding the forging of reshaped émigré identities liberated by the New World.[23]

In terms of the more conventional stories of aspiration and accomplishment, the Hollywood film studios are undoubtedly the best-documented cultural industry associated with California and the enormous influence of European-American émigrés. Architectural and interior design figures such as the Viennese-trained Hungarian Paul László, who was dubbed "architect to the celebrities," built their reputations on an association with glamorous Hollywood clients. But the region also generated a distinctive breed of industrial designers whose avant-garde ideas impacted the 1930s West Coast, most notably in the work of early influential émigrés Neutra and Schindler, both of whom had worked with Frank Lloyd Wright. Juxtaposed with myriad other cultural influences—Asian, Latin American, Arts and Crafts, Native American—their work set the West Coast apart from its Eastern counterpart.

A handful of designers succeeded in harnessing the otherwise incongruous aesthetic mix of European modernism, American "moderne" with the "sunshine" lifestyle of California. The early twentieth-century émigrés—Austrian American shopping mall pioneer Victor Gruen, Viennese Paul T. Frankl, and the Berliner Kem Weber—operating studios in 1930s and 1940s Los Angeles and the Bay Area exemplify this phenomenon. Historian Christopher Long identifies Frankl and Weber, along with similar émigré designers prominent in the region, as being central to the making of modern American design more generally. Their contribution was forged through a "persistent quest for Americanness" precisely because they were "not American at all." In other words, they were forced to seek out and invent their own Americanness as émigrés.[24]

Through his engagement with the Southern California environment, Frankl in particular (already famed for his New York-inspired modernistic "Skyscraper" furniture) generated a style of interior that Long asserts "voice[d] a truth about American life, or, at least, the vision many Americans held at the time about what their lives could be."[25] In the closing years of the Depression, these quintessentially Californian interiors proffered a vision of prosperity, leisurely living, casual conviviality: "a place to withdraw but also a site for active engagement with the pursuit of pleasure," which according to Long "somehow still [fell] within the scope of most people's aspirations."[26] This description of aspirational "loosened-up" interior design perfectly defines the range of designs Papanek experimented with upon his arrival on the West Coast.

As a means of reconciling this transition, a niche Southern Californian lifestyle publication, *La Mer*, proved particularly useful to Papanek upon his arrival in the Golden State. In July 1950, the magazine ran a back-cover advertisement for a portable "Transite Table." Made of asbestos, the "burn-proof, liquor-proof, scratch-resistant," "almost indestructable [*sic*]" coffee table—customizable through its interchangeable surface parts of "painted Transite, glass, panels of abstract painting, rare woods, photomurals, metals, plastics, textiles, etc."—keyed into the relaxed conviviality of Californian living (see figure 3.6). An illustrative sketch explained how the interchangeable surface insets offered a "MINIMUM of 39,062,500 of possible variations in colour, texture & pattern without raising basic MFG.cost."

The name "Transite" derived from the portability and adaptability of the design and the branded cement-asbestos of the same name, from which it was molded. Yet, as Papanek's fellow Viennese Sigmund Freud might have observed, "Transite" also alluded (if only subconsciously) to the "transitory" state of its designer. By 1953, when the indestructible coffee table reappeared in the lifestyle pages of another Californian consumer publication, the widely known and broadly distributed *Sunset: The Magazine for Western Living*, Victor and his family had relocated from the Bay Area, going further south toward Los Angeles. Advertising his services as a "contemporary interior and furniture designer" in Laguna Beach, the *Sunset* inclusion offered a breakthrough not least because the magazine was a staple of the Californian lifestyle scene, renowned for its vividly color-saturated sunshine-filled celebratory depictions of modern living, an "arbiter of western values and taste."[27] The table's promotional tag, "light enough to be picked and carried," pitched to the contemporary regional sensibility of "easy" living devoid of monumental furnishings, and notably preempted Papanek's later *Nomadic Furniture* (1973–1974) design initiative, which further pushed the theme of contemporary living as an ephemeral and transitional phenomenon.[28]

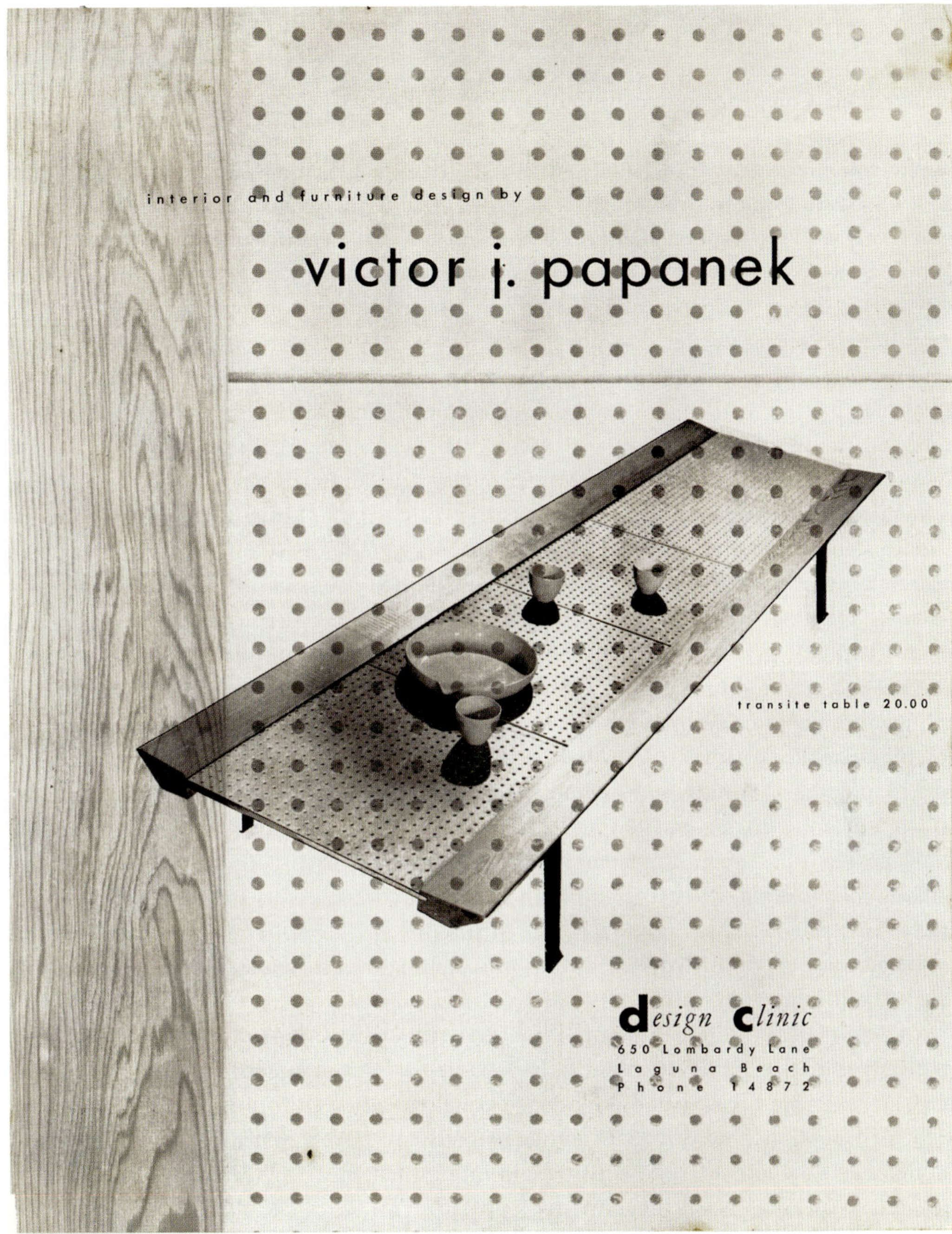

Figure 3.6

La Mer, August 1950, verso cover, "Transite Table 20.00, interior and furniture design by Victor J. Papanek: Design Clinic, 650 Lombardy Lane, Laguna Beach." Courtesy Victor J. Papanek Foundation, University of Applied Arts Vienna

Goodbye, Mr. Chippendale: The Birth of a Design Critic

Despite the minor success of having his funkily designed cigarette-burn and alcohol-spill-proof low table featured in California's premier lifestyle magazine, Papanek's career break-through came in the form of a commissioned article of design criticism rather than furniture or interior design. The August edition of *La Mer* featured his first foray into design criticism with an essay titled "Why Contemporary?" exploring the quandaries of mid-century modern design for the magazine that had advertised his Transite Table in its previous issue.[29] Featured beside an advertisement for a "Surf and Sail" shop in Laguna Beach, the commentary offered fascinating insight into the hazards of furnishing the contemporary middle-class home. Written in a polemical tone that would become his hallmark style, "Why Contemporary?" predates *Design for the Real World* by two decades but is instantly recognizable as its literary predecessor by its breathless, unabated hyperbole. Lamenting the fact that the American public found themselves "confronted by evermore bewildering array of so-called modern furniture, an array as diverse as it is unsound," Papanek offered the consumer a lesson in avoiding "the chintz-and-knack school of inferior desecration [*sic*] the cobbler's bench ('antiques made to order, $19.99 up')" and modern furniture formed from "birch bean-poles and chromium grilles."[30] Stranded between condemnation of nostalgic kitsch and denouncement of European modernism, besides offering recommendations for built-in closets and bookcases, readers must have been perplexed as to what exactly might constitute contemporary design by Papanek's definition: "Le Corbousier's [*sic*] dictum that 'the house is a machine for living in' and all the fascist negation of living which that statement implies is now proven a lie," he growled, "[b]ut neither is there any excuse to drag all the seedy and sentimental pap dating back to grandmother into our contemporary living space."[31] Viewed in the context of Papanek's own hybridized interior and furnishing designs, this acerbic prose belies his confusion and frustration in searching for an American and, more specifically, a Californian design style.

With the Design Clinic transplanted from New York City to San Francisco paired up with his gallery initiative Studio 44, Papanek had wholeheartedly embraced a freshly honed taste for the "California Look," a loosely defined conglomeration of forms, colors, objects, and aesthetic aspirations that designers and consumers collectively and intuitively understood through exposure to endless advertising and interior features. The October 1951 edition of the *Los Angeles Times' Home* magazine took up the challenge of defining the style once and for all. Under the title "What Makes the California Look?," the issue's front cover illustration of abstractedly arranged handcrafted vessels, patio furnishings (from planters to braziers), ethnic objects, "glowing colors" of terracotta and azure blue, and informal furnishings suited to indoor and outdoor living alike offered, at last, something akin to a definitive answer.[32]

Many of these clichéd visions of West Coast living, from rattan-and-cord furnishings to the permeability of outdoor and indoor, celebrated an iconography of "unslackening" rather than representing a practical lived reality. Designs such as the fiberglass and plastic Eames chair depicted in the *Los Angeles Times* feature—shifted nonchalantly between living room and patio, as the piece suggested—would never have withstood exposure to intense Californian sunlight. Likewise, the use of large expanses of plate-glass windows created endless problems for protecting the interior of West Coast homes (and their inhabitants) from over-exposure to sunlight and heat.[33] Nevertheless, the emergence of specific genres of room settings and interior configuration, from the hearthside conversation ensemble to low-slung seating arrangements, undeniably transformed the bodily, as much as the visual, experience of the modern lived interior of mid-century California.

In the early 1950s, Papanek experimented enthusiastically with his own version of the genre, making a series of exuberant watercolor renderings embracing a flamboyant aesthetic far removed from the pared-down socially responsible design for which he gained worldwide renown by the 1960s and early 1970s. The feature hearth, sunken den, coffee table or cocktail table, mobile sculpture, ethnic ornament, animal-skin area rug, and "floating" lamp emerged as his tropes of informal modern living. One such interior design, "Made Exclusively for Wini (Laguna Beach) by Vic Papanek," has as its focal point a low rectangular table surfaced with cherry wood and rush matting squares, with a base of splayed black wrought-iron legs (see figure 3.7). Papanek's design for "Wini" diligently checks all of the boxes required of mid-century West Coast living: an overflowing metal fruit bowl perches on spindly legs, a Japanese rice bowl is poised at the table's edge, a twisted piece of driftwood entangled with vegetation is illuminated by a floating Calder-style sculptural wall lamp, the table and its contents accessorized by an animal-skin area rug beneath, and a vivid red "hearthside stool" at its side. The "client" for his California contemporary flourish was none other than his new bride, Winifred.

To some extent, then, the unshackled environs of the West Coast exerted a pressure on designers to push the boundaries of convention, turning design into an predominantly expressive rather than functional practice. In this vein, Papanek supplemented his speculative design commissions during the early 1950s with work as a part-time instructor in commercial art: "*Creative Design Workshops* are led by Victor J. Papanek (author, editor, designer Knoll Associates, T. H. Robsjohn-Gibbings, Helena Rubenstein, Container Corp., etc.)," read the Art League of California's promotional brochure.[34] The pamphlet hints at the emergence of Papanek's less conventional "creative" approach to industrial design, as well as his propensity for self-aggrandizement, overemphasizing as it does his prestigious but tenuous connections.

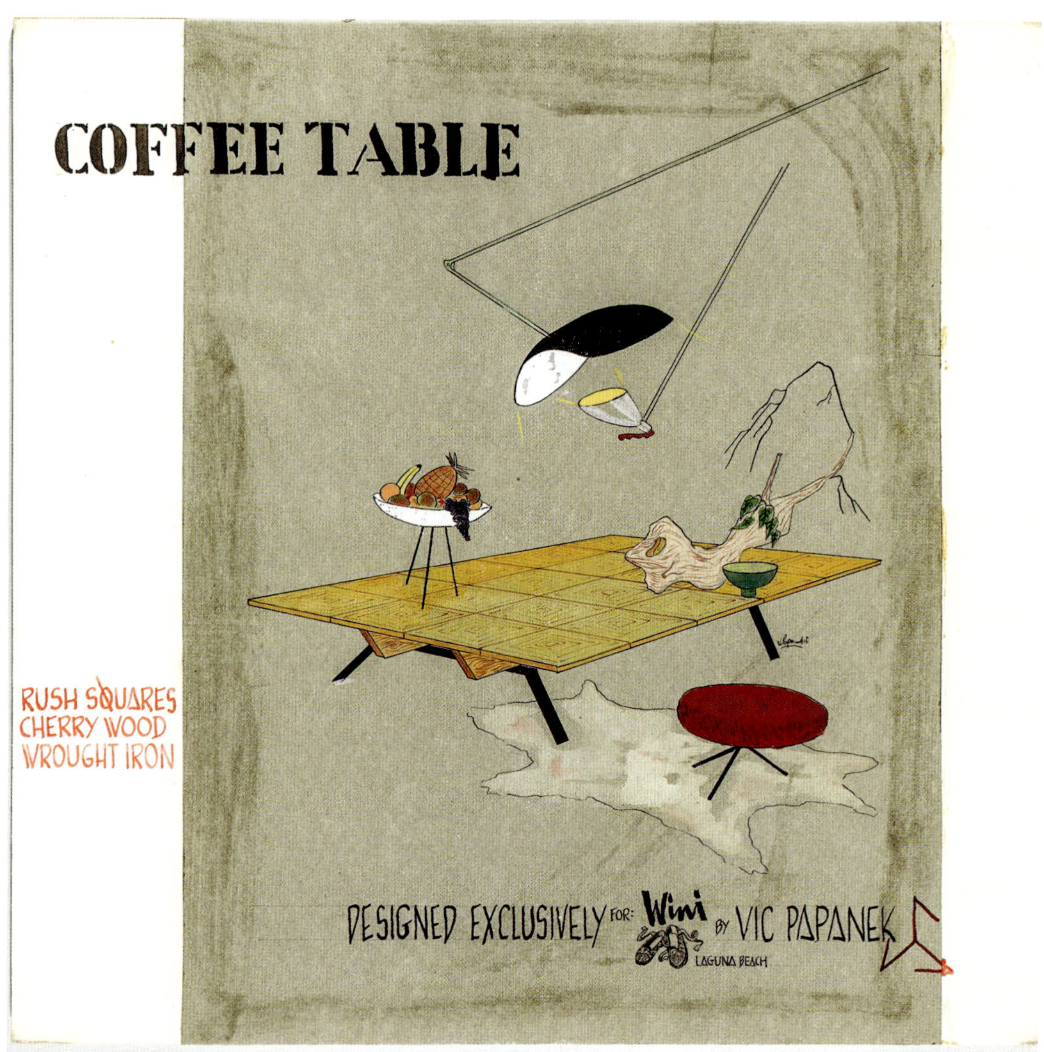

Figure 3.7
"Coffee Table—Designed Exclusively for Wini, by Vic Papanek," c. 1952–1953. © Victor J. Papanek Foundation, University of Applied Arts Vienna

In this respect, Robsjohn-Gibbings (1905–1976), a British-born US furniture designer, is a notable addition to Papanek's network of real and imagined connections. Creator of the biomorphic Mesa table (1951), Robsjohn-Gibbings garnered a reputation as an arbiter of taste from the early 1940s to mid-1950s, applying his classic European education to the creation of high-end restrained contemporary pieces, which eschewed the harsh-angled modernism of the Bauhaus and its followers. The *La Mer* think piece "Why Contemporary" that inadvertently launched Papanek's career as a design critic, alluded to Robsjohn-Gibbings's "classical" contemporary style when it condemned "modernistic," "chinese [*sic*] modern," "pacific moderne" interior furnishings, stating instead that "True contemporary design is truly traditional!"[35]

As with Knoll Associates, Helena Rubenstein, and the Container Corp., there is no evidence to suggest Victor Papanek had at this point actually been commissioned by the British American émigré Robsjohn-Gibbings (himself commissioned by Rubenstein). However, an undated "scrapbook" portfolio of one of his earliest designs for a biomorphic "3/4 inch glass-top table" states clearly, "for T. H. Robsjohn-Gibbings." The design, set amid a luridly colored interior, bears more than a passing resemblance to the Noguchi Table produced by Herman Miller from 1947 onward, which also involved a biomorphic shape, albeit with unpainted wooden base and plate glass top.[36] There is every possibility, as with his designs "for" Paul Bry, that Papanek's rendering was speculative, rather than evidence of a successfully secured commission with, or for, Robsjohn-Gibbings.

Under the name of their main designer Robsjohn-Gibbings, the United States Widdicomb Furniture Company produced a heavy biomorphic glass coffee table identical to that featured in Papanek's portfolio and renderings in the early 1950s; the only slight difference in design detail was that Papanek's version had a painted base rather than a natural wood one. If interpreted generously, Papanek's design portfolio may have acted as a source file, rather than an accurate record of his commissions. Conversely, he may have been attempting to pass off Robsjohn-Gibbings's biomorphic table as his own, merely illustrating Gibbings's work with minor adaptations suited to a colorful Californian interior.

Whether Robsjohn-Gibbings's Mesa table was semiplagiarized by Papanek or merely inspirational to him, it is surely indisputable that Robsjohn-Gibbings's best-selling book *Goodbye, Mr. Chippendale* (1944), a treatise on the societal ills borne of "bad" reproduction design and a design industry "rotten to the core," was the direct inspiration behind Papanek's "Why Contemporary?" article of 1950 and would form the basis of his tone as a design critic. Written in a brusque and sarcastic way, the lighthearted, jauntily illustrated *Goodbye, Mr. Chippendale* is quite possibly the nearest to an ur-source for the politically provocative *Design for the Real World* as one could find. One reviewer of Robsjohn-Gibbings's book compared the conspiratorial tone to that of the lead character and architect Howard Roark in Ayn Rand's *The*

Fountainhead (1943), making Papanek's own adoption of Roark's renegade persona as a student at Cooper Union all the more intriguing.[37] With its critique of an unwitting public and an irresponsible and negligent design industry peddling outmoded impractical objects unfit for contemporary living, *Goodbye, Mr. Chippendale* targeted a postwar audience eager to embrace the contemporary as a new expression of "graceful living and ease."[38] Significantly, Robsjohn-Gibbings (like other figures in his network—Loewy, Nelson, and Rudofsky) offered Papanek an exemplary model of the successful "designer-author," and one who also worked coast to coast.

If the stylistic tastes advocated by Robsjohn-Gibbings proved inspirational for Papanek's emerging brand of design criticism, so too did *Good Taste Costs No More* (1951)—penned by none other than Richard Gump, purveyor of exotica and owner of the famed department store, and released the year of Papanek's arrival in San Francisco.[39] The best-selling book provided a catalogue of selected products from the store, interspersed with sage advice on achieving taste through the canny selection and juxtaposition of items rather than of costly inappropriate interior choices. Together with *Goodbye, Mr. Chippendale,* these publications helped set the tone of the writing on popular design that Papanek fostered in his campaign for socially responsible design several decades on.

Likewise, Papanek's tutoring at the Art League of California, San Francisco, launched his academic career, framing an expanded transdicisplinary approach to "creative design," rather than the brand of "industrial design as styling" promoted by his forebears. The four-semester workshops Papanek delivered there between 1951 and 1953 appear startlingly innovative even by contemporary standards, and reveal the beginning of his integrative approach to media in design pedagogy—courses involving the "investigation of advanced media: light optics, film (various hand-drawn and animated films), perception (the Princeton-Dartmouth University experiments by Ames and Ittselson); space (stage sets for an avant-garde ballet); empirical work regarding the psychological, physiological, anthropological, sociological and medical aspects of design."[40] Practical design projects centered on product development, furniture, film animation for television, display and exhibition work, but recounting the themes of his slide lectures and seminars at the Art League—a highly visual teaching methodology aimed at creating a historical, critical, and social science background to his students' work—becomes apparent as the prospectus reveals: "Among the many slide lectures, films, seminars etc. were: comic books, Semantics, Jungian Archetypes, the Kabuki and Noh Plays of Japan, Balinese and Javanese dance, Clothing, City Planning, Surrealism, Dada and the Art of the Insane, Ikebana, Minoan-Cretan Symbolism, The German Bauhaus Movement, Bionics, etc."

An avid reader of San Francisco's influential *City Lights Journal* during this period, which by the 1960s went on to feature leading writers and cultural commentators including famed beat poet Allen Ginsberg, Papanek claimed to have cut his teeth as a critic there in the early

1950s. Despite persistent references in his résumé to authored articles on comic books and popular culture having been published in the ground-breaking journal, in actual fact Papanek's first crack at art criticism appeared in 1952, in the pages of the San Francisco *Chronicle*, the daily newspaper in which Papanek had announced his daughter Nicolette's birth the previous year.[41] He published two further reviews dealing with modern architecture publications that summer, one of which berated the author for having omitted Frank Lloyd Wright's Taliesin, Spring Green, and V. C. Morris building from her survey, while the other review praised the innovation of the Bauhaus project and fellow émigrés Herbert Bayer and Walter and Ise Gropius.[42]

Legitimate and competent entrées into art and design criticism, the *Chronicle* reviews built on Papanek's earlier lifestyle piece in *La Mer* magazine. Yet the meticulousness with which Papanek tweaked and manipulated the facts on his résumé (associating himself with the acclaimed *City Lights Journal*, for example) to maximize his accomplishments and professional connections often to the point of deception, reveals the extent of his unadulterated ambition and desperation to carve out a living for himself and his family. When viewed alongside the systematic obfuscation of his personal life history, his compulsion to engineer an enhanced version of himself, personally and professionally, reveals the extent to which he viewed himself as an outsider desperately clawing his way into the cultural design milieu of America. Ultimately his failure, despite his best efforts, to make an impact or commercial success in San Francisco spurred him onward to explore the promise and potential of the Los Angeles area.

Moving South, Heading North: From Los Angeles to Toronto

In the early 1950s Papanek had just managed to keep afloat both the Design Clinic and the gallery enterprise Studio 44, but by late 1953, admitting defeat, he abandoned the Studio 44 retail gallery outlet on Fillmore, then quitting San Francisco altogether, moved with his family to start a fresh life further south along the coast in Laguna Beach. By 1954, his new home was registered as being in the Altadena district, north of Pasadena.[43] Occasionally tutoring at the Chouinard Art Institute (which by 1961 became the basis for the formation of CalArts at which he would eventually take on a senior design role in 1970), he also embarked during this period in Southern California on an intriguing design for the entrance lobby of the Gallery of Living Arts in Corona Del Mar, Newport Beach (see figure 3.8).

Prior to leaving New York, Papanek had encountered the Museum of Living Art at New York University (1927–1943), the small-scale enterprise of A. E. Gallatin that in its promotion of contemporary arts predated the Museum of Modern Art, New York, the Whitney Museum of American Art, and the Solomon R. Guggenheim Museum of Non-Objective Art.[44] Into the

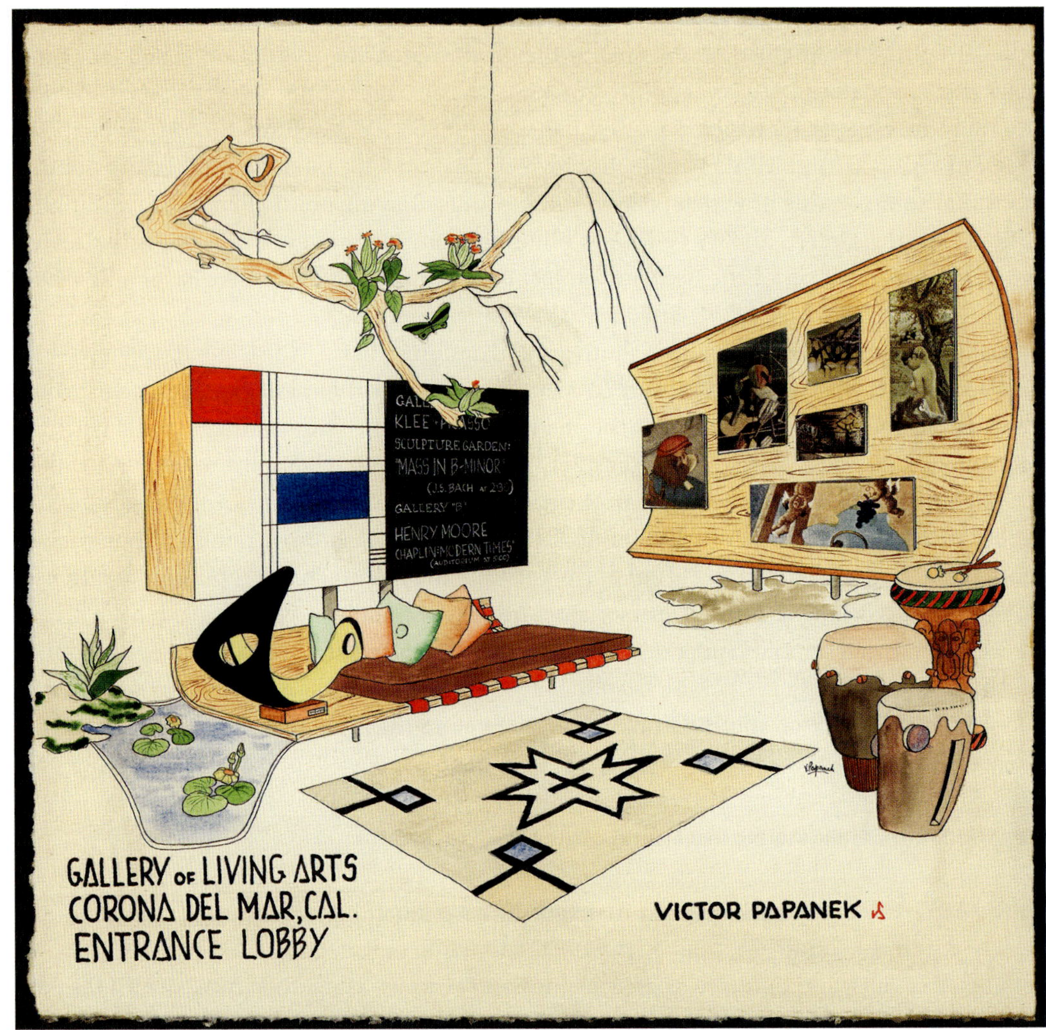

Figure 3.8
"Gallery of Living Arts in Corona Del Mar, Entrance Lobby," interior design by Victor Papanek, c. 1950–1952. © Victor J. Papanek Foundation, University of Applied Arts Vienna

postwar period New York continued to be the hub of avant-garde art in the United States, with California struggling to gain equivalent stature.

Modern art occupied a precarious position in the climate of Cold War politics and McCarthyism, with the House Un-American Activities Committee and the Los Angeles County Council, specifically, condemning contemporary art as Communist propaganda. Art historians have noted how "the Californian art scene was plagued by a shortage of exhibition space and a lack of interest in contemporary art, and local art alike."[45] Along with his ventures into art and design criticism, Papanek's design for a gallery space in the seemingly incongruous setting of the "neo-Mediterranean Seaside community" Corona Del Mar, in Newport Beach, offers insight into his broader ambitions to belong to the arts establishment in an expanding region in 1950s, and perhaps a politically motivated attempt to transfer his progressive New York arts-and-design scene knowledge to the West Coast and its developing counterculture.[46]

The Laguna Beach Museum of Art, which in 1951 doubled in gallery size through donor funding, offered a comparable institutional model to the Corona Del Mar proposal and may indeed have inspired the project. The commission was most likely connected to Papanek's friendship with a character named Briffault, who also relocated from New York to California in the early 1950s. Papanek designed an interior for a hunting lodge for an "A. E. Briffault" shortly after his design for the Gallery of Living Arts in Corona Del Mar; the patron may have been a wealthy figure keen to support the arts in the region (see figure 3.9).[47]

While the origins of the Gallery of Living Arts project remain unclear, ultimately it was unrealized. With no other substantial commissions in the offing, at the age of thirty-one, Papanek still failed to muster a consistent means of financial support for his family. At this point Helene, his elderly mother, continued to provide an essential support system for her son. She followed him loyally from location to location, as he looked for a secure setting for a design career, until her death in 1967. She served up homemade Viennese comfort food, provided childcare for his family, and remained steadfastly supportive through the ups and downs of his often-complex personal life. His abiding confidante in all of his relationships and friendships, she alone surely knew of the details of his first marriages to young Jewish women in New York, and may even have lent a hand in arranging these liaisons to ensure his citizenship in the United States. Between marriages and girlfriends, she remained his close companion until her death, emotionally supporting him and financially supplementing his livelihood, taking paid domestic work as a nanny to help support her son as he forged a life and career in design.[48]

Despite the move from the Bay Area further south to be closer to the Los Angeles design scene, Papanek's opportunities soon dwindled, and he realized the possibilities of forging a design profile in California were fast waning. Perhaps surprisingly, the single most valuable

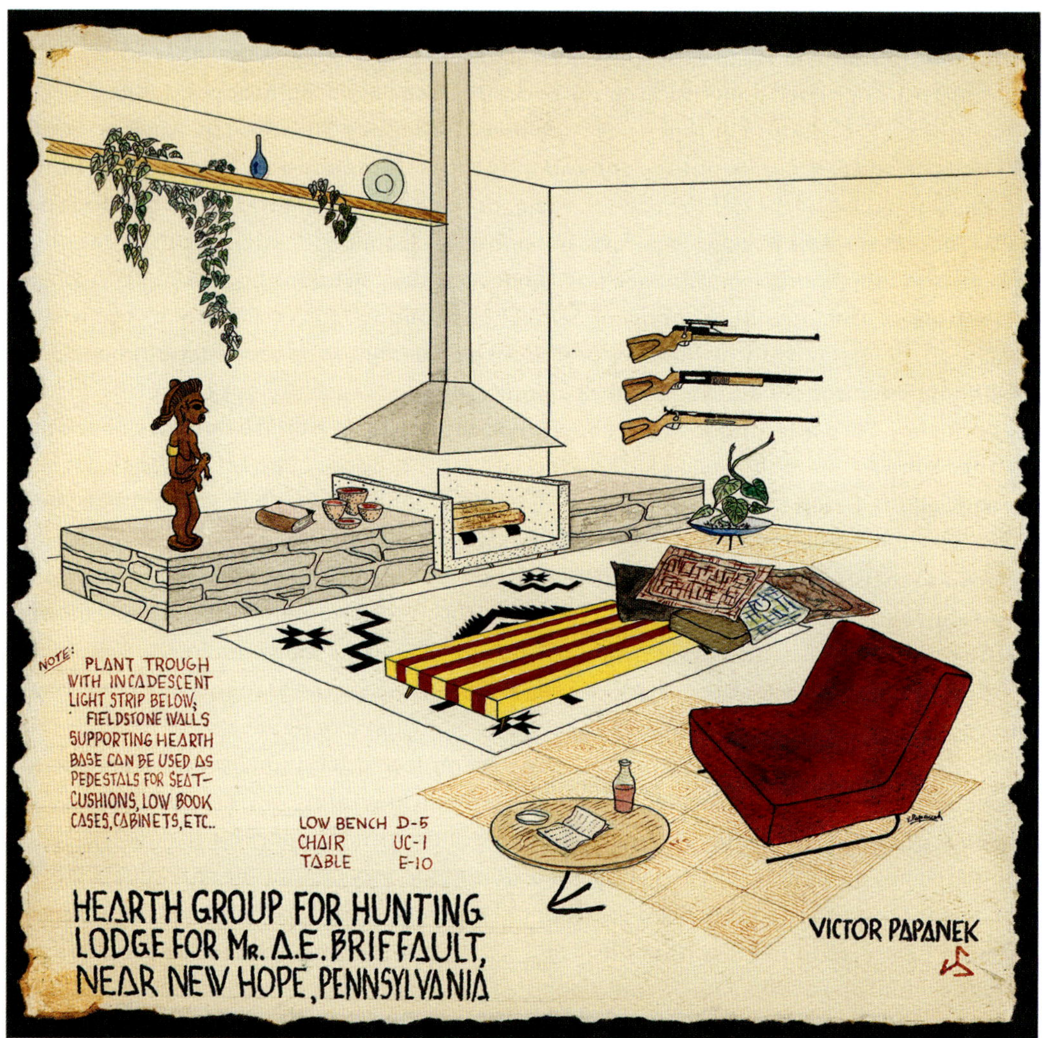

Figure 3.9
"Hearth Group for Hunting Lodge for Mr. A. E. Briffault, Near New Hope, Pennsylvania," interior design by Victor Papanek, c. 1950–1951. © Victor J. Papanek Foundation, University of Applied Arts Vienna

asset gained from his stint on the golden coast turned out to be his part-time teaching at the Art League of California, San Francisco. The Creative Design Workshops he initiated there, combining art, humanities, film, and experiment, struck a chord with the direction that postwar design training more generally was heading in the Cold War period. Inconsequential as it might have first seemed, this part-time tutoring position set him up as a transdiciplinary designer and, unbeknownst to Papanek at the time, helped secure his future as a vanguard design pedagogue. Where the upbeat, carefree optimism of the Golden State had failed to deliver, the snow-filled Canadian state of Ontario unexpectedly succeeded, introducing him to the groundbreaking work of anthropologist Edmund Snow Carpenter and philosopher and media theorist Marshall McLuhan, as well as their emerging intellectual circle theorizing the societal influence of new media technologies. Described by Carpenter as a bleak and uncosmopolitan city best compared to the relentless drabness of 1950s Communist Moscow, Toronto became Victor Papanek's new home in the mid-1950s. He secured his first full-time academic teaching post at Ontario College of Art, launching his shift from furniture designer to industrial designer and theorist.[49] For, the mid-century move northward to Canada marked the true beginning of his trajectory to becoming a pioneering experimental designer and the twentieth century's most prominent design critic and activist.

4 The Comprehensive Designer: Toward a New Politics of Design

From the mid-1950s onward, US anxiety surrounding the Soviet Union's increasingly innovative technological strategy led to a major North American governmental initiative to "humanize" design and engineering across colleges and universities. This was a concerted effort to unite art and humanities in the formulation of an all-powerful science and technology paradigm. From the height of the 1950s Red Scare through the late 1960s disruptive Cold War politics, the definition and role of art took on a new urgency. According to historian Matthew Wisnioski, during this period art constituted "a range of shifting meanings among scientists and engineers" paving the way for the rise of an interdisciplinary field dubbed the "technological arts."[1] In a similar vein, the appetite for freshly configured transdisciplinary understanding of design and engineering gained momentum, and with it changing understandings of the Cold War potential of industrial design, a field formerly viewed as a technical branch of engineering or the mere practice of aesthetic styling under the rubric of the "commercial arts."

That prominent universities, including the archetypal Cold War institutional edifices of Harvard and Stanford and the Massachusetts Institute of Technology (MIT), were bolstered and shaped by corporate contracts and federal government sponsorship is an established fact of Cold War history; lesser known is the broader impact of these policies on art and design institutions.[2] In early 1954, boosted by the broad-scale shift toward technological arts and the phenomenon of technology's "humanization," the Ontario College of Art (OCA; known today as Ontario College of Art and Design University), Toronto, set about enlarging its faculty and facilities to keep pace with an increasing postwar demand for industrial design training. Impressed by the unconventional bent of Victor Papanek's approach, following his submission of a speculative résumé and the internal recommendation within the art faculty, OCA offered him a full-time position as an experimental design instructor, commencing that same academic year. The art college had operated a successful small-scale industrial design department since 1950, the only one of its kind in Canada, which by 1952 was training a maximum of a dozen students annually. Their vision, and the pretext of Papanek's appointment, was to respond to a growing postwar demand for product designers by enhancing the "maturity and diversification of its courses and instructors" and to "include courses in advanced technical theory, economics, business, and industrial administration acceptable to the various universities and industrial design organizations in the country."[3]

A report made to the National Industrial Design Committee by Canada's five schools of architecture, concerning the growth potential of industrial design at the University of Toronto, observed, "There seem[s] to be no reason why [OCA] could not be looked upon the same way as the Chicago Institute of Design, now incorporated in the Illinois Institute of Technology." With this impetus, the leading faculty member of Ontario College of Art, a painter named Lawrence A. C. Panton (1894–1954), recommended Papanek as an ideal addition to the faculty. An advocate of the experimental design approach that Papanek had showcased in his tutoring at the Art League of California, Panton died prior to Papanek taking up his position in Toronto, and therefore did not live to see his vision for a new interdisciplinary design department come to fruition.[4] Panton's demise may also have influenced the position shifting from an experimental to the more generalist teaching approach Papanek eventually took up. His résumé for this period covered industrial design, design psychology, product diversification, design economy, transportation design, market analysis techniques, and, by 1956, advertising design and packaging design. Nevertheless—a testament to the innovative design ambitions of OCA—to prepare Papanek for the groundbreaking nature of his first full-time teaching position, in 1954 the college registered him in a highly innovative MIT summer seminar program pioneering an industrial design approach that combined "the technical skills of engineering with [a] more comprehensive human-centered approach."[5]

Creative Engineering: The Shifting Cold War Politics of Design

As America's premier postwar technological institution, MIT stood at the forefront of an increasingly politicized US higher education agenda that not only promoted the coming together of the arts and technology, but also openly instrumentalized humanistic education as a means of culturally reinforcing "America's claims to moral superiority."[6] As much as the humanistic education policy seemed an effective means of bolstering innovative thinking, its principal aim was to produce critically attuned, well-rounded US scientists to be pitted against what were perceived as their passive, brainwashed technical counterparts in the USSR. American designers and technologists were envisaged as the hypercharged ammunition of a free-thinking democracy: borne of a critical engagement with a technological future, ideologically (as much as practically) these individuals stood in opposition to the disengaged robot-like operatives manufactured by an unquestioning, Soviet totalitarian state. For winning the "hearts and minds [of college students] became a national security imperative," historian Peter Justin Kizilos-Clift argues, "particularly important in the case of science and engineering students, upon whose knowledge and skills the nation relied to maintain political, economic, and military superiority over the Soviet Union."[7] Therefore, policy makers persuaded them by shaping their "allegiance and subjectivity" that they were involved in a collective cause that was both "socially and morally legitimate."[8]

MIT's highly unconventional Creative Engineering Laboratory (CEL; with its spin-off summer seminar program) perfectly encapsulated the humanistic education targeted at reiterating the free-thinking, independent-minded, individualistic model on which America's broader claims to moral superiority relied. The principle figure of this movement was John E. Arnold, MIT Professor in Engineering Mechanics, whose "teaching methods and philosophy were designed to help students overcome cultural inertia and defy the gravitational forces that pulled hearts and minds into conventional orbits."[9] Although markedly quirky, the creative engineering genre coined by Arnold offered a kind of free-association methodology. Having encountered the creative engineering mentor firsthand at MIT's summer seminar programs prior to his commencement as tutor at OCA in 1954, and then a year later in his capacity as a fully fledged industrial design tutor, Papanek incorporated Arnold's methodology as a principal facet of his design pedagogic approach. Notably, Arnold coupled his conventional teaching with broadcasts on design thinking for MIT's public educational TV channel, joining the prominent industrial designer George Nelson, whose sardonic short film *How to Kill People* (CBS, 1960) later tackled design's integral role in fueling the Cold War arms race. Arnold and Nelson, along with other media broadcasters in the field of culture and design, inspired Papanek later in the decade to use television and radio shows as a mainstay of his design proselytizing.

In fact, attending the Creative Engineering Laboratory summer seminar at MIT in the mid-1950s would set Papanek on the single most significant track of his design career—a track that proved as profoundly instrumental in shaping his design approach as his 1940s connection to the renowned architect Frank Lloyd Wright, or the critical politics of racial and social inclusion espoused by social psychiatrist and public intellectual Fredric Wertham. Opened up to interested parties outside the institution through the summer seminar, the "one-man" MIT initiative led by Arnold offered the freshly appointed industrial design instructor the opportunity to mingle with a diverse set of designers, engineers, and military and consumer manufacturers, in a series of formal lectures and informal cocktail parties. Here Papanek was deeply immersed in the cutting-edge research and design thinking of a postwar university system, reinvented to address the increasing demands of postwar consumer culture and the direct objectives of the US military-industrial complex. During these two-week-long events, he met and befriended the instigator of creative engineering himself, design engineer turned educational philosopher Arnold, whose extraordinary combination of philosophical and futuristic design pedagogy attracted loyal advocates ranging from Buckminster Fuller to Henry Dreyfuss (both of whom lectured in the CEL seminar programs). Papanek even took part in a number of immersive experiments and was even subjected to a Minnesota Multiphasic Personality Inventory (MMPI) as part of his studies; the MMPI was one of the earliest psychometric tests, first devised in 1943 by psychologist Strake R. Hathaway and neurologist J. C. McKinley, faculty members of the University of Minnesota. The test offered intriguing results, with Papanek's reading for "hypomania"—which signaled a tendency toward unstoppable, excitable flights of ideas, irritability, grandiosity, and egocentricity—peaking well above the normal range.

Critically, Arnold founded and directed CEL from 1953 to 1957 before consolidating his ideas in the volume *Creative Engineering: Promoting Innovation by Thinking Differently* (1959) at the height of Cold War politicking over the role of industrial design in boosting the American economy.[10] His annual creative engineering seminars had arisen from a broader mid-century expansion of the application of product design to the fields of science, commerce, and planning. As the abundance of services and products swelled, according to historian Arthur Pulos, design, as a discipline, was inflated to address the immensity of the postwar consumer project: "Businessmen had become more and more distant from the product and services to which they were responsible," and this, according to Pulos, made them more receptive to a range of specialists offering their services to relieve the burden.[11] The increased prominence of industrial design in the popular media, and its rising profile among professionals outside the immediate areas of art and design, led to a fresh breed of experts who "believed that intuition, imagination, and talents could be rationalized, measured, and dispensed to the uncreative."[12] Arnold's move away from conventional mechanical engineering formed part and parcel of this postwar design boom, the reiteration of design practice, and design's cultural and political role, more generally. His classes at MIT were conducted in seminar format, focusing on design aesthetics, the problems of mass production, marketing, and consumer analysis. Sessions revolved around product-design case study assignments, judged by external expert juries briefed to scrutinize results based on originality, quality of presentation, and what Arnold described as the "fluency level" of the engineering approach applied to a project. Students produced renderings, full-size models, and engineering drawings for these juried presentations. One typical first-year project at MIT had included a "Dental Mobilmeter," an electronic device to measure teeth mobility. In contrast with a standard engineering assignment, the students were expected to look beyond the electric circuitry of the object, to the broader aesthetic and cultural questions of how the components might be chosen in relation to the packaging of the product, and to its user's expectations.

The success of this teaching method was then rolled out to a broader professional audience in the guise of the CEL summer seminars, which incorporated the case study assignments of Arnold's students as an illustrative example of the application of human creativity to problem solving and borrowed the format of design school "crits," the individual projects subjected to public, collective critique as a part of a reflective, creative process. The group dynamics at the summer seminars must have been intriguing, to say the least, given the potent mix of attendees, the majority of whom hailed from the upper echelons and middle ranks of the US military, armament research, nuclear research, automotive industry (General Motors), consumer corporations (Whirlpool, Corning glassware), chemical industries (Dow Chemical Company, DuPont), and telecommunications companies, along with a smattering of progressive product design educators and freelance product designers. Aside from the exceptional "accompanying wife" (Victor's partner Winifred accompanied him in 1955), the seminars were resolutely male, and even the opening speeches given by Arnold as chief organizer read as if addressing a rather smug and exclusive Gentleman's Club—which they essentially were. According to the "Roster of Registrants" of 1955, a lone female mathematician named Miss Aida Kalish, from the Polytechnic Institute of Brooklyn, courageously enrolled herself into the all-male proceedings. Although there was no formal policy on the exclusion of attendees on the grounds of gender or ethnicity, the events were overtly targeted toward a homogenous group of leaders within the military-industrial complex (defense, manufacturing, and consumer representatives).[13]

In one such opening speech, Arnold set out specifically to define the creative process for a potentially skeptical audience. Creativity, a highly regarded facet of design process, far from being intuitive, he argued, is measurable, and as such can be engineered:

> Gentlemen, I think that it is appropriate that we start out the two-week seminar with a Short history lesson. . . . What is a creative problem? We believe the creative process is one that involves mental activity in which one combines and recombines past experience with present experience in such a fashion that you end up with new patterns and configurations that better satisfy some need of man. These needs may be an expressed need or an implied need. In every case however you must end up with some tangible evidence of a better solution to that need . . . I probably don't have to emphasize the word tangible to you men, but unfortunately many of my students seem to think that all a creative person has to do is to get an idea and the rest takes care of itself. They quickly learn, however, just as you men have, there is a great deal of work between conception of an idea and its final birth in a new product.[14]

As product design shifted further away from the art school notion of creativity, to be repackaged as a hybridized offshoot of "design thinking" geared to servicing wartime technological and security agendas, its masculinization steadily accelerated. Mid-century science and engineering, notorious for its instrumental and cultural protectionism, as one historian observes, utilized the designation of "scientific authority as the province of university-educated men" to precipitate and legitimate women's exclusion.[15] The structural elements of exclusion—namely, educational access, ethnicity, and social class—that ensured only the tiniest minority of outsiders gained entry, were further perpetuated through the professional culture of science and engineering, expanding to include "styles of dress, modes of interaction, hierarchies of practices and values [which] visibly marked women as outsiders."[16]

The contemporary figure of the omnipotent, technocratic, polymath male designer, besuited and bespectacled, a style Papanek himself enthusiastically adopted from the mid-1950s onward, inferred a kind of visionary genius. Whether heading a corporate branding initiative, or proselytizing a comprehensive planetary resources system theory, this uniform bolstered the profession as an expression of strictly white-male authority. Although situated at polar ends of the design theoretical spectrum, Buckminster Fuller and Henry Dreyfuss, high-profile designers and self-professed visionaries of the mid-twentieth century, epitomized this profile. They shared the privilege of occupying the upper echelons of US design discourse, both collaborating broadly with federal agencies and corporations and promulgating global ambitions for their ideas. Their projected identities were perfectly in tune with the paradigm of the Cold War application of design, technology, and engineering to the military-industrial complex.

In 1955, Fuller and Dreyfuss (who that same year authored the highly influential manifesto *Designing for People*, advocating the social and aesthetic benefits of ergonomics) delivered lectures at Arnold's creative engineering seminar, presenting under the titles "Comprehensive Design" and "Creative Collaborators," respectively, with Papanek avidly taking notes in the audience. Although *Designing for People* reemerged in the late twentieth century as part of the canon of inclusive design, and undoubtedly influenced Papanek's early ideas around the social responsibility of the designer, the book is better understood as promulgating the Cold War paradigm as Dreyfuss himself made plain: "The goal in [military projects] is a contribution to morale, the intangible force that impels soldiers to have confidence and pride in their weapons and therefore in themselves and that, in the long pull, wins battles and wars."[17] As historian David Serlin observes, the easy application of a commercial design practice to the objectives of the military-industrial complex meant that Dreyfuss's shift from prosthesis design for wartime veterans (with consultation work for the National Research Council's Advanced Council on Artificial Limbs) to the designing of Howitzer army rifles, "missile launchers as well as the ergonomic interiors of the M46 and M95 tanks for the army" exemplified "the collaborative symbiosis between government and industry that so marks the Cold War period."[18] Rather than "designing for people," under the auspices of ergonomics Dreyfuss streamlined and aesthetized the ravages of war.

The Design Mavericks: Victor Papanek Meets Buckminster Fuller

Exposure to the creative engineering approach, fueled by his ties to the expanding ambitions of the design faculty at OCA, seems to be Papanek's breakthrough moment as a contemporary designer and design thinker. It proved doubly significant as it marked the beginning of an enduring collegial relationship with the nation's renowned maverick designer, R. Buckminster Fuller, a relationship that would have a momentous impact on Papanek's global profile, his design approach, and his life's work.

Papanek's first encounter with the design visionary had been from an admiring distance as a member of the audience at a lecture Fuller delivered at Syracuse University, New York, in the spring of 1955. But it was one of the maverick designer's legendary four-hour lectures, presented on the topic of comprehensive design at the MIT CEL seminar during his first year as tutor of industrial design at OCA, that irrevocably fired up Papanek's imagination. It is entirely feasible that the two men met informally over cocktails, or during the networking banquet sessions. Whatever their preexisting connection, immediately following the CEL summer seminar Papanek—in a typically audacious manner—penned Fuller a letter introducing himself as a designer with a shared mission to transform the twentieth-century material world, brazenly volunteering to assist Fuller's research into "Dymaxion formulations and Geodesic structures." On August 3, 1955, Papanek received a much-anticipated reply from Fuller via the great man's assistant, John Dixon.[19] Containing copies of newspaper cuttings pertaining to Fuller's geodesic initiatives designed "to house the listening devices of the Air Force's Distant Early Warning Line, the 3,000-mile strip of radio installations along the northern edge of Canada and Alaska," the missive reassured Papanek his interest in Fuller's work had been duly noted: "Should it occur that [Fuller] would need the assistance of an individual with your training and background he will communicate with you to explain what aid is required." Whether or not Papanek's letter had ever actually passed under the personal gaze of his design hero, Dixon offered further reassurance of the Fuller's recognition of Papanek's admiration: "Mr. Fuller is pleased that you found yourself in concord with his Dymaxion formulations and Geodesic structures and principles." The message also contained a reprint of the Alfred Korzybski Memorial Lecture that Fuller had delivered that year, which inspired Papanek (who was familiar with Korzybski's work in semantics from the late 1940s onward) to attend a workshop at the Institute of General Semantics (IGS) a year later, in 1956. Fuller taught intermittently at the IGS in Connecticut, following the death in 1950 of its founder, whose dictum "the map is not the territory" summarized a theoretical approach to human knowledge as an abstraction.[20] Along with Fuller, Korzybski offered Papanek further models of the complex relationship between technological innovation and the human condition.

A Polish American émigré trained as an engineer, Korzybski left his native country after experiencing firsthand the violence of World War I. Initially influenced by the early US technocracy movement, he went on to pursue the question "What makes humans human?" and championed the application of engineering and mathematical methodologies to the human condition and its evaluation. Korzybski led a new field of empirical science he termed "general semantics," which was promulgated through his early works *Manhood and Humanity: The Science and Art of Human Engineering* (1921) and *Science and Sanity: An Introduction to Non-Aristotelian Systems and General Semantics* (1933).

Papanek's attendance of the workshop in general semantics, most likely inspired by Korzybski's link with Fuller and their pursuits of comprehensive knowledge and universal design principles, respectively, was just one of numerous mildly eccentric educative escapades the budding designer sampled in the early stages of his career. Papanek's résumé offers a taster of the sheer breadth of pedagogical and intellectual pursuits that he partook of during the late 1950s: "Experimental Psychology, Human Engineering Factors, Perception, Anthropometry, Gestalt Theory, Cultural Morphology, Anthropology, Oriental Art, Neurophysiology, etc."[21] Despite this unbridled intellectual curiosity, his admiration for Fuller's ideas began to dominate his thinking. Where in the 1940s he had constructed Frank Lloyd Wright as the pivotal reference in his growing (real and imagined) network, the renaissance figure of Buckminster Fuller served as a role model extraordinaire throughout the late 1950s and early 1960s.

Astonishingly, following their initial meeting at MIT in the mid-1950s, Fuller and Papanek, construed as North America's most forward-thinking and controversial designers, went on to share platforms together, with equal billing, at high-profile international design events, from Helsinki, Finland, to Aspen, Colorado. As well as facilitating this career-changing connection with Fuller, the MIT seminars of 1954 and 1955 offered Papanek an expanded model of design thinking and pedagogy, one he would utilize in perpetuity through the borrowing of Buckminster Fuller's notion of "the Comprehensive Designer" and Arnold's creative engineering methods.

Their collegial relationship culminated in the ultimate public endorsement, when Fuller contributed a foreword to Papanek's first book, *Design for the Real World: Human Ecology and Social Change*. Fuller's words perfectly encapsulate the momentous shift that had taken place in design since their first meeting during the Cold War technological drive, in the mid-1950s:

> At M.I.T. there are buildings full of rooms, and rooms full of yesterday's top priority machinery that is now utterly obsolete. They have a vast graveyard of technology. The students don't want to take classes in mechanical engineering anymore because they have heard that what they learn is going to be obsolete before they graduate, its evolutionary events cover all phases of technology and physical sciences. Victor Papanek's book conducts a mass funeral service for a whole segment of now obsolete professionals.[22]

Following his encounters with Fuller, Papanek began experimenting with botanical and biological forms, searching for a model of universal design structure that might match that of his new role model. These experiments, which began at the height of Cold War design experimentation, continued into his later career fluctuating dichotomously between military and humanitarian application. One particular biodynamic experiment, showcased in an annual exhibition of OCA student work at the Art Gallery of Toronto in 1955, caught the bemused attention of the local press. "Project Mapleseed" revolved around workshop experiments whereby "Papanek gave every one of his students in industrial design a mapleseed, made him or her toss it in the air, watch its spiral flight to the classroom floor, and the assignment was to come up with a practical design."[23] The winning design and model, a "'sustained flair' for aircraft in distress," would be taken to the International Design Conference in Aspen, Colorado, "thence to Harvard school of design." By 1970, the Mapleseed Project and similar biological experiments he conducted at Purdue University in the 1960s, made their way into Papanek's *Design for the Real World* as an essential part of the thinking designer's repertoire:

> The very simplest growth characteristics of almost any plants may provide the solution to imaginative design problems. Thus the growth of an ordinary green pea may become instructive. If the pea is permitted to "go to seed," at one growth stage a string at the back of the peapod ceases to grow. As the rest of the pod continues to enlarge, within a few days it very slowly opens, and the pea seeds are slowly raised above the level of the pod. A manufacturer of suppositories for children was persuaded to adopt this concept in his package.[24]

Tied to the discourse of design and development peddled by US corporate designers as part of a 1950s governmental campaign to use "design as a political force," it was never entirely clear at this point whether Papanek himself envisaged such experiments (for example, imitation of the dynamics of an exploding seed pod unit) as being weaponized for military use or elaborated for humanitarian purposes. Like Fuller, whose libertarianism "appealed to both political conservatives and the counterculture alike," allowing him to move between governmental military agencies and radical cultural forums with ease, Papanek courted both ends of the political design spectrum.[25] Ultimately, like Fuller, he too would be called out for his apolitical posturing more than a decade later.

From Tiger Underwear to *Domus*: The Contradictions of a Social Designer

Ironically, parallel to his search for highbrow design theoretical paradigms with the potential to bridge large-scale human concerns and the ethical dilemmas of everyday material and social life, much of Papanek's work in Ontario transpired to be the most profoundly conventional of his career. In spite of his forceful advocacy of experimental design and creative engineering, the commissions he undertook during this period were notably mundane: the graphic branding of a men's underwear label named "Tiger," interior design for local pharmaceutical stores, and speculative packaging for Fourth Dimension, a new Helena Rubinstein perfume. Running counter to his increasingly urgent pedagogic pleas for a socially valiant design culture, and the tone of his freshly printed business stationery, which listed "Industrial Design: Product Development: Basic Design Research: Creative Engineering" among its specialisms, Papanek's real-world design practice was far removed from the rigors and debates of cutting-edge Cold War design.[26] With a tinge of further irony, in the same year Papanek succeeded in forging a tenuous alliance with North America's most controversial, experimental designer-architect, Buckminster Fuller, his furniture design was featured in the mainstream design press celebrating a quintessential mid-century lifestyle. The Italian magazine *Domus,* in April 1955, ran a spread exploring the "Canadian" design of a Victor J. Papanek; it included the Transite table, an updated version of his early portable "E-1: Carry About Table," a black aluminum chair, and a fluffy hearth stool.[27] By the 1950s and 1960s, *Domus*, first established in the late 1920s by Gio Ponti, was undergoing a major revival, heightening its cultural prominence as a principle forum for the burgeoning architecture and design scene. Inclusion in *Domus,* then, was clearly Papanek's greatest mainstream design accolade to date, and particularly prestigious for an American designer, for whom representation in the much-vaunted pages of the premier European design monthly conferred the ultimate stamp of approval. Yet, bestowed on an increasingly vociferous critic of contemporary design, with a self-professed mission to expose the nefarious role of designers and their perpetuation of modish consumer culture, the accolade proved to be a double-edged sword. Following closely on from this representation of him as a fashionable furniture designer, Papanek's work appeared in yet another high-profile design journal, the widely distributed US magazine *Interiors*. Under the feature title "Fresh Sculptural Approaches," Papanek's Uchiwa and Samisen chairs sat jauntily side by side with other up-and-coming lifestyle designs. The editor, apparently aware of the potentially derivative aesthetic of Papanek's furniture, pointedly drew attention to the fact that the Samisen design in particular "invites comparison" with a 1948 chair design by influential Italian designer Carlo Mollino.

Aside from the contradictory elements of his design initiatives, with the steadily expanding design program of OCA underpinning his career ventures, Victor Papanek's star was steadfastly on the rise. A testament to this fact: the ambitious design tutor attended the 1956 International Design Conference in Aspen (IDCA) in Colorado, the highlight of the US design industry's calendar. Under the conference theme of "Ideas of the Future of Man and Design," Papanek made a number of provocative interventions during panel discussions, transcripts of the proceedings reveal. One such panel, "Management and Design," debated the extent of the designer's role beyond commerce—provoking Papanek to propose a new economic model of industrial design that might also facilitate handicrafts. Another, "Education and Design," discussed the importance of an all-around liberal education in making a "cultured man" fit to become a designer or architect, whereby Papanek questioned the move toward overly technical specialism within design.[28] His interventions, as a conference attendee, reveal a previously novice émigré designer now fully versed in critical engagement and able to hold his own in the upper echelons of the US design fraternity. No longer on the periphery watching his creative heroes with awe, Papanek had remade himself in their composite likeness.

"My client retains but one right when he hires me and that is to fire me," asserted a revitalized Victor J. Papanek in a feature in the publication *Canadian Packaging*, photographed in an iconic butterfly chair, ever-present cigarette dangling nonchalantly from his hand—cocksure in his tailored black suit and dark, heavy-rimmed spectacles.[29] Rather than the illustrious pages of the *Domus* or *Interiors* design journals, the full magnitude of the up-and-coming designer's newfound confidence first emerged in this obscure trade magazine. Under the title "How to Work with a Designer," the illuminating 1957 interview depicts an impassioned Papanek in a range of animatedly posed black-and-white photographs, debating topics including hand soap design, cigarette packaging, clothes-washing detergents, upgrading consumer taste, and the thorny issue of designed obsolescence. Within a few paragraphs, Papanek conjures the specter of Raymond Loewy: "I would like to jump in here with a story," he interjects, before continuing, "I once worked for Raymond Loewy in New York. Loewy had two words on the wall over every desk and the two words were 'Maya' and 'Tanya.' 'Maya' meant most advanced yet acceptable and 'Tanya' meant too advanced, not yet acceptable. He felt . . . that the designer had to be somehow aware that he must be advanced but that this stuff must still be acceptable. There comes a point of advancement when you are leaving the public behind you."[30]

Any notion of experimental design fades into the background, as Papanek the consumer protagonist trained by America's premier branding stylist comes to the fore. Social responsibility of design is touched on briefly in the piece, but only to reveal Papanek's frustrated ambivalence toward the impact of public health issues and design psychology on good design: "The Phillip Morris cigarette package [was] to my mind a classic design because it reflected tobacco. But now it no longer reflects the tobacco because people started worrying about lung cancer. Because olive drab and yellow suggested lung cancer, it had to be changed to new colors and consequently new, though inferior design."[31]

In less than a decade Papanek would become a lynchpin of the wider 1960s anticonsumerist movement. All the more striking, then, that he wholeheartedly embraced designers as active accelerators of consumption, drawing on public sociologist David Riesman's theory of consumer competence, popularized in Reisman's bestselling book *The Lonely Crowd: A*

Study of the Changing American Character (1950). "I wonder," Papanek reflected in the *Canadian Packaging* piece, "when we're talking about over-designing and under-designing, if we shouldn't remember that in North America one thing people are trying to get at is what Riesman calls competent consumership." Papanek continued, "they want to acquire methods of spending more and more money. Isn't it necessary therefore to provide people with stepping-stones so that if they buy wine at 75 cents per bottle, the bottle should look 75 cents so that it is a relief when they graduate to one at $1.29 and that they feel a sense of accomplishment when they are buying $2 wine."[32]

Comparatively obscure though the *Canadian Packaging* journal might be, as a contemporary source it provides an exceptional insight into Papanek's thinking at a crucial point in his journey toward becoming the champion of a socially responsible design movement. The trade journal reveals the contradictions and tautologies of a designer simultaneously searching for an original theoretical standpoint and striving to earn a living as a practitioner. Later in his career, the conventional design press acted largely as a vehicle in promoting his inflammatory, oftentimes two-dimensional rhetoric and a regurgitated potted biography, which relied on a repertoire of name checks, from Frank Lloyd Wright to Buckminster Fuller. Rarely did the reporters offer, though, the kind of critical insight into his thinking, or the fraught process involved in his staking a viable position in the complex field of postwar contemporary design politics, allowed by this journalistic debate over the power of design packaging. Papanek was himself an avid consumer of the mainstream professional design press, from the earliest emergence of his career in the 1940s through his latter days as a prominent design pedagogue. In particular, the mouthpiece of North America's design establishment, the US journal *Industrial Design (I.D.)*, kept him abreast of major developments in the field.

The growing Cold War discourse surrounding design's intervention into so-called developing countries, featured prominently in the editorials of *I.D.* in the late 1950s, had a profound impact on Papanek's future semi-anthropological design approach. During this period, designers were asked to take up a role intriguingly close to that of the conventional anthropologist, by embedding themselves as ethnographic informants within the cultures of "nonindustrial peoples," albeit with the sole purpose of serving the US government and its burgeoning development agenda.

Design and Development: The Origins of Design Anthropology

"When a designer turns up these days in an oriental or Caribbean marketplace, camera and notebook in hand," ran an *I.D.* editorial of 1957, "there is a good chance that he is more than a souvenir-seeking tourist."[33] Instead, continued the feature, "He may be an important member of a foreign economic aid program that began nine years ago with the Marshall Plan to reconstruct war-torn Europe, and that today seeks to help nations around the globe with the struggles of their peacetime economies."[34] Leading US industrial designers were commandeered in the postwar task of developing local and global strategies for industrial expansion, making detailed on-the-ground field surveys of regional labor, production, and craft methods with the purpose of transforming developing nations into emerging capitalist economies. Unambiguously titled "Design as a Political Force," the piece explained to the North American design establishment how the bulk of the Mutual Security Program's "$3,776,570,000 budget this year" was earmarked for military equipment; but a tenth of the sum had been specifically set aside for the economic development of "underprivileged nations that the USA hopes to keep on its side of the political friends by encouraging stable and prosperous future for them."[35]

A select, all-white group of male corporate designers, bankrolled by the International Cooperation Administration (ICA), acted as expert advisors pitting themselves against the Soviet Union by generating design strategies aimed, for example, at enhancing the Western-market desirability of Indigenous handicrafts, and expanding the US appetite for them. A large-scale project with global ambitions, the objective was to assess the extent of the over-haul and infrastructure required, as made clear in the article's comment that "as anthropologist Margaret Mead points out, it is usually more effective to introduce entirely new elements and even to change the culture as a whole, than it is to make piecemeal modifications."[36] The project was part of a broader objective to utilize design to "raise the standard of living of the man on the street in Pakistan or Jamaica, give him a better chance of living a productive life free of the another of communist ideology," a proposition that transformed the designer into an ersatz ethnographer and an invaluable Cold War weapon.[37]

The call to action positioned American designers as a bulwark against the threat of spreading communism. As such, it was part of a targeted initiative to use Western designers to colonize emerging markets, particularly those situated perilously close to politically sensitive geopolitical areas. The "Design as a Political Force" piece had been preceded a year earlier by a similar feature exploring the efforts of leading US industrial designer, Russel Wright, contracted by the ICA to conduct a survey regarding the commercialization of South East Asian craft products and their suitability for adaptation to Western tastes. In a highly orchestrated US mapping of the Cold War geopolitics, individual designers were each assigned a cluster of developing nations to investigate. Walter Dorwin Teague, for example, a principal choreographer of the corporate design spectacle that was the 1939 New York World's Fair, oversaw Greece, Lebanon, and Jordan. Dave Chapman, whose company garnered a reputation as specialists in the readaptation of World War II companies to peacetime production, took on the potent ensemble of Pakistan, Afghanistan, Mexico, Suriname, El Salvador, Jamaica, and

Costa Rica.[38] Despite its unapologetic campaign to utilize design as a global political tool in the service of Western power, the article's author stressed the importance of sensitivity toward local traditions, warning that "otherwise the [design] critic runs the danger of bringing in standards from America or imposing a universal criterion of good design."[39]

The intrinsic dilemma regarding design's political potential as a means of generating increased standard of living and social betterment, yet conversely its potential to destroy Indigenous cultures through homogenization, is what forms the core of Papanek's future treatise *Design for the Real World: Social Ecology and Human Change,* which followed his decade-long engagement with the problematics raised in these late 1950s pages of *I.D.* But his lifelong dedication to the humanitarian potential of design, which drove Papanek's curiosity into the Third World application of design, and non-Western artifact and design cultures, also found its origins in a distinctly Canadian intellectual discourse of the Cold War period.

The Anthropologist, the Media Theorist, and the "Angry Young Design Instructor"

Mid-1950s Toronto had a reputation as an austere and markedly noncosmopolitan city, yet its university would emerge as a hotbed of mid-century intellectual debate, with its theories around culture, media, and communication leaving an indelible mark on late twentieth-century thinking. In 1953, a year before the city opened its first subway, University of Toronto anthropologist Edmund Snow Carpenter and philosopher Marshall McLuhan, author of the groundbreaking *The Mechanical Bride: Folklore of Industrial Man* (1951), cowrote a grant proposal exploring the potential of interdisciplinary media research, titled "Changing Patterns: Language and Behavior and the Media of Communications." The bid a success, the resulting "Culture and Communication" seminar (1953–1959), with its highly influential spin-off academic journal *Explorations: Studies in Culture and Communication,* took place just a few blocks walk from Papanek's office at OCA. His in-depth familiarity with McLuhan's work from an early date suggests that Papanek attended a number of these Culture and Communication research debates. His connection with McLuhan might also explain why the media theorist included an otherwise incongruously obfuscated art historical article penned by Papanek, titled "A Bridge in Time," in a 1957 edition of *Explorations,* alongside luminaries such as economic anthropologist Karl Polyani.[40]

Just as the industrial design programs headed by Papanek at Ontario College of Art had substantially benefited from enhanced governmental funding, the interdisciplinary research project that gave rise to the Culture and Communication seminars was sponsored by the Ford Foundation, bastion of back-door ideological Cold War funding. Both the Ford and Rockefeller Foundations maintained well-documented Cold War connections to the CIA, and were used to backchannel cultural funding in the cause of broader patriotic and propaganda causes.[41] This sponsorship coincided directly with Canada's heightened prominence within Cold War North American defense strategies, and the perceived Soviet threat coming from above the Arctic Circle. The nation and its cutting-edge researchers were placed center stage on the geopolitical map when, in the mid-1950s, the construction of sixty-three integrated radar and communication stations stretching three thousand miles across the arctic regions of Canada, Northern Alaska, and Greenland was initiated as a means of intercepting Soviet nuclear missiles. The Distant Early Warning (DEW) line, the result of a carefully negotiated United States and Canadian cooperation, marked the pristine arctic landscape with Buckminster Fuller high-modernist geodesic domes, signaling the dawn of a communication and cybernetic network revolution. By its completion at the end of the 1950s, swaths of Indigenous Inuit populations had been displaced from the Arctic Circle region in the process, the political consequence of which boosted a renewed anthropological interest in the communities (particularly taken up by Carpenter).

The formative significance of this Canadian-US edifice of the Cold War era revealed itself in McLuhan's later work and ideation: "I think of art, at its most significant, as a DEW line, a Distant Early Warning system that can always be relied on to tell the old culture what is beginning to happen to it," he declared in his groundbreaking 1964 exploration of information politics, *Understanding Media: The Extensions of Man*.[42] In 1969, at the peak of "McLuhan-mania" and the height of his popularity, he produced a pack of "Distant Early Warning" playing cards released to subscribers of his *DEW-Line Newsletter* to nurture creative problem-solving and thinking; a clear precursor to the "gamification" processes of twenty-first-century design thinking. As founders of the "Toronto School of Communication," the work of McLuhan and Carpenter, however academic and ideological in endeavor, was never fully uncoupled from the broader discourses and machinations of Cold War politics and realities. Underpinned by a web of funding bodies keen to support innovative thinking in emerging fields of communication technology, and with the backdrop of Canada's increased prominence in Cold War politicking, the seminars of McLuhan and his anthropologist colleague Carpenter became an epicenter of a new critical thinking about culture. Papanek's arrival in mid-1950s Toronto at the moment of the city's accidental emergence as a hotbed of critique concerning contemporary society, media, and technology was certainly serendipitous. Now embedded in Canadian cultural politics, it would not have escaped Papanek's attention that the impact of the military geodesic structures Fuller's assistant John Dixon had drawn his attention to in his 1955 letter, had become a stark geopolitical reality rather than a technocratic-aesthetic exercise.

Politically, the ideas of Edmund Carpenter, retrospectively hailed as a pioneering media ecologist through his early introduction of media theory to anthropology, were of particular significance to Papanek. Because he was a specialist in the anthropology of Inuit cultures,

Carpenter's ethnographic research focused on the pending destruction of Indigenous culture and visual media's efficacy in expediting this process: "We use media to destroy cultures, but we first use media to create a false record of what we are about to destroy."[43] Either through coincidence, or on account of exposure to Carpenter's work, Papanek collected and utilized Indigenous Inuit objects in his own design teaching, and read widely from the 1950s onward on aspects of pre-literate Inuit life and material culture. An inspiring role model for public intellectual pedagogy, Carpenter not only explored the impact of media consumption and distribution on contemporary and Indigenous societies, but he also utilized it himself through the making of numerous anthropological films. Notably, under the title *Explorations* (after his publication with McLuhan), he was also the first professional anthropologist to present a weekly Canadian TV program. Together, McLuhan and Carpenter generated an "unorthodox academic agenda"—until 1957, when Carpenter rejected the conventional anthropological department of the University of Toronto, moving to America's West Coast to set up an experimental program in art and anthropology (and ethnographic film) in California.[44]

Although Papanek's presence at the University of Toronto predated the impact of McLuhan's renowned Program in Culture and Technology, which started in 1963 as the Centre for Culture and Technology, his exposure to McLuhan and Carpenter's early theoretical ruminations over media and humanistic experimental visual anthropology (with its nuanced observations and recordings of the erosion and transformation of Indigenous worlds) sharpened his critical awareness of the comparable role that design played in eroding authentic cultural tropes and social relations through its principal mechanism of obsolescence. Opening up even further his ambition to be a comprehensive designer, Toronto's hotbed of Cold War cultural debates inspired Papanek to become vice president of the Canadian Society of Comprehensive Designers (CSCD) between 1955 and 1956 and, by 1956, its president—or so he claimed on a (possibly puffed-up) résumé some years following his departure from Ontario.[45] Whether or not the CSCD ever actually existed, by the late 1950s Papanek's profile as a critic of mainstream design began to gel. In the spring of 1959, he took a crucial step toward public prominence with his first sole-authored contribution to North America's most influential and widely read design journal, *I.D.*, published under the heading "Student Project: An Angry Young Instructor Attacks Teaching Practices and Problems."[46] Based in part on a talk he delivered at the 1956 International Design Conference in Aspen, the article appeared to attack—in an apparent about-face—the expanded, experimental, transdisciplinary mode of design that Papanek himself had sought to pioneer: "During the last two decades the field of design education has been broadened and enriched through additional knowledge and new points of view. But the product of this educational process, the graduate industrial designer, has become narrower and less inventive, less aware of aesthetic values in his approach."[47] As if reciting the teaching offers he boasted as part of his own résumé, he recounted the sheer breadth of theory and subject matter contemporary industrial design students were now expected to familiarize themselves with: "Moholy-Nagy and the Bauhaus, Gestalt psychology, human engineering, Ittleson's and Ames's perception tests, motivation research, Osborn's 'brainstorming,' semantics and cybernetics—all this, together with excursions into such disciplines as sociology, automation, bioeconomics and configurational anthropology, has been presented to the designer/teacher as 'the truth.'"[48] While he was an "angry young instructor," Papanek's main objection was not to the ambition of broadening the designer's education per se, but rather to the degeneration of interdisciplinarity to the point of faddish relativism whereby one previously fashionable contextual subject became obsolete, another merely emerging to take its place. Rather than creating a "comprehensive" designer, the exposure to a tsunami of "fashionable dogma, clichés, rules and regulations without any real insight into the creative process" riled Papanek. Further, he thought design education was busy producing a generation of "neat, prepackaged graduate[s], possessed of highly technical skill but totally incapable of any true creative thinking any basic original insight."[49]

Under the subtitle "Creativity v. Conformity," Papanek asserted a marked ambivalence toward art school education, an ambivalence displayed repeatedly in the course of his peripatetic pedagogic career. In his opinion, the iconoclastic power of the designer was fast being eroded by education's innate "culture-preserving mechanism." By way of contrast, initiatives such as John Arnold's Creative Engineering at MIT (an approach Papanek pioneered with his third-year design graduates at OCA) "recognized that discovery, invention, original thought are culture-smashing activities."[50] Beyond the polemic, the most intriguing aspect of the *I.D.* piece is the insight it offers into the actual research and design projects with which Victor Papanek's students engaged. These included a "Design for Braille Writing Instrument" by George Filipowski and a "hand sculpture combined with steel springs" intended to serve as a "projection mechanism for women in labor" designed by Susie Takahashi and shown with Papanek's side comment in parentheses: "One difficulty with this assignment was the trouble of the students had in getting a product tested by actual use."[51]

Design Dimensions: Proselytizing Design on Cold War Television

The same year *Industrial Design* published the "Angry Young Instructor Attacks Teaching Practices and Problems" intervention, Papanek departed Ontario College of Art at the University of Toronto to take up the position of associate professor at the College of Education, State University of New York (SUNY) at Buffalo.[52] "Industrial designers occupy a position shared only with architects and city planners in designing directly from the physical environment

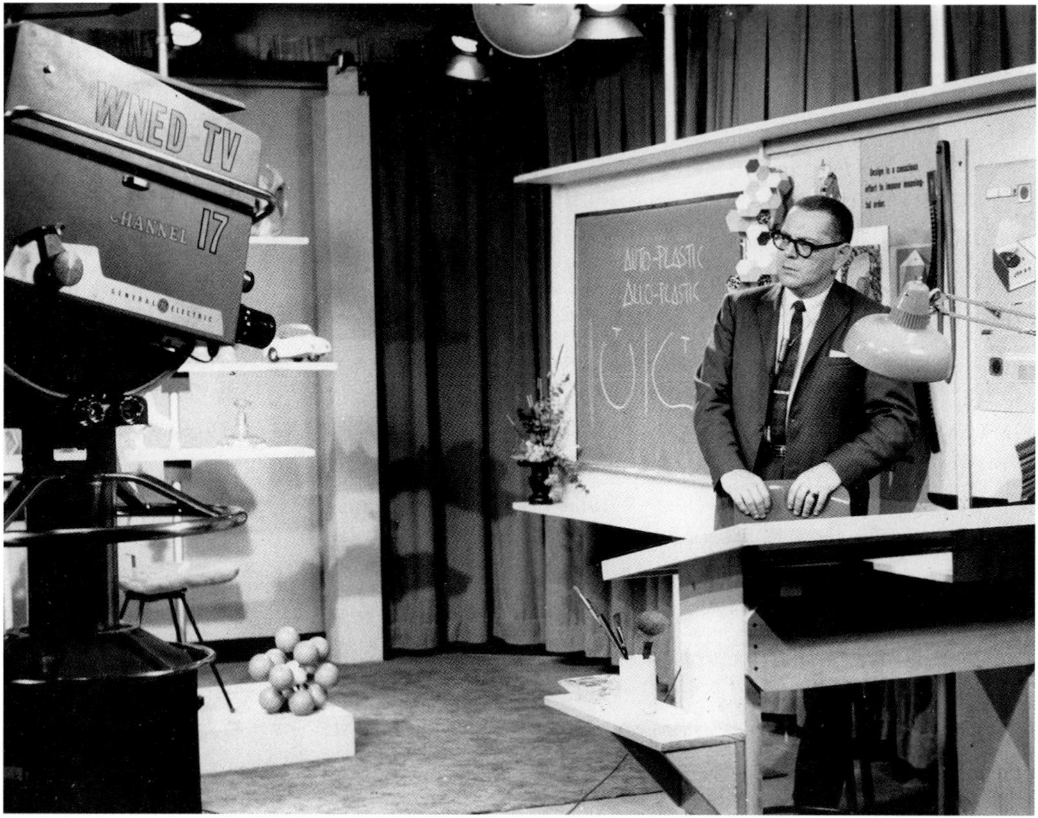

Figure 4.1
Design Dimensions TV show presented by Victor Papanek. Courtesy Victor J. Papanek Foundation, University of Applied Arts Vienna

of human species," he declared in the March 1959 article that coincided with the end of his work with the Ontario design students, and the beginning of his engagement with SUNY, adding that "automation and mass production gives the product of the industrial designer a mass impact comparable only to that of film and television."[53]

Upon his arrival in Buffalo, following in the footsteps of Carpenter and buoyed by McLuhan's theories on the power of media in society, Papanek launched his own broadcasting and public intellectual career, hosting a WNED educational series of television shows titled *Design Dimensions* (see figure 4.1). An ersatz Buckminster Fuller, with sharp suit-and-tie combo, bespectacled and authoritative, he slipped readily into the uniform and authoritative demeanor of the omnipotent male designer.

In 1958, the year after the Soviet Union launched the first satellite, Sputnik, anxiety over communist technological and educative advantage was further stoked. In response to the perceived threat, the Eisenhower government set out the National Defense Education Act (NDEA), increasing funding for educational initiatives on multiple levels. This gave emerging disciplinary areas, such as the audiovisual instruction movement and the communication arts, an unparalleled boost as well as underpinning a concerted new educational broadcasting scheme, and an interest in films about "other cultures."[54]

When it opened up media as a serious domain of academic inquiry in 1957, McLuhan and Carpenter's coedited journal, *Explorations*, had been received with rapturous enthusiasm by a broad range of individuals, corporations, and institutions in addition to the field of communication arts more generally.[55] Prior to this, Michigan State University launched the College of Communication Arts and Sciences in 1955, explicitly poised to educate a new generation in the potential as well as danger of a communication culture overshadowed by the specter of the Cold War, as explicated by the editorial of the university's newsletter: "Like the unleashed power of the atom, the unleashed power of communication may enrich or destroy the human race. Hydrogen bombs, deadly germs and gases all wrapped and waiting. And men communicating with men had better devise ways of keeping them there, or else."[56] A growing fear of television and radio as means of propagandization made the otherwise purely academic pursuits of Marshall and Carpenter all the more salient. As the newsletter continued, "Undoubtedly communications knowledge—like nuclear knowledge—can be dangerous. To give it to only a few would leave the many to be hopelessly propagandized against their own best interests. The consumers of mass media must learn how to react to this potent element in their environment. Professional communicators must develop a code of ethics along with his new insights and skills."[57]

When in 1959 the Western New York Public Broadcasting Association WNED-TV (under the slogan "Better Television") commissioned Papanek to present a series of educational shows on design, the Cold War context of their distribution would not have been remotely obvious to viewers (who hailed from Buffalo and from across the border in his former home city of Toronto) (see figure 4.2). But *Design Dimensions* arose directly from the funding made available by the NDEA and a renewed interest in consumer culture and communication as politically salient areas, with WNED-TV describing its channel as "the largest TV classroom in the country." Significantly, *Design Dimensions* premiered the same year that the famed "Nixon-Khrushchev Kitchen Debate" aired on TV, showing the US President Richard Nixon and the Soviet Union First Secretary Nikita Khrushchev publicly arguing for the relative merits of the capitalist and communist standards of living at the American National Exhibition in Moscow in 1959 through the example of the average US kitchen. Nixon placed the

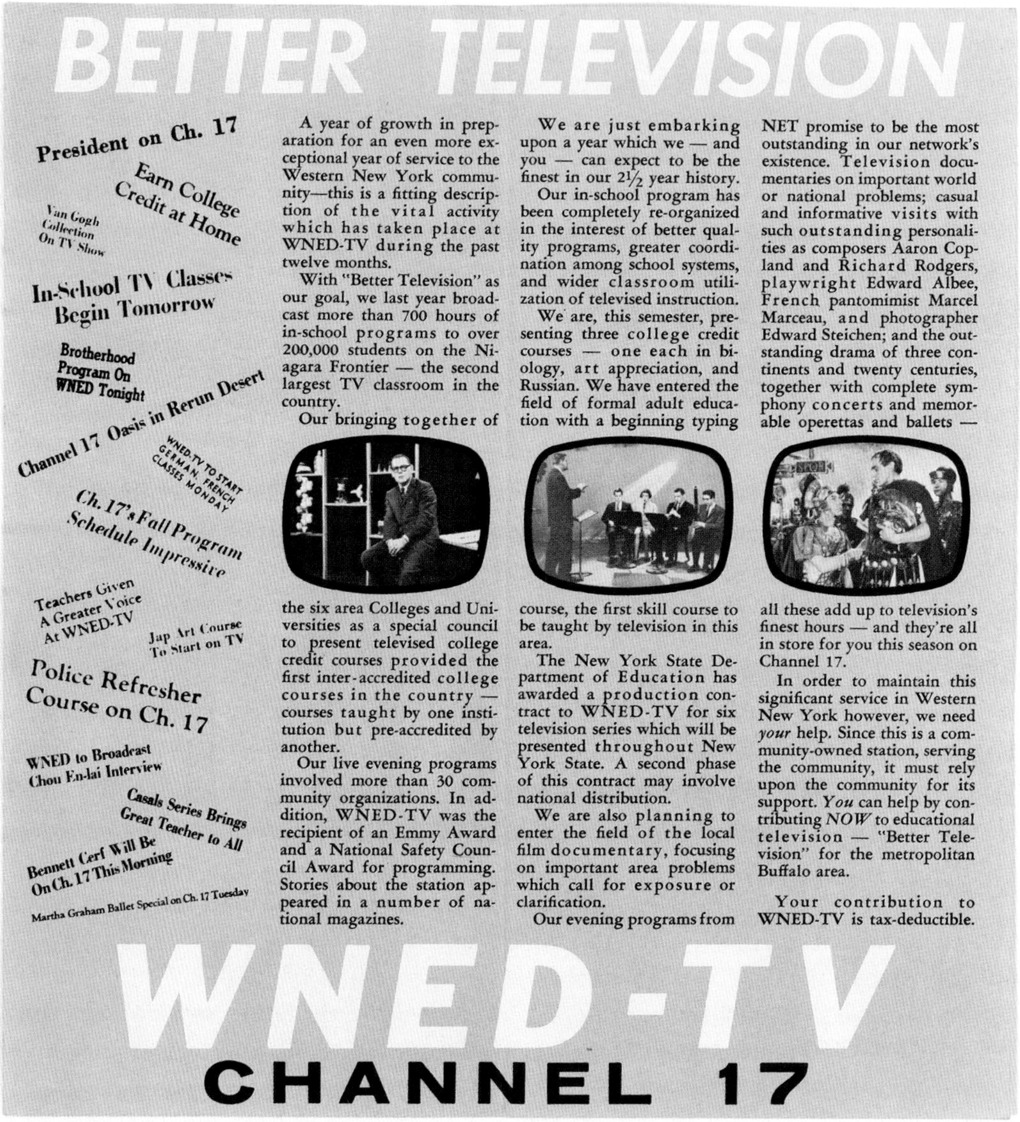

A year of growth in preparation for an even more exceptional year of service to the Western New York community—this is a fitting description of the vital activity which has taken place at WNED-TV during the past twelve months.

With "Better Television" as our goal, we last year broadcast more than 700 hours of in-school programs to over 200,000 students on the Niagara Frontier — the second largest TV classroom in the country.

Our bringing together of the six area Colleges and Universities as a special council to present televised college credit courses provided the first inter-accredited college courses in the country — courses taught by one institution but pre-accredited by another.

Our live evening programs involved more than 30 community organizations. In addition, WNED-TV was the recipient of an Emmy Award and a National Safety Council Award for programming. Stories about the station appeared in a number of national magazines.

We are just embarking upon a year which we — and you — can expect to be the finest in our 2½ year history.

Our in-school program has been completely re-organized in the interest of better quality programs, greater coordination among school systems, and wider classroom utilization of televised instruction.

We are, this semester, presenting three college credit courses — one each in biology, art appreciation, and Russian. We have entered the field of formal adult education with a beginning typing course, the first skill course to be taught by television in this area.

The New York State Department of Education has awarded a production contract to WNED-TV for six television series which will be presented throughout New York State. A second phase of this contract may involve national distribution.

We are also planning to enter the field of the local film documentary, focusing on important area problems which call for exposure or clarification.

Our evening programs from NET promise to be the most outstanding in our network's existence. Television documentaries on important world or national problems; casual and informative visits with such outstanding personalities as composers Aaron Copland and Richard Rodgers, playwright Edward Albee, French pantomimist Marcel Marceau, and photographer Edward Steichen; and the outstanding drama of three continents and twenty centuries, together with complete symphony concerts and memorable operettas and ballets —

all these add up to television's finest hours — and they're all in store for you this season on Channel 17.

In order to maintain this significant service in Western New York however, we need *your* help. Since this is a community-owned station, serving the community, it must rely upon the community for its support. *You* can help by contributing *NOW* to educational television — "Better Television" for the metropolitan Buffalo area.

Your contribution to WNED-TV is tax-deductible.

Figure 4.2
WNED-TV, New York State Department of Education, promotional pamphlet. Courtesy Victor J. Papanek Foundation, University of Applied Arts Vienna

exemplary quality of US industrial design (from refrigerators to washing machines) at the forefront of formal Cold War politics. Papanek certainly tried his utmost in the promotional materials for his new series to capture the viewer's imagination and feed into a newfound national and political interest in the merits of well-designed domestic goods and industrial design. But, as one newspaper clipping reveals, he steered dangerously close to voicing an anticapitalist rhetoric. "If you think this program sounds too specialized possibly dull, listen to its aims in the words of Papanek," pronounced the journalist, before going on to quote the provocative designer verbatim: "In this age of the squeeze bottle-martini, the stereophonic babysitter and charge-a-plate divorce, what could be legitimate areas of design for the industrial designer? While the mink-covered golf tee and the hula hoop holds sway; while the smug consumer relaxes evermore comfortably in his bathtub of lukewarm Farina, enjoying his 'music from easy listening,' and we add another tailfin to a moon probe rocket, it is still design for money instead of for the many."[58]

The provocative and sardonic episode titles of the *Design Dimension* series, which ran sporadically during his time in Buffalo, from 1959 to 1962, sketched out the themes that would drive Papanek as a designer and critic into the next decade: "Prescription for Rebellion," "Do It Yourself Murder," "The Chrome-Plated Marshmallow," "The Mechanized Bride," "The Air-Conditioned Nightmare," "Design for Easy Glistening," "The Split-Level Conscience," "Our Kleenex Culture," "Survival Through Design" (see figure 4.3). Recycling the titles of works produced by those he admired, including psychoanalyst Robert Mitchell Linder, media theorist Marshall McLuhan, American novelist Henry Miller, and Austrian American émigré architect Richard Neutra, Papanek set himself up as part of the canon of critical thinkers and public intellectuals.

Branding a New Persona: Polyhedra and the Papanek Power Couple

The 1956 Aspen conference that proceeded Papanek's arrival in Buffalo had proved a watershed for Papanek in several ways: it offered him an entrée into the upper echelons of the North American design world, raised his profile as a design critic, and provided a platform for a talk that would become the basis of his groundbreaking article in *I.D.* in 1959. He attended this conference alongside a native German-speaking colleague and friend, graphic designer Hans Kleefeld, who, emigrating to Canada in 1952, would later be renowned as one of the nation's leading corporate designers.[59] It was Kleefeld who had almost certainly sent graphic design and packaging work Papanek's way in his earliest years in Ontario as the young designer found his way professionally and financially in Toronto. This would account for his creative detour into underwear branding and perfume packaging for Helena Rubenstein in

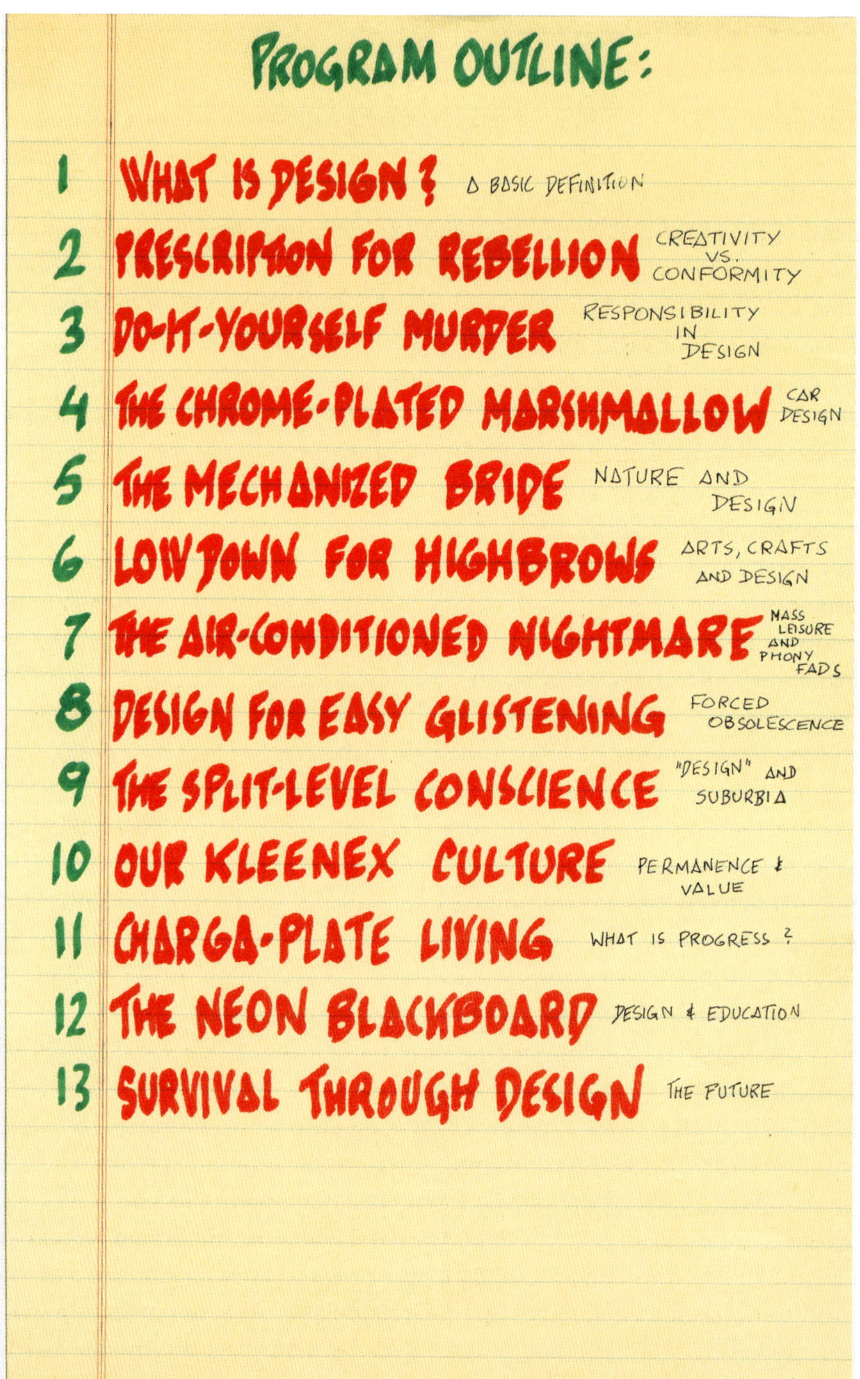

Figure 4.3

Design Dimensions program list, conceived and handwritten by Victor Papanek. ©Victor J. Papanek Foundation, University of Applied Arts Vienna

the mid-1950s. More significantly, though, Hans seems to have introduced Papanek to the woman who would go on to become his fourth wife, Diana Jean Kleefeld; at the time Papanek met her she was already married to Hans. This might explain Papanek's sudden departure from an otherwise promising full-time position as vice chairman of industrial design at a thriving design institution. Whatever the actual circumstances behind his quitting OCA, by 1959 Diana Jean accompanied Papanek on his new venture to Buffalo, where they cohabited until their marriage in 1962, following Papanek's divorce from Winifred Nelson.[60]

Newly installed in the Western New York area, and relishing the local fame generated by his new TV show, Papanek set about reinventing himself. Offering a unique snapshot of this transition to a northerly, less than cosmopolitan city, a local news feature exploring the background of SUNY Buffalo's latest faculty member appeared in the *Buffalo Evening News* in the fall of 1959. Under the title "Design Professor's Home a Place of Individuality," candid shots showed "Mr and Mrs Papanek" in situ with their cherished objects and interiors (see figure 4.4). Diana Jean Kleefeld (captioned as "Mrs Papanek") was photographed clutching an exotic musical instrument "designed by one of her husband's pupils . . . to duplicate the sounds normally created by using kitchen implements." The couple were shown surrounded by "examples of new and individual design"; these included Papanek's scratch-proof, alcohol-proof, cigarette-proof Transite table (originating from his 1940s Design Clinic studio) and a "series of polyhedral, three-dimensional geometric forms" hanging from the ceiling. Alluding to Buckminster Fuller, Papanek explained, "When these polyhedra have fourteen sides they can be fitted together. A new type of architecture, maybe suitable for an apartment on the moon could be built from a group of these clinging together." "After all," he continued, "we are going to have to face the problem of extraterrestrial architecture soon and will need to be truly creative as we face the problem of building foundations that have to swim with the dust."[61]

No longer dabbling in brand design for the likes of Tiger underwear, Papanek reveled in his newly made Bohemian credentials, regaling the local news reporter with size of his art book collection, which he boasted was worth the equivalent of "three Cadillacs." Young, attractive and fashionably dressed, Diana Jean slotted perfectly into place in the contemporary domestic tableau of an emerging charismatic American designer.

A couple of years after Victor and Diana Jean had left New York State for North Carolina, Papanek commissioned a hand-drawn Christmas card to announce his new position, and recent bride, to friends and colleagues (see figure 4.5). "Happy Yule, You'All!" exclaimed the card, mimicking the region's southern dialect and using Papanek's distinctive graphic handwriting style. Seated in an Eames-style easy chair, cool as cucumber, cigarette in hand, the budding designer was flanked by an exaggeratedly curvaceous depiction of Diana Jean, who, donning a skin-tight black evening dress with a risqué thigh-high split, towers over him in

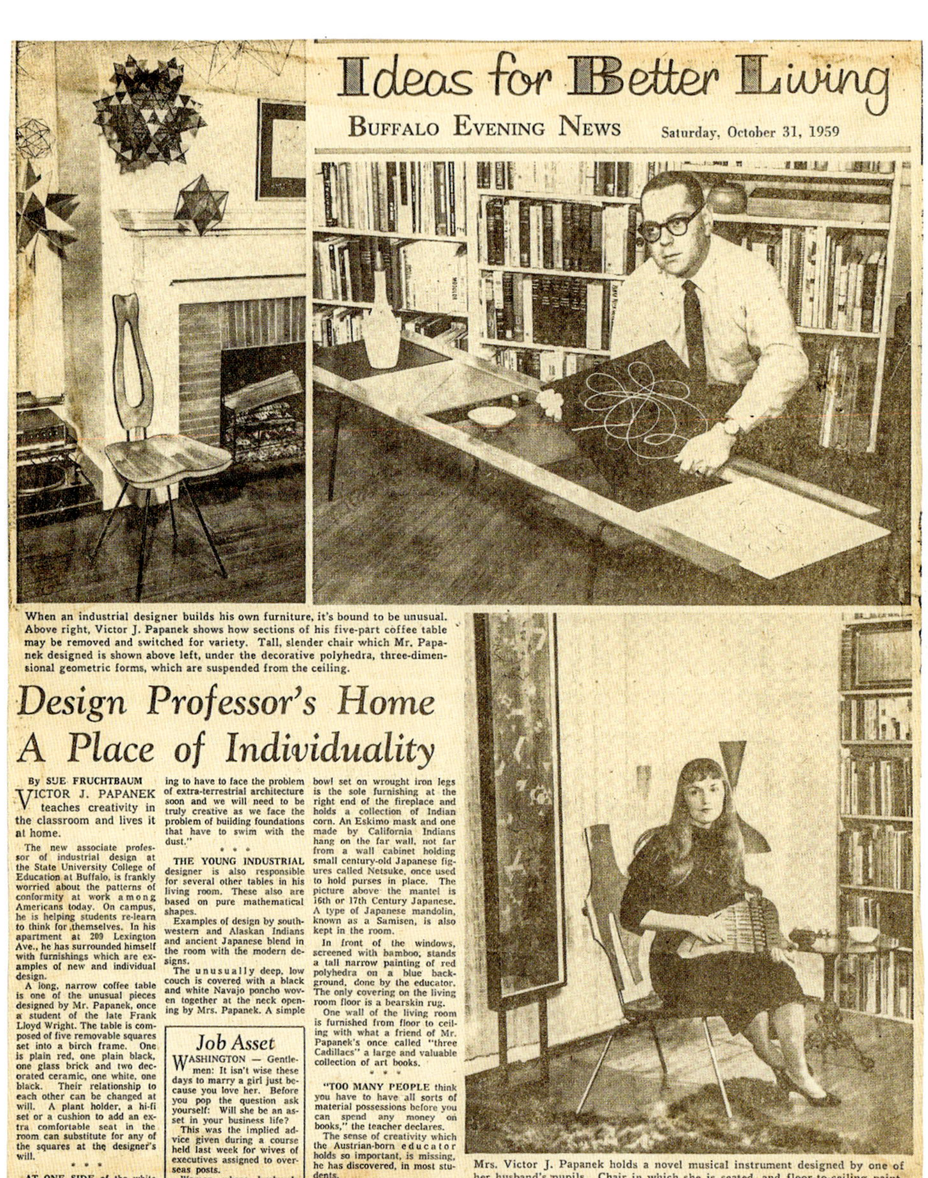

Ideas for Better Living

BUFFALO EVENING NEWS   Saturday, October 31, 1959

When an industrial designer builds his own furniture, it's bound to be unusual. Above right, Victor J. Papanek shows how sections of his five-part coffee table may be removed and switched for variety. Tall, slender chair which Mr. Papanek designed is shown above left, under the decorative polyhedra, three-dimensional geometric forms, which are suspended from the ceiling.

Design Professor's Home A Place of Individuality

By SUE FRUCHTBAUM

VICTOR J. PAPANEK teaches creativity in the classroom and lives it at home.

The new associate professor of industrial design at the State University College of Education at Buffalo, is frankly worried about the patterns of conformity at work among Americans today. On campus, he is helping students re-learn to think for themselves. In his apartment at 209 Lexington Ave., he has surrounded himself with furnishings which are examples of new and individual design.

A long, narrow coffee table is one of the unusual pieces designed by Mr. Papanek, once a student of the late Frank Lloyd Wright. The table is composed of five removable squares set into a birch frame. One is plain red, one plain black, one glass brick and two decorated ceramic, one white, one black. Their relationship to each other can be changed at will. A plant holder, a hi-fi set or a cushion to add an extra comfortable seat in the room can substitute for any of the squares at the designer's will.

AT ONE SIDE of the white fireplace is a narrow-backed chair also designed by Mr. Papanek, a chair that looks stiff to the observer but which is unusually comfortable seating.

Hanging from the ceiling above the chair are a series of polyhedra, three-dimensional geometric forms, made by former students of Mr. Papanek.

"When these polyhedra have 14 sides," he explains, "they can be fitted together. A new-type of architecture, maybe suitable for an apartment on the moon, could be built-up from a group of these clinging together. After all, we are going to have to face the problem of extra-terrestrial architecture soon and we will need to be truly creative as we face the problem of building foundations that have to swim with the dust."

* * *

THE YOUNG INDUSTRIAL designer is also responsible for several other tables in his living room. These also are based on pure mathematical shapes.

Examples of design by southwestern and Alaskan Indians and ancient Japanese blend in the room with the modern designs.

The unusually deep, low couch is covered with a black and white Navajo poncho woven together at the neck opening by Mrs. Papanek. A simple bowl set on wrought iron legs is the sole furnishing at the right end of the fireplace and holds a collection of Indian corn. An Eskimo mask and one made by California Indians hang on the far wall, not far from a wall cabinet holding small century-old Japanese figures called Netsuke, once used to hold purses in place. The picture above the mantel is 16th or 17th Century Japanese. A type of Japanese mandolin, known as a Samisen, is also kept in the room.

In front of the windows, screened with bamboo, stands a tall narrow painting of red polyhedra on a blue background, done by the educator. The only covering on the living room floor is a bearskin rug.

One wall of the living room is furnished from floor to ceiling with what a friend of Mr. Papanek's once called "three Cadillacs" a large and valuable collection of art books.

* * *

"TOO MANY PEOPLE think you have to have all sorts of material possessions before you can spend any money on books," the teacher declares.

The sense of creativity which the Austrian-born educator holds so important, is missing, he has discovered, in most students.

"Apparently youngsters are losing their sense of creativity at the age of 6 or 7. Even the beatniks in beards and sandals are looking about for others like themselves and are searching for something on a superficial level. The one way to get back to a student's creativity, is to give him a problem totally unrelated to any previous experience of his," he notes.

Then, by way of explanation, he pulls from a bookshelf an unusual new musical instrument designed by a student who had been given the problem of developing an instrument which could duplicate sounds normally created by using kitchen implements.

Job Asset

WASHINGTON — Gentlemen: It isn't wise these days to marry a girl just because you love her. Before you pop the question ask yourself: Will she be an asset in your business life?

This was the implied advice given during a course held last week for wives of executives assigned to overseas posts.

Women whose husbands are opening branch offices for their companies in foreign lands were told just how important they are to the success or failure of hubby's new job.

"The conduct and attitude of the wife during the overseas assignment may be decisive in the success or failure of the man's mission," these attractive "students" were told. The School for Wives was conducted by the Business Council for International Understanding and the School of International Service of American University.

Mrs. Victor J. Papanek holds a novel musical instrument designed by one of her husband's pupils. Chair in which she is seated, and floor-to-ceiling painting back of the chair are by Mr. Papanek.

Figure 4.4

Sue Fruchtbaum, "Design Professor's Home a Place of Individuality," newspaper clipping, *Buffalo Evening News*, October 1959, Ideas for Better Living, 31. © 1959 The Buffalo News; courtesy Victor J. Papanek Foundation, University of Applied Arts Vienna

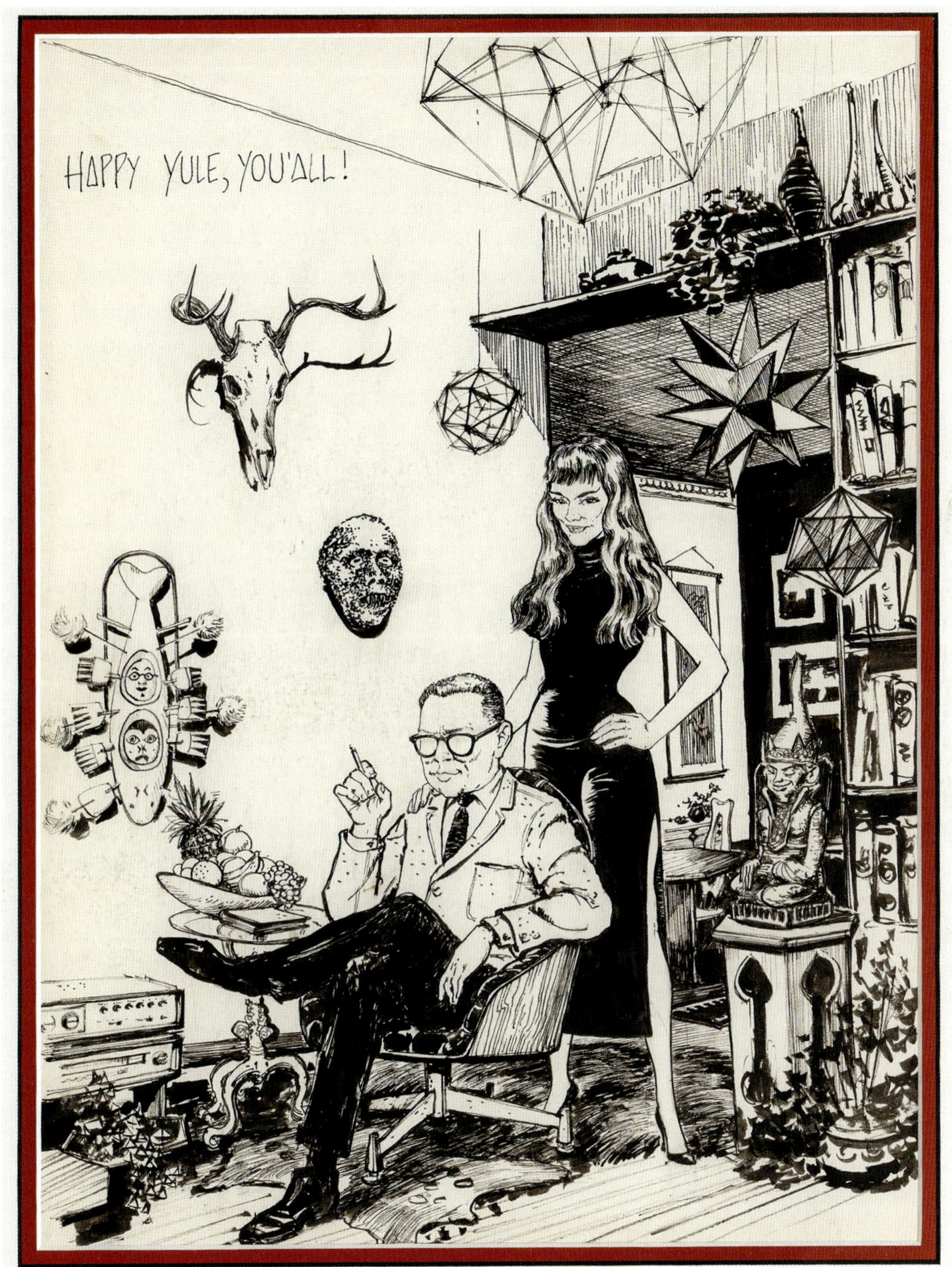

Figure 4.5
"Happy Yule, You'All!" Christmas Card depicting Victor Papanek and Diana Jean Papanek, hand drawn by Victor Papanek, c. 1962. Courtesy Victor J. Papanek Foundation, University of Applied Arts Vienna

her high-heeled stilettos. The ultimate stereotype of a 1960s trophy wife, with a curvaceous figure, long dark hair, and thick, short-cut bangs, she shows an uncanny resemblance (at least in Papanek's depiction) to his pin-up idol, fetish model Bettie Page. The finely detailed line drawing depicts the power couple set against the backdrop of Victor's nomadic interior, which acted as a kind of portable rendering of his cultural capital, a materialized shorthand of his design credo. Its components (including Inuit masks and artifacts, Navajo weavings, Buddhist figures, polyhedral structures, and early furniture designs) recurred time and time again in press profiles of the designer following the interior's first appearance in the *Buffalo Evening News*.[62]

Southward to North Carolina

With his establishment of an experimental studio at SUNY Buffalo, Papanek's time there launched the designer firmly into the stratosphere of Cold War design politics (see figures 4.6 and 4.7). Poised on the edge of a turbulent new decade that would ask radically different questions of design's societal role, Papanek took a leap of faith. He had become familiar with the design department of North Carolina State University (NCSU) through the work of his newfound ally there, Buckminster Fuller. Papanek decided to quit Buffalo, and his educational broadcasting career, and move south to a university that had recently benefited from huge federal sponsorship following the 1960 Black civil rights protests in nearby Greensboro. On May 16, 1962, Paul G. Bulger, writing on behalf of the Office of the President of State University of New York at Buffalo, accepted Papanek's letter of resignation, signing off: "Best wishes to you in your new position as Chairman of the Industrial Design Department in North Carolina."[63] In the southern state, Papanek would be exposed to an entirely new level of political discourse, which tied a critique of Cold War technocratic machinations to broader issues of anticolonialism and African American rights.

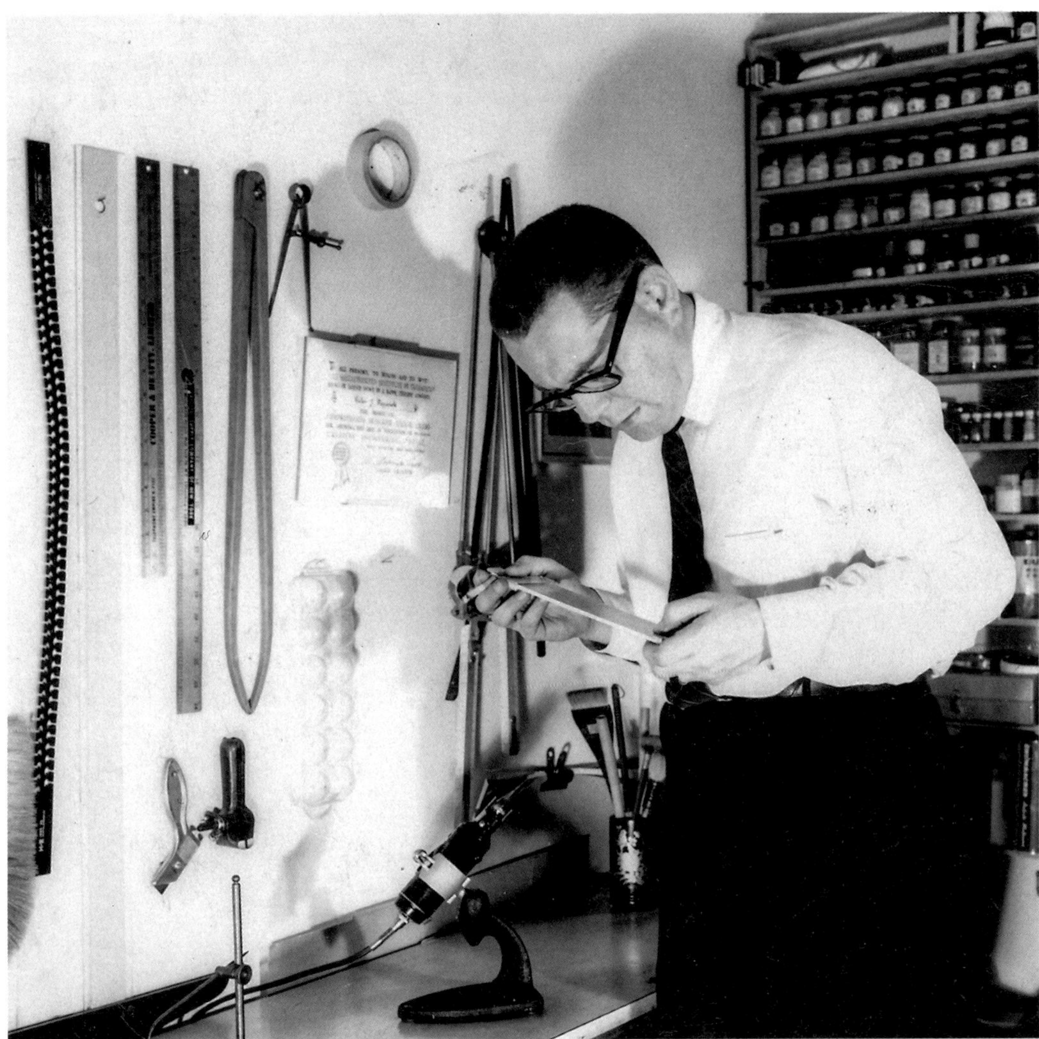

Figure 4.6
Victor Papanek in his experimental design studio, SUNY Buffalo, 1958. Courtesy Victor J. Papanek Foundation, University of Applied Arts Vienna

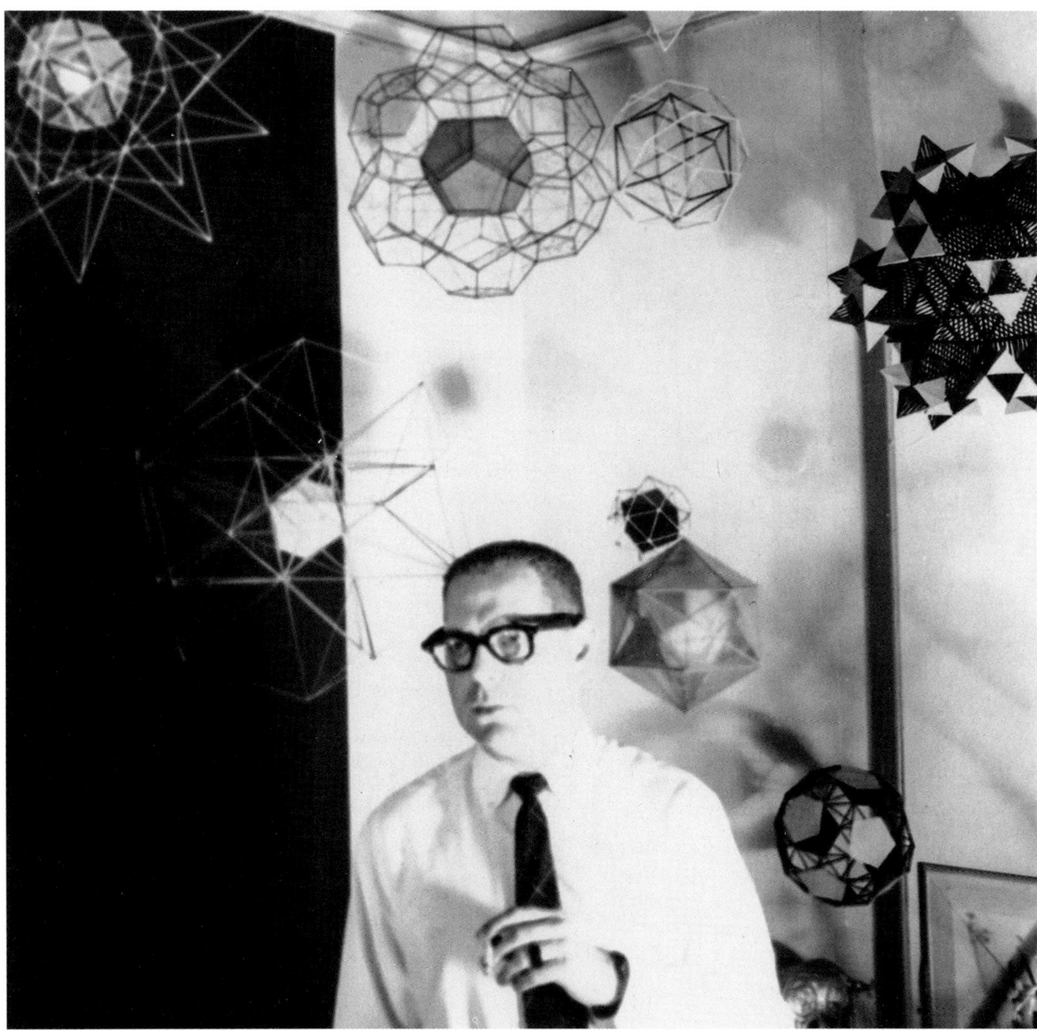

Figure 4.7
Victor Papanek with his collection of polyhedral hanging sculptures, Buffalo, 1959. Courtesy Victor J. Papanek Foundation, University of Applied Arts Vienna

5 From Bionics to Racial Politics: The Making of a Cold War Designer

In the spring of 1964, just months after the assassination of America's progressive leader President John F. Kennedy had shattered the political landscape of the nation, a young faculty member of Germany's famed Ulm School of Design, Gui Bonsiepe, set off on a seemingly incongruous journey. His mission was to meet in person a forward-thinking social designer recently installed in the design school of a public university in the conservative southern state of North Carolina. Touching down at Raleigh-Durham airport, Bonsiepe was greeted by his enthusiastic host, a dashingly dressed fellow native-German speaker and design professor named Victor Papanek. Duly whisking Bonsiepe away in his startlingly conspicuous, vermillion-red Volvo sports car, Papanek introduced his young design colleague to the local political climate by revealing a hidden revolver stashed in the side-pocket of his convertible: "It was amazing to me," Bonsiepe later reflected, "that a private person, a university professor, would consider it necessary to have a weapon in his car to protect himself."[1] For the newly arrived visitor, this unsettling encounter, coupled with the shockingly routine nature of the racism visible in the passing scenes of everyday life in Raleigh, from segregation at drinking fountains to reports of violent protests against African Americans in the local press, undermined the model of US cultural democracy the idealistic European designer had nurtured in his imagination.

This 1964 encounter between two intellectually ambitious designers is the earliest evidence of the ways in which Papanek's ideas, born of a distinctive North American Cold War politics of design, came to impact innovative European design thinking. A decade after this pivotal meeting in Raleigh, Bonsiepe and Papanek emerged as intellectual adversaries fervently pitted against one another with their vastly different stances toward design for the "Third World": favoring a rational, systems-theory approach to what he termed peripheral economies, Bonsiepe accused Papanek of advocating an anthropological and holistic approach that was overly utopian. Their polarized positions would eventually be aired in the upper echelons of the Italian design press, when in 1974 Bonsiepe put forth a bitingly critical and widely read review of the Italian translation of Papanek's seminal work *Design for the Real World* (*Progettare per il mondo reale*).[2] But at this early point of contact, there was every indication that Bonsiepe and Papanek shared not only the same native language, but also a shared belief in industrial design as the driver of a new paradigm of postwar democracy.

Forging a Career in Cold War Design: The Rise of User-Centered Research

A protégé of leading design theorist Tomás Maldonado, and his coeditor for the influential design journal *Ulm: Quarterly Bulletin of the Hochschule für Gestaltung*, Bonsiepe at the time of his first meeting with Papanek had begun to garner his own reputation as a cutting-edge designer-theorist at the Ulm School of Design. Brought over to the States by an internationally minded US colleague, William S. Huff, to act as guest tutor for the highly regarded Industrial Design (I.D.) program at Carnegie Institute of Technology (CIT) in Pittsburgh, Bonsiepe arrived in the winter semester of 1963–1964.[3] Huff, himself a guest instructor of the "Basic Design" course at Ulm School of Design between 1963 and 1968, occupied a unique position as interlocutor, introducing the "Ulm Model" of design pedagogy to US industrial design students.[4] It was in his capacity as a Euro-American bridge builder between the industrial design traditions of Germany and the United States, and because of the creative methods of the Ulm School of Design in particular (aspects of which ran openly against the grain of conventional vocational US design pedagogy), that Huff invited Bonsiepe to shake up the I.D. department. Despite the creative influence that Bauhaus émigré Josef Albers had exerted in the CIT architecture department, there persisted among the product design faculty a pronounced antipathy toward European-style design approaches. This sentiment echoed across a number of mainstream, postwar US design education institutions of this period, with one leading design figure summarizing this ambivalence in a proclamation that such modes were "quite unsuited to our [US] type of student who is anxious to reach economic self-sufficiency in a minimum amount of time."[5] Bonsiepe's intervention into American design pedagogy, however, proved remarkably successful; he was described by one alumnus as injecting "electricity through the department" and motivating the student cohort "like a shot of adrenaline" with his perfectionism, unmitigated internationalism, and theoretically ambitious approach.[6]

The studio environment Bonsiepe encountered in Pittsburgh contrasted starkly with that of the Ulm School. He set about immersing himself in the prevalent national design debates. Familiarizing himself with the ways in which the priorities of American design practice and education differed from those of the more progressive, research-based contingent at the Ulm School, he came across what he described as the "anti-establishment" work of Papanek, whose approach immediately set him apart "as an outsider in the North American design environment."[7]

One lengthy feature article investigating the future of design pedagogy particularly spiked Bonsiepe's interest. In late 1963, a special edition of the *Industrial Design* journal titled "Education of a Designer I" published a state-of-the-art overview of cutting-edge student work across the United States framed by debates emanating from the Industrial Design Education Association (I.D.E.A.) national forums. The editorial targeted the general complaint that there was still "no real evaluation of schools of design from either an admission or a graduating point of view," a situation deemed "deplorable for a profession which is directly involved in visual excellence, industrial quality control, and the advancement of good design."[8] The editorial protested, "high standards begin with well-educated human-beings," before advocating an adherence to minimum standards in industrial design education, with the ideal curriculum making a measured balance between humanities and professional training.[9]

The special education-focused edition of *I.D.* showcased innovative student projects, interspersing them with the voices of leading design educators opining on the shifts in pedagogic discourse and the future trajectories of design. The bombastic and unabashedly forthright words of Papanek, head of product design at North Carolina State University's School of Design (now known as the College of Design), took the lead page: "Certainly it is plain that there are two main directions in the field of product design today, which have to be fought: one is appearance design, that is styling, 'streamlining,' and the whole attempt of design toward forced obsolescence. The other is the 'artsy-craftsy' approach, the 'I-hear-America-singing' method of hand-chewing pretty statements out of silver, plastics, and wood. Naturally, both of these are to be rejected."[10]

The tone echoed uncannily Papanek's earliest foray into design criticism in his early 1950s article "Why Contemporary?," penned for a local Californian lifestyle magazine. A decade and a half on, flushed with the success of his students who had been showcased as recipients of the journal's prize for "design for the underdeveloped" (their projects featured in the same edition), Papanek openly exploited the publication as his mouthpiece, zealously rehearsing his objection to conventional product design. In fact, the *I.D.* spread was nothing short of a personal coup for Papanek, for within a year of taking up the reins as head of product design, and associate professor specializing in "Design Analysis, Creative Engineering and Bionics," he had firmly placed the school and North Carolina on the map for innovative, future-forward design. Pointing out the "State's unusual emphasis upon projects for backward, underdeveloped countries," *I.D.* specifically identified his product design program at the forefront of a sea change in pedagogy, placing it alongside the design divisions of renowned institutions such as the Cranbrook Academy of Art, Illinois Institute of Technology, University of Illinois at Carbondale, Stanford University, and UCLA.[11]

From the commencement of his appointment, Papanek had honed a user-based pedagogical approach with self-imposed parameters of subject matter, which included "design for backward and underdeveloped areas, medical design and experimental breakthroughs in areas that appear to have dead ends."[12] This shift in design education, the feature maintained, pivoted on design's unique capacity for addressing human issues and its application to the understanding of the cultural specificity of design's value beyond the purely commercial imperative. Although warning against an "overdose of sociology," considering its standing as the profession's most widely read journal, the *I.D.* feature made a markedly provocative flourish by openly challenging the modernist conceit of design as a universalizing force: "Educators," the editorial beseeched its readers, "needed to serve a balanced diet of values and direction to students. Not all peoples need the same type of products and the most convincing contemporary U.S. values may be the worst for Thailand or even for Harlem."[13]

This passing reference to Harlem under the umbrella rhetoric of so-called underdevelopment applied simultaneously to the Southeast Asian military-governed nation of Thailand (a key ally in the war against communist Vietnam subjected to a modernization program bankrolled by the US government) betrays the subtext of much of US development discourse of this period, as it applied concurrently to the internal national African American population and external geopolitics of the "developing world."

This conflation of "African American" and "underdevelopment" held particular prescience in a southern state in which segregation and racial violence persisted to an extent that outsiders like Bonsiepe were shocked to find such conditions existed in 1960s America. The extreme racial violence inflicted on the Black community of the region acted as an unspoken subtext for many of Papanek's development projects during his time in North Carolina: the development of Africa became a metaphor for a broader domestic agenda of top-down "racial" control.

From the outset the School of Design (originally founded as the School of Architecture and Landscape Design, and renamed the School of Design in 1948) was perceived as a haven of progressive thinking in the otherwise conservative region dubbed the Tar Heel State in recognition of its blue-collar industrial past. Papanek's position at NCSU was buoyed by both state and private funding, offering well-established links to industry and untold potential for an ambitious product designer set on challenging the status quo of the profession. The school also thrived on the legitimacy lent by a succession of eminent (exclusively white and male) creative figures—among them Lewis Mumford, Frank Lloyd Wright, Charles Eames, George Nelson, Mies van der Rohe, and Marcel Breuer—who were incorporated into the institution through lectures as well as studio and workshop activities enabled by copiously funded distinguished visitors' program. Exceptional for its period, this program broadened the school's significance beyond region and nation, making it a prominent institution on an international level, albeit in the largely conventional terms of showcasing its affiliations with great design and architectural masters.

Prior to winning a position at NCSU, Papanek had been aware of Buckminster Fuller's integral role in the school's success, as well as his ties to the region as an early faculty member of North Carolina's experimental enterprise, the famed Black Mountain College (BMC). The maverick inventor-designer had been part of NCSU's distinguished visiting faculty in the first seven years following the launch of the ambitious school, receiving an honorary doctorate in recognition of his services in 1954. Fuller's was not a cursory commitment; he led month-long design projects with students on variants of the geodesic dome; NCSU faculty member Duncan Stuart devised mathematical calculations critical to the structure and exerted an influence on related curricula and methods.[14] Fuller's close relationship with the US military, members of which regularly mingled with students at design project briefings, revealed the deep entrenchment of postwar industrial designers within the configuration of the growing military-industrial complex. Perhaps not coincidentally, the honorary doctorate was bestowed on Fuller during the same year that he realized the first full-scale version of the geodesic dome created for the US military to house the Distant Early Warning (DEW) line.

Also that same year, 1954, Papanek had taken the bold step of initiating personal contact with the inventor while attending a Creative Engineering seminar at Massachusetts Institute of Technology, where Fuller delivered a keynote address. Papanek maintained sporadic correspondence with the Fuller's office from that date onward in an attempt to win his patronage. His doggedness paid off in the mid-1960s when Papanek secured a position alongside Fuller on the speakers' platform at progressive design conferences across Northern Europe. While it is unclear exactly how Papanek came to the attention of the NCSU hiring committee, it is highly conceivable that Fuller recommended Papanek for the position, or at the very least vouched for the validity of the up and coming designer's work. Papanek's experimentations on the periphery of conventional industrial design certainly matched the institution's mission in carving out a unique profile in the late Cold War era as United States policy shifted emphasis from postwar European reconstruction enshrined in the Marshall Plan to development of the Global South by the 1960s through the embracing of a design-for-development paradigm.[15]

The Contradictions of North Carolina as a Hub of Progressive Design: Black Mountain College, Penland School of Crafts, and the Race Question

According to design historian Russell Flinchum, by the latter part of the twentieth century North Carolina's design culture, in which Papanek's new product design program was embedded, had benefited "since the Great Depression" from "waves of progressivism" washing over it.[16] The innovative thrust of the School of Design sprang specifically from its inception as an equivalent to Walter Gropius's Graduate Design School at Harvard; it was launched in 1948 as an amalgamation of the Department of Architectural Engineering and the Department of Landscape Architecture. Backed by unprecedented governmental investment, its infrastructural strategy was to boost design's alliance with regional and national industries and, most significantly, the US military.[17]

Some viewed the success of NCSU's School of Design as part of the region's broader legacy in the field of arts and design experimentation, albeit of a more alternative bent. Just over two hundred miles away, near the city of Asheville, stood North Carolina's famed independent liberal arts institute Black Mountain College, an "unaccredited college in rural Appalachia"' that, while offering "little more than train fare and a bed for its faculty," in the words of historian Eva Diaz, "became a vital hub of cultural innovation."[18] Established in 1933 as an experimental hothouse of liberal arts and design pedagogy, it presented a nonhierarchical interdisciplinary environment giving space to celebrated alumni and faculty such as Ruth Asawa, Gropius, Merce Cunningham, and Fuller. Bauhaus émigrés Josef and Anni Albers famously led its visual arts program. John A. Rice, its founder, hailed from an upper-class, educationally privileged background, and BMC was molded on his personal mission to "remap a liberal arts college education" according to the democratic principle of self-governance, with creativity set at the core of a new model of pedagogical community.[19]

Although the Black Mountain College pedagogical experiment had folded five years prior to Papanek taking up his post in North Carolina, its unique reputation as creative refuge famed for hosting numerous émigré designers and artists would have been thoroughly familiar to him as a student at the progressive Cooper Union in New York City in the 1940s. Indeed, the BMC's transdisciplinary and experimental approach to art and design echoed that of Papanek's own teaching as far back as 1952 at the San Francisco Art League, where the design curriculum he devised centered on performance, filmmaking, and critical theory practices. Perhaps more significantly, though, as a hub of creative experimentation, BMC lent a cosmopolitan cultural profile to this corner of the southern state otherwise defined by its industrial history and political conservatism.

By the early 1960s, this progressive cultural cachet accrued through North Carolina's affiliation with experimental arts education stood in stark contrast to the state's notoriety for violent responses to civil rights protests, resistance to desegregation, and the rampant racial violence inflicted on African Americans through the institutional racism of the Jim Crow segregation laws. That such an endemically reactionary region of rural North Carolina emerged as home to two notable design initiatives in the first half of the twentieth century—Black Mountain College (dissolved through ideological disputes and bankruptcy by 1957) and Penland School of Crafts (1929–present)—seems wildly contradictory.[20] They were diametrically opposed in terms of the social-class backgrounds and intellectual impetus of their respective founders and their respective pedagogic emphasis, with Black Mountain College embracing an avant-garde and internationalist outlook, and Penland drawing on the local craft traditions of the Appalachian region. Yet, as Penland archivist Carey Hedlund contends, the basic fact "that both schools were located in Appalachia, barely 50 miles and a mountain range apart, begs us to look for a common denominator."[21] Certainly both shared a remoteness and immersion in the distinctive landscape offered by the awe-inspiring North Carolina Blue Ridge mountains; each school embraced its location as a type of muse, an anchor used to ground initiatives that were otherwise sustained through diverse external networks of intellectual and practical relations. Both schools exerted a considerable influence on Papanek; BMC in terms of the legacy of its interdisciplinary design approach and avant-gardist émigré heritage, and Penland as a school at which Papanek taught during the summers of his tenure at NCSU. Somewhat incongruously, perhaps, it was this traditional, conservative craft school that fostered the expansion of his profile as a purely technologically driven "comprehensive designer" to that of an advocate of the vernacular.

Black Mountain College remained inextricably linked to the cosmopolitanism of New York City, and the transatlantic émigré connections of its Bauhaus alumni, swiftly accruing the status as a crucible of creativity that would forge the "Who's Who of Postwar American Art."[22] Conversely, Penland drew its instructors from the Midwest and consciously embedded itself in the Appalachian craft traditions, at least in part due to the fact that the enterprise had been initiated by a locally trained teacher and craft reform advocate, Lucy Morgan, who recognized the fast disappearance of regional craft knowledge.

Each of the schools regularly hosted external art, craft, and design faculty, drawing instructors and visiting experts to an otherwise remote region through their reputations for innovation and immersive, hands-on pedagogies. Yet with the notable exception of a couple of tenuous links, including a visiting lecture made by Anni Albers at Penland in 1945 under the auspices of a meeting of the Southern Highland Crafts Guild, the schools remained bafflingly unconnected.[23] That Josef and Anni Albers purportedly held in disdain "non-intellectual" craft practices such as pottery, which they "associated with hobbies, Nazi kitsch, and therapy," may go some way to explaining the paradox of the schools' shared geographical proximity yet studied indifference toward each other's enterprises. Despite their marked differences, they coexisted in the adverse sociopolitical environments of overt racism and opposition to desegregation, which at least on the surface would appear antithetical to their respective creative initiatives.[24]

Black Mountain College and Penland School of Crafts were both detached from the urban diversity of the state's towns and cities, celebrating their immersion in a rural landscape and the creative particularity this remoteness afforded them. Yet their faculties adopted vastly different approaches to the negotiation of regional racial politics and its impact on their respective pedagogical projects.[25]

Design Experimentalism Negotiates Civil Rights

The mythology surrounding Black Mountain College's radical and progressive approach to pedagogy as a holistic pursuit (influenced by the ideas of educational reformer John Dewey, author of *Democracy and Education* [1916]) has been furthered still by a burgeoning scholarship placing the school at the forefront of pre–civil rights antisegregationist movements. The presence of African American New York–based artists Jacob Lawrence and Gwendolyn Knight Lawrence as faculty members in its summer school in 1946 has, for example, been the focus of scholarship reiterating the college's status as a socially progressive as well as creative hub.[26] Similarly, a growing number of historians have staked a claim for BMC's progressiveness "as one of the first, if not the very first, white post-secondary institution[s] in the South to admit African-American students."[27]

The incorporation of African American culture into the BMC curriculum was far more than a tokenistic gesture; in fact, according to historian Micah Wilkins it constituted a manifestly political act: by "exposing students to African American culture, engaging faculty in an ongoing conversation on African American culture and race relations, and serving as a model of what southern independent schools might become, Black Mountain College entered a prescient socio-political conversation on the dignity of the human person."[28]

Despite the claims for its radicalism and advanced approach to racial inclusivity, Black Mountain College, described pointedly by locals as "a Yankee island in a Southern sea" was ostensibly an elite, white Euro-American concern, which in the course of its twenty-four-year history saw only 1,200 students pass through its doors.[29] Interpretations of its gestures toward racial inclusivity—from the teaching of "Negro Gospel Music" and invitations to celebrated Black singers such as the tenor Ronald Hayes, if viewed more skeptically through the paradigm of cultural appropriation—might be seen as serving the purpose of boosting the international cultural capital of the core white avant-garde faculty as much as advancing the cause of African American civil rights and liberty.

Yet the startling contrast of Black Mountain College's approach to negotiating the conditions of extreme racism in the South are lent a renewed significance when considered alongside those of its neighboring arts' institution, the Penland School of Crafts, and the reactionary origins of its own institutional history, and more specifically, Papanek's decision to teach there in the 1960s at the pinnacle of his engagement with design for the socially and ethnically excluded.

Racial Integration—"A Bitter Pill to Take": Penland and the Troubled History of Craft

Since the post–Civil War period, applied craft had formed a component of education for formerly enslaved peoples. During the 1930s, Penland School of Handicrafts (as it was called then) trained white instructors in manual activities, such as weaving, as a means of equipping them to become instructors to African American "freedmen" at the Hampton Institute in Virginia, an institution originally founded under the auspices of the American Missionary Association.[30] This postcolonial application of craft, and craft teaching, extended at least into the mid-1940s, when Penland's founder Morgan recorded that "missionaries from Africa . . . have come to Penland to get information and experience to take back to their missions."[31] In the same report Morgan describes how "for some years, instructors from the Penland staff have gone to Fort Valley State College of Negroes at Fort Valley, Georgia, to conduct a special class in crafts. Their class has been attended by outstanding young Negroes from High Schools throughout the state."[32]

These snapshots of Penland's naturalized role in shoring up of elements of America's practical and ideological infrastructure of white dominance culminated by 1959, with the school's unmitigated support of racial segregation and its proactive attempts to exclude Black students and visiting instructors entirely. Board of trustees records from that year show that acting director Bonnie Willis Ford (covering for the year-long sabbatical leave of Penland's founder) actively changed the admissions procedure "to work out a way on the application blank of screening future applicants as to race, and to reject any applications from negroes."[33]

A full quarter of a century after the establishment of BMC and its incorporation of African American students and culture into the school, the incendiary theme of segregation still dominated the South as a form of intimate terrorism overshadowing everyday lives, actions, and relations. Willis's invidious adaptation of the Penland admission form (just four years prior to Papanek's arrival as a visiting tutor there) was orchestrated in order to surreptitiously circumvent the inclusion of a "blunt statement in the catalog that we are not integrated" and was purportedly motivated by a desire to avoid the "disruption" that had arisen on the occasion of an African American student's visit to the school in the summer of 1959.[34]

An accomplished occupational therapist practicing in a mental health hospital in Chicago, Novella Machin had been recommended by a fellow Penland student and was admitted (apparently under duress) on the grounds that she was vouched for as "a high type individual completely dedicated to her work in serving humanity."[35] Despite concurring that the "Negro woman" (the entire report fails to mention Novella Machin by name) had lived up to the expectations of the faculty in terms of her respectability and earnestness, Penland's acting director objected vehemently: "Her presence created problems and special planning which I feel we should not be called upon to deal with again under present conditions."[36]

The ensuing detail of the report, focusing on the management of the occupational therapist's visit, offers excruciatingly vivid insight into the everyday inhumanity involved in the choreography of racism during this period in the South.

Machin was made to reside offsite, in a staff member's home, but was permitted to eat onsite in the main dining room and to attend classes with fellow students. Just prior to the African American student's arrival, a delegation of disenchanted Penland staff approached the acting director "to register their protests and regrets" at her attendance at the school. Willis apparently responded with a plea for their "tolerance and support," but nevertheless went on to describe the extent to which the "Negro Woman's" visit had been an affront to rights of its white faculty members: "We have teachers and workers on our summer staff from the South, the North, and the middle West, and I do not think there is a one of them from any part of the country who approved of integration at Penland, and for some of the fine, loyal instructors from the Southern states, it was a bitter pill to have to take."[37] Certain faculty members took few, if any, pains to disguise their animosity toward the visitor from Chicago, and likewise their manager made no attempt to stymie their overt racism: "They did not accept her socially, nor did I ask them to, and this meant that sightseeing trips, picnics, and other social events either had to be cancelled or she had to be taken with individuals who volunteered to do it. The usual trip to the Lily [textile] Mills for that session was cancelled because it was felt that we should not subject our hosts there to any embarrassment."[38] Further damning details provided by Willis are couched in terms of the discomfort inflicted upon the Penland faculty rather than betraying any empathy for the hostility their guest had been subjected to under their care: "She was taken in a private car to the Craftsman's Fair but after they got there, no eating place in Gatlinburg could be found that would serve her lunch, and so food had to be taken to her in the exhibit building. One of our students from South Carolina—a gentleman of the Old South—who has been coming to Penland every year for several years cancelled his reservations and left."[39]

The emotive report filed by Willis, as acting director to her superior, should perhaps be viewed with a modicum of skepticism in terms of its validity as an objective account of the events that unfolded around Novella Machin's visit. Although details of the racism Machin encountered during her short stay at Penland are borne out by factual histories of the period, the apocryphal report is also knowingly constructed as a means of defending Willis's opportunistic instigation of a prosegregation policy at Penland during the absence of its director: "Our teachers and workers are fine, conscientious people with honest scruples who object not to the Negro as an individual," she concludes, "but to the philosophy of integration."[40] This statement, made in December 1959, came five years after the US Supreme Court had ruled in *Brown v. Board of Education* that segregation in public schools was unconstitutional, and despite the fact that it had been legally permissible (at least since the mid-1940s) for

African Americans to be admitted into private institutions of higher education in North Carolina.

What Is Design? The 1960s Politics of Popular Consumption and Design

The regional specificity of racial and social politics, seen through the examples of Black Mountain College and Penland, cast light on the ways in which Papanek's designs for the "underdeveloped" as head of product design at NCSU, were intertwined on a microlevel with the region's racial politics, and on a macrolevel with those of broader US Cold War development strategies for which industrial design had been overtly weaponized.

Besides Fuller, there is no record of Papanek having made relationships with alumni of the Black Mountain College, despite his dual profile as an émigré designer and experimental teacher. Strangely it proved to be the more traditional environs of Penland that exerted an influence on his design thinking, compelling him to become a visiting tutor in "comprehensive design" at the handicraft college in the summers of 1962 through 1967.

Deeply incongruously, given Papanek's technological approach to product design and bionics at NCSU, Penland School of Crafts (with its anti-progressive stance on racial segregation and its homespun design approach) emerged as his second pedagogical home during his time in North Carolina, and even after his later move to Indiana.[41] A counterpoint to his technologically driven medical and development design experimentation, its craft traditions became part of his holistic design approach.[42] Tutoring there considerably impacted the key tenets of his design treatise: when later addressing the issues of design's capacity to generate social inclusion in his seminal work *Design for the Real World*, Papanek specifically referenced "uneducated rural poor people" as potential beneficiaries of a shift toward socially responsible design.[43] He also featured handcrafted Southern Appalachian wooden candlesticks prototyped in the workshops at Penland as a means of illustrating how designers might harness cottage industries in an attempt to see beyond corporate "managerial decision-makers and consultants" to an alternative model of the integrated design team.[44]

The appeal of craft industries, Indigenous making, and vernacular form seems moderately less anachronistic when viewed from the perspective of Papanek's critical engagement with popular culture and his broader anticonsumer culture rhetoric coined in the early 1960s. As head of product design, one of his first public initiatives had been to build on his previous broadcasting experience in the late 1950s presenting and writing the series *Design Dimensions*, inspired by his connections with media theorist Marshall McLuhan. In the summer of 1962, shortly after his arrival in Raleigh, he launched a twelve-part series on the University of North Carolina public television (WUNC-TV) titled *Pop Culture* (see figure 5.1). Each program in the series promised an "essay in mass media," inducting the viewer into edified popular culture consumption, understanding the "drawbacks" of certain cultural forms as well as their "good attributes."[45] The analytical tone Papanek invoked is startlingly contemporary in tone, predating media studies and cultural studies as discrete disciplines by several decades. Take, for example, his treatise on the soap opera, a popular form he identified (with a nod to émigré cultural theorists Adorno and Horkheimer) as having already been "studied in the thirties and forties" but whose "sterilized, homogenized, pasturized [*sic*] television daytime counterpart" now stood as "even more interesting phenomenon."[46] Echoing the voice of the influential critic of mass cultural comics and media Frederic Wertham (with whom Papanek had worked as a young refugee in New York City), other essays dealt with "comics," "girlie magazines," and "mystery confession magazines," with further "essays in mass media" exposing the workings of "image building and soft sell advertising in America."[47] According to the account of one enthusiastic journalist covering the success of Papanek's television debut in North Carolina, the series had provoked unprecedented response from viewers, prompting over fifty letters and phone calls to the station either congratulating or disputing the presenter's pronouncements. Deemed an enormous success, the regional newspaper article mooted the possibility of a spin-off series covering topics such as "Henry Miller and architect Frank Lloyd Wright each of whom Papanek knew personally."[48]

The TV broadcasts built on a growing interest in design first spawned in the mid-century and, by the late 1950s, a didactic literature that had emerged within the design profession itself led by celebrity industrial designers such as George Nelson (*Problems of Design*, 1957) and Henry Dreyfuss (*The Human Factor in Design*, 1960). Nelson's own film, a sardonic twenty-minute commentary on design's complicity within the Cold War arms race through analysis of the design evolution of weaponry, had been broadcast in 1960 on CBS under the provocative title *A Problem of Design: How to Kill People*. But the discursive approach of Papanek's *Pop Culture*, with its self-conscious mélange of classical and pop references, also spoke to the emergence of a new cultural, anthropological, and discursive form of design and architectural theory.

In the first semester of Papanek's teaching at Raleigh, a US-based French writer and design theorist named Paul Jacques Grillo visited him, inscribing his latest publication as a gift with the following words: "To Victor Papanek, to one among the very few—most cordially, Paul J. Grillo."[49] Grillo's book *What Is Design?*, first published in 1960, was indicative of a broader theoretical shift within design and architectural analysis of the material world as an interrelated set of relations, in which vernacular and historical variations of form and action, were seen as constituting an overarching structure of meaning. By the mid-1960s, underpinned by this broadly structuralist paradigm, a gamut of design publications focused attention on the act of making, as well as the vernacular, and the ad hoc, as antidotes to the formal rationalism of modernism and its heroic narrative of progressive technological determinism.

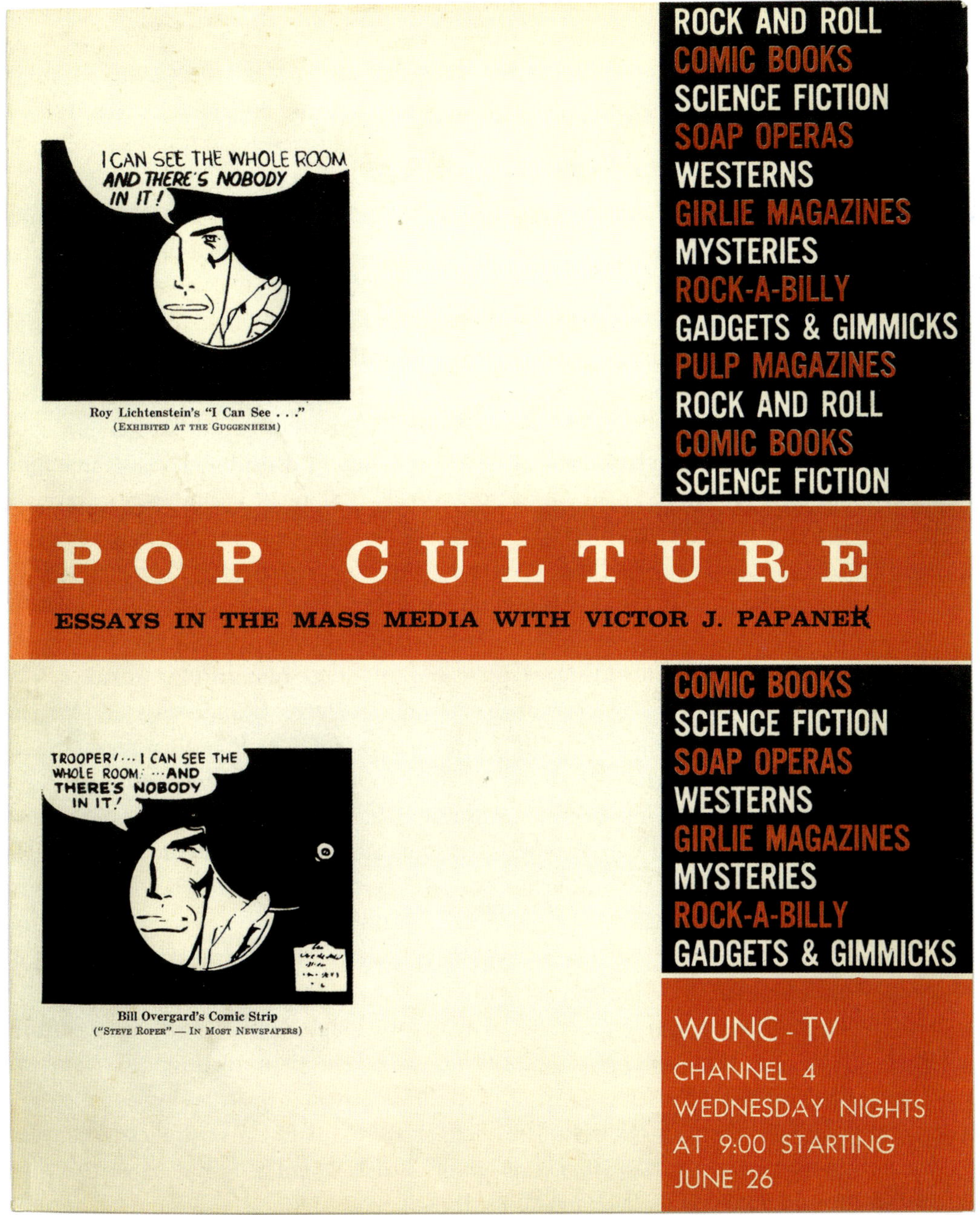

Figure 5.1

Pop Culture WUNC-TV promotional pamphlet, 1955. Courtesy Victor J. Papanek Foundation, University of Applied Arts Vienna

Grillo's publication embraced a turn to informal modes of design specifically to address the design-consuming public, adopting a stirring rhetoric, redolent of that used by Papanek himself a decade later with the publication of *Design for the Real World*. With a flourish decidedly similar to Papanek's typical pronouncements, the inner dust-jacket flap of *What Is Design?* offers this quote from Grillo: "This book is addressed to the public—you and me—abused day in and day out by all the bad design we have to put up with and suffer from in our lives. It is not only written in defense of the public against bad design, but also as a guide to the appreciation of good design, stripped of all high-pressure publicity or prefabricated smiles, cleaned from the dust of conventional thinking."[50]

While Grillo's 1960 philosophical exploration focused on the entwined historical, anthropological, and natural gestations of design "shown in its athletic and wholesome simplicity," that same year a markedly more mainstream exploration of design hit the shelves when bestselling author and social critic Vance Packard released his seminal work on design obsolescence, *The Waste Makers*. An exposé of the inner workings of consumer product manufacturing companies, it proved to be an enormously timely critique of US industry's perceived conspiracy to generate artificial and planned style obsolescence and built on the success of his enormously popular exposé of the advertising industry, *The Hidden Persuaders* (1957). Lifting the lid off the murky world of marketing and design, *The Waste Makers* offered its readers insight into the cynical methods applied to make unwitting customers boost the profits of corporations. These included the application of functionless product features, such as faux automatic push buttons, to encourage ever-rapid turnover in model styles. One particularly hyperbolic chapter, titled "How to Outmode a $4,000 Vehicle in Two Years," traced the Machiavellian strategies devised in the manipulation of consumer consciousness in order to generate "dynamic obsolescence." Consumers were targeted and lured in by "sham" products: once-durable items (such as refrigerators, automobiles) reconfigured through the application of ephemeral and useless fashion details. According to Packard, even everyday consumer appliances were subjected to this regime of bogus styling: "I counted thirty-five buttons and dials on one Hotpoint gas stove," he lamented, noting that "the dials had no connection underneath the cover. They were dummies."[51]

Worse still, according to the popular social critic, the emergence of a "Kleenex culture" of disposable design, concocted through the machinations of marketing psychologists, undermined an American character born of the Puritan ethos of thrift and rationality, and it was one figure in particular whom Packard held accountable for this unprecedented shift to consumer hedonism: Austrian American émigré Ernest Dichter. President of the Institute for Motivational Research, Dichter (dubbed the "Freud of Madison Avenue" due to his introducing psychoanalytic theories to the behemoths of the advertising industry) had been openly accused by Packard of steering a generation of Americans toward a "new mood of self-indulgence" replacing old-fashioned tastes with whims and desires. Dichter's own provocative book, *The Strategy of Desire*, published the same year as Packard's *The Waste Makers*, ran as a counter-thesis to the growing condemnation of postwar consumerism.[52] "Far from being objects of delusion, everyday products from food mixers to vacuum cleaners are objects of 'psycho-economic value,'" Dichter argued. "Objects have a soul," he continued. "People on the one hand, and products, goods, and commodities on the other, entertain a dynamic relationship of constant interaction."[53] In direct contradiction to Packard's approach, which interpreted the 1960s boom in consumerism as being indicative of the growing alienation of modern life, Dichter posited the exact opposite: objects instead offered "a new and revolutionary way of discovering the soul of man."[54]

Although a decade and half older, Dichter shared much in common with Papanek in terms of his émigré background. Like Papanek, he escaped Vienna for the United States perilously late following the Nazi regime's *Anschluss* of Austria, and was born into a family with a small Viennese business. But in terms of his celebratory approach to American consumer culture, he bore more in common with Viennese American émigré Victor Gruen, the pioneering designer of the US shopping mall, both figures adhering to the model of what historian Daniel Horowitz identifies as "the émigré celebrant of American consumer culture."[55] Papanek was undoubtedly aware of the writings and popular discourse around these two high-profile adversaries Packard and Dichter, and certainly drew on both the intellectual posturing on design of Grillo and the core critiques of *The Waste Makers*. Condemnation of planned obsolescence and the detrimental effects of the automobile industry remained salient themes throughout Papanek's career. But by affiliating himself self-consciously with the great cultural characters of America, namely Frank Lloyd Wright and Henry Miller, and by avoiding the avant-gardist and modernist émigré network, Papanek set himself apart. He was certainly not celebrating consumer culture, but neither did he refute the potential benefits of design for the masses.

Besides, there were sources closer to home, in the fields of architecture and design, that proved far more pertinent. One example, *God's Own Junkyard: The Planned Deterioration of the America's Landscape* (1964) by architect Peter Blake, exposed the inherent relationship between the "uglification" of the American landscape and the country's reliance on unbridled capitalistic expansionism. In a condemnatory tone, not dissimilar to that adopted later in Papanek's own writing, Blake identified how the structural elements (speculative development, unmitigated waste production, and dismemberment of the city through the prioritization of the automobile) underpinning a contemporary model of American progress had become culturally naturalized to the long-term detriment of the collective landscape.[56]

But the most visible challenge to heroic modernist formalism emerged in Bernhard Rudofsky's renowned New York Museum of Modern Art exhibition and eponymous book

Architecture without Architects (1964), which constituted a critique of capitalism's impact on architecture by instrumentalizing the comparison of "the serenity of architecture in so-called underdeveloped countries with the progressive chaos and blight of our urbs and suburbs."[57] For a generation of progressive designers and architects weary of the homogenizing constraints and hierarchical politics of Western modernism, Indigenous craft, and vernacular style, the building and material culture of nonindustrialized countries stood as counterpoints to capitalist mass-consumer culture and its discontents. As the dedication in the frontispiece of Grillo's *What Is Design?* and the extensive collection of Rudofsky's writings kept in his personal library attest, Papanek actively engaged with the emergence of this design theory discourse and its identification of nonmainstream Western cultures as an ur-source.

Rather than the bourgeois Bohemian avant-gardist and émigré legacy of Black Mountain College, it was the rural working-class southern setting of Penland coupled with the firsthand observation of racial segregation in urban Raleigh that ultimately helped shape Papanek's design rhetoric; it was a rhetoric that conflated rural and working-class poverty, racial exclusion with the peripheralization of the Global South and the detrimental effects of unbridled consumer culture. Although emanating from a liberal, progressive standpoint, Papanek's ideas intersected with popular and intellectual critiques of consumer culture and a newly honed late Cold War culture of industrial design. This conflated group came to constitute a homogenous entity known as underdeveloped peoples, a problematized mass, the improvement of which would determine the successful expansion of the US economy and of democracy as a global project.[58]

"Design for the Underdeveloped Peoples"[59]

Under the auspices of the design for development rubric at NCSU, Papanek devised his most iconic, enduring, and yet deeply contradictory work: polemical designs that upheld the institutional politics of US expansionism while, dichotomously, appealing to the counter-cultural trope of "hippie modernism" through their upending of core assumptions around the designer's role in perpetuating capitalist consumer culture.[60] Although openly engaged in developing a codesign approach geared toward diverse social groups with users at the helm, Papanek's tutoring took place in an exclusively white, American design program led by male instructors in an equally nondiverse university typical of the period.[61]

Yet from the onset of his career at North Carolina State University's College of Design, Papanek positioned himself on the moral high ground, proffering a more democratic, socially imbued mode of design. The secondary headline of a feature titled the "New Professor's Approach" in a regional newspaper summarized his take-home philosophy on contemporary consumption: "Designing for Need, Instead of Want Makes Products More Usable."[62]

Pictured with Diana Jean, by this point officially his wife, Papanek holds his early design for the organic Shibumi wooden chair. Mirroring the words of the new professor, the article begins, "Industrial design is one of the phoniest professions in the world. It could be one of the most important."[63] Foregrounding his work in bionics, which is described as "the application of nature's principles to industrial design," the journalist (clearly prompted by his interviewee) draws parallels between the projects of Fuller in architecture and those of Papanek in design. But most revealingly, the piece shows how Papanek perceived no inherent contradiction in his advocacy for "the basic needs and wants of undeveloped countries" and the creation of a bionic, self-repairing "marauder suit" for the US Army. The design enabled "an infantry man to have the striking power of a battalion and to leap over obstacles" and, modeled from "encapsulated plastics," enabled the wearer "to blend into the background as chameleons do."[64] The quite aptly named "marauder suit" also featured a "wire launcher on the back for launching missiles," the design for which was based on bionic experiments in the design studio, and "adapted from the pump weed which throws seeds 29 feet."[65] By the release of the first edition of *Design for the Real World* in Sweden in 1970 (the English edition would arrive in 1971), these overt and celebratory references to military design were all but eradicated from Papanek's prose and his résumé, replaced by a vision of socially responsible design geared to the needs of the disabled, aged, socially excluded, and peripheralized. Of all his media coverage, this small newspaper clipping puts the single most powerful spotlight on the true nature of the bionic experimentation Papanek engaged in with his student cohort and the extent to which it blurred military and humanitarian concerns in the broader interests of the US military-industrial complex.

The *All-Terrain Vehicle,* the *Mini-Camp,* and the *Tin Can Radio*

At first glance, Papanek's codesigns with students during this period seem uniquely progressive: *Mini Haul,* a 1962–1963 collective project focused on a brief to devise an inexpensive and versatile *All-Terrain Vehicle* (with a detachable trailer) (see figure 5.2). The *Mini Haul* was conceived with the potential to "be manufactured from materials native to most of the countries included in the list of underdeveloped nations aided by the United Nations."[66] Molded from fiberglass, with a "uni-stick accelerator-brake combination" and an engine allowing for a top speed of thirty miles per hour, the semiamphibious, all-terrain, low-slung vehicle designed "to carry a driver and three stretcher cases, or driver, two passengers and cargo, or a driver and 1,000 pounds of cargo" boasted an ability to "negotiate roadless terrain, sand dunes, snow, ice, swamps, 45° inclines, and small streams" (see figure 5.3).[67] The design's practicability remained unproven, but as a conceptual design the *Mini Haul* project was featured in the university and regional press, as well as the highly regarded *I.D.* journal,

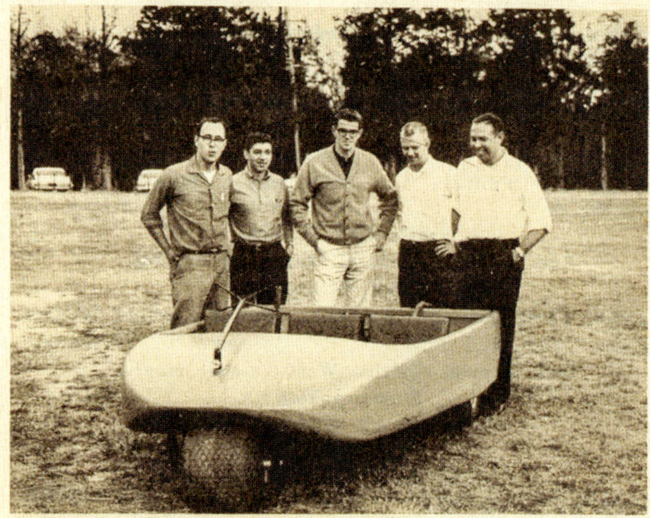

U Haul It In 'Mini Haul'

By Bill Fishburne

Project "Mini Haul" has been completed.

Assigned last year as a project for fourth-year product design students, the vehicle is designed to go anywhere in undeveloped countries.

The vehicle has a 7½-horse power engine, "terra" tires that allow it to have no springs, fiberglass construction and a uni-stick accelerator-brake combination. It will carry five people seated or two people seated plus two more on stretchers; top speed is 30 miles per hour. It can be steered from the center seat or from in front.

Water holds no threats to this vehicle. It floats, and the tires act as paddle wheels to power it. The body is contoured to allow the vehicle to enter the water from a bank as steep as 45 degrees without submerging itself. Speaking about 45 degree banks, it can traverse them in any direction and remain stable, even with a full load of people.

It can be manufactured from materials native to most of the countries included in the list of under-developed areas aided by the United Nations.

Students participating in the project were Bert Olivari, Bill Phifer, George Heeden, Bruce Auld, Bill Huntly, and Don Peeler. The project was under the direction of Clark Macomber and Victor J. Papanek.

Project "Mini Haul" is being admired by five of its creators after its first test run. They are (from left) Bruce Auld, George Heeden, Don Peeler, Billy Phifer and Bill Huntly.

(Photo by Holden)

Figure 5.2
Bill Fishburne, "U Haul It in 'Mini Haul,'" newspaper clipping, North Carolina c. 1962–1963. Photographer: Holden. Courtesy Victor J. Papanek Foundation, University of Applied Arts Vienna

emerging as a significant component of Papanek's repertoire of designs promoted as political tools of self-empowerment. The principal role of the contemporary designer was, according to Papanek, to harness and mediate an interdisciplinary knowledge set, to produce precisely such objects: "By directing the student towards design work that has the social good in mind and exposing him to the interrelation of such skills as bionics, anthropology, and advanced psychology, he will be able to fulfill a more valuable function in our society."[68] "Certainly a stiff order but worthwhile," commented the *I.D.* journal editorial, belying a decidedly more skeptical take on the socially transformative potential of design.[69]

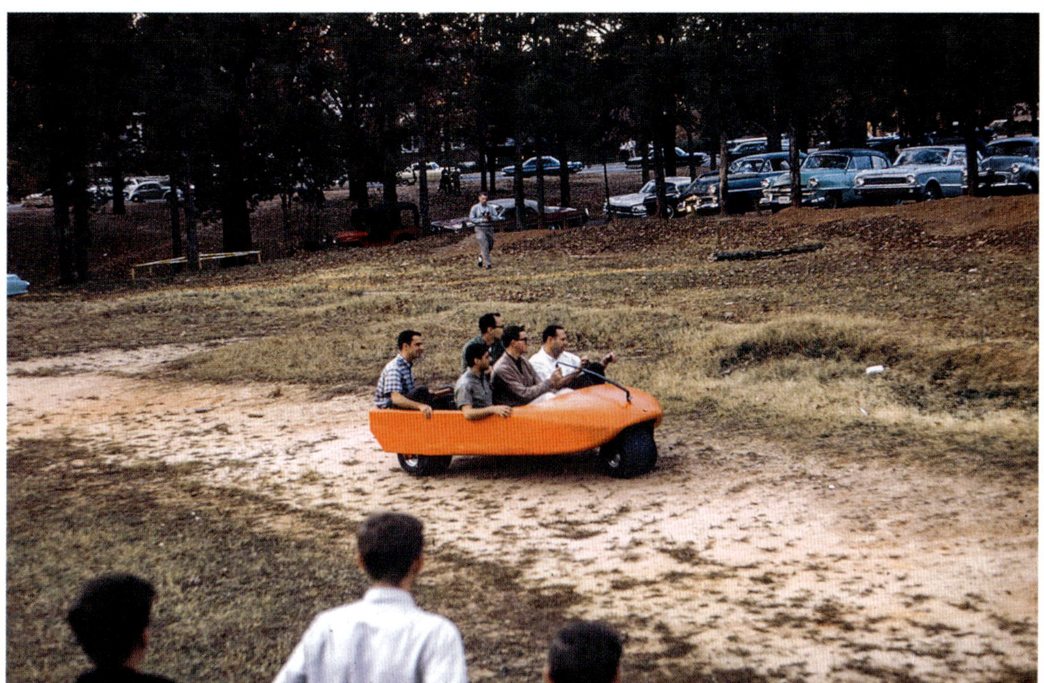

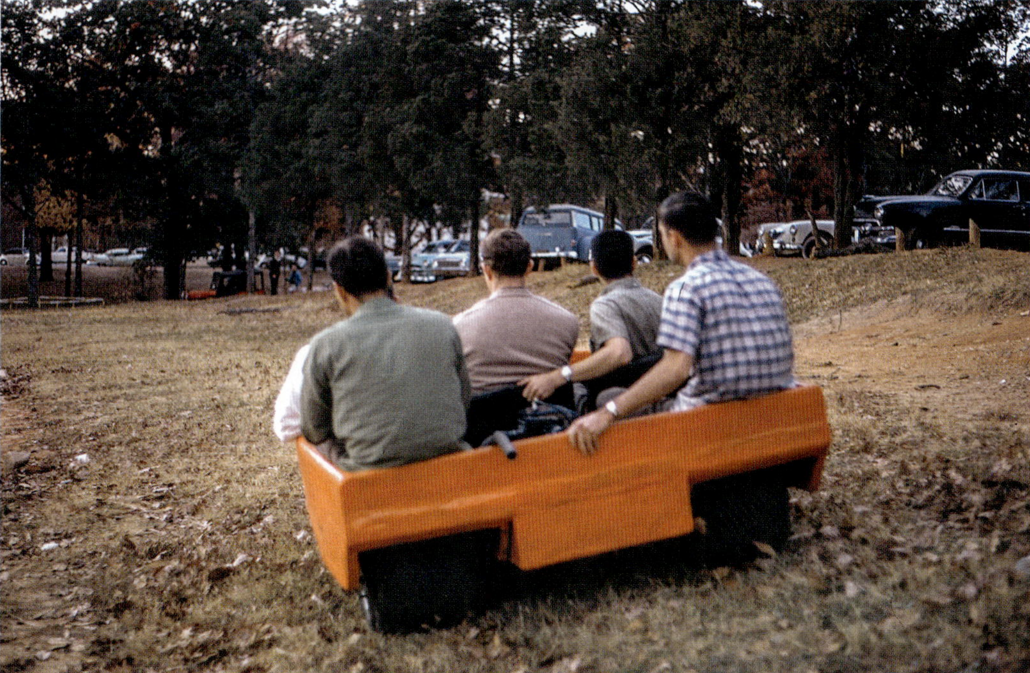

Figure 5.3
"Mini Haul/All-Terrain Vehicle" 1962–1963 color slide, Victor Papanek. © Victor J. Papanek Foundation, University of Applied Arts Vienna

Alongside the national design press's interest in the project, the number of press images released with the launch of the *All-Terrain Vehicle* testifies to the enthusiasm that the North Carolina design college management harbored for the project as a showcase of the school's unique output. An evocative representation of the futuristic design shows a clean-cut, all-male, all-white posse of industrial design students accompanied by their dapperly dressed professor, squeezed into a luminous orange prototype pod set against the anachronistic backdrop of conventional gas-guzzling 1960s automobiles in a nearby university parking lot. Despite its mention of all-terrain capabilities, the *Mini Haul* was principally aimed at developing countries on the African continent. Formally, the project is indelibly marked by what historian Felicity D. Scott describes as the construction of a "neocolonial form of dependency" generated through the 1960s shift of US attention from "the reconstruction of Europe through the postwar Marshall Plan to focus on the Global South."[70] Yet for a design intended for local appropriation and adaptation, the micropolitics of its creation, as well as the proximity of the project and its makers to the highly visible racial politics of the South, suggests a gesture (albeit an inadvertent one) toward the contemporary understandings of self-determinism and pan-Africanism.

Significantly, the design school at Raleigh sat in close proximity to some of the most significant civil and racial rights struggles of the early 1960s, with North Carolina witness to major violent protests. Indeed, in 1960, the historic African American Greensboro sit-in had taken place just an hour's drive from the school campus. African American students from the North Carolina Agricultural and Technical College (later dubbed the Greensboro Four) famously occupied the "whites-only" Woolworth's lunch counter in defiance of segregationist policies, acting as the catalyst for similar organized protests in nearby southern communities and nationwide. Papanek's arrival in 1962 coincided with a succession of high-profile, violent antisegregation protests in North Carolina drawing on the impact of the Greensboro sit-in.[71] The Pullen Park Pool integration protest of August 1962 in Raleigh, heavily reported in the local press, could not have escaped his attention; four African American young men accompanied by their two white friends jumped in the all-white pool, prompting the sounding of a bullhorn alarm and the immediate removal of the group by the employees and the temporary closure of the pool.[72]

The year 1962 also proved a watershed for the pan-African cause, when Algeria's secession to independence from France led to the instatement of an enthusiastic advocate of pan-Africanism as its president, further invigorating the idea of a unified political bond between people of African heritage that had first emerged with the 1957 independence of Ghana, which itself had offered "a beacon to a new generation of African Americans" in terms of defining "new diaspora identities."[73]

On a personal level, as a white European émigré Papanek experienced an early (albeit limited) exposure to debates around African American rights and empowerment through his association with the Lafargue Clinic for Black Americans in the 1940s (where, as a newly arrived Jewish refugee to New York City, he acted as a lay clinician, carrying out Rorschach tests). The Lafargue Mental Health Clinic in Harlem had garnered a reputation for its extraordinarily progressive and practical response to what its supporters identified as the mentally debilitating effects of culturally institutionalized racism. The institution benefited from the support of high-profile individuals such as the famed actor and baritone Paul Robeson, a key figure of pan-Africanism and Black internationalism whose controversial passport cancellation in 1950 by the FBI on account of his politics being viewed as a threat to American security became a cause célèbre.[74] By the early 1950s, the radically progressive ideas around African American youth and working-class rights that Papanek had witnessed during his work with the Lafargue Clinic were being systematically quashed by the stringent political censorship of early Cold War ideologies. The pivotal Asian-African Bandung Conference of 1955 had been a self-determining attempt of the Third World to extricate itself from the polarized politics of the Cold War—and distance the continents from the vestiges of colonial power. Sections of the African American press hailed the conference, to which neither the United States nor the Soviet Union was invited, as "a turning point in history" and a "clear challenge to white supremacy," revealing that "the majority of the world's people think there is an alternative to following blindly the lead of either Russia or the United States."[75] Robeson and leading African American activist W. E. B. DuBois had both been prevented by the US government from traveling to the Bandung Conference, perceived as a gathering of disruptive voices set on jeopardizing the handling of the inflammatory "color question," which was viewed by the Eisenhower administration as one of the weakest links in America's defense against Communism.[76] This is not to suggest that the intricately complex and opposing debates around Black political empowerment and anticolonialism—which many critics viewed as being eroded rather than advanced through the compromise of liberal, pro-civil rights legislation of the Kennedy administration—occupied the curriculum of students engaged in design for development initiatives in 1960s North Carolina. Yet the broader geopolitics of America's "color question" unquestionably acts as a set of previously unexamined indices by which the projects of Papanek and his codesigners could be located. Anthropologist Arturo Escobar, in his study of the intertwining of design and the Western project of development, flags the prescience of Papanek's definitive work, *Design for the Real World*, as an intervention made "as industrialism and US cultural, military and economic hegemony were coming to their peak."[77]

Papanek's designs for the Global South and the socially excluded emerged concurrently with hybridized and inherently dichotomous material manifestations of liberal countercultural progressivism and Cold War political and military expansionism. Certainly, the concepts that underpinned his work never shied away, for example, from the military application of an expanded, late-postwar definition of industrial design. In a 1963 issue of *I.D.* Papanek stated:

> Born as a gimmick laden sales device in the Depression, design is now beginning to assume its true status: to provide meaningful design of all the objects that mass production and an expanded economy needs. Such fields as military hardware, farm implements, appliances, furniture, toys, surgical and medical tools, all come within the range of the product designer. In spite of the tawdry fads and gimcracks, the field has become the logical tool with which mankind can transform its environment into a more meaningful one.[78]

One such project, the *Mini-Camp*, a lightweight inflatable tent trailer prototype of 1964, won the Alcoa Student Design Merit Award from the Aluminum Company of America. Weighing around two hundred pounds, the "umbrella tight tent" fixed to a trailer with a built-in "compressed air supply" could "be erected in about two minutes through the normal device of inflated nylon tubes that form part of the tent wall" (see figure 5.4).[79] A versatile contraption for either humanitarian or outdoor-leisure use, it did not take a great stretch of the imagination to envisage its adaptability to military ground operations.

The *Tin Can Radio*, a humanitarian object pitted against "tawdry fads and gimcracks," evolved as the single most iconic design of Papanek's career (see figure 5.5). A "one-transistor radio, using no batteries or current," formed from a recycled can filled with paraffin wax and wick, and powered with rising heat to be "converted into enough energy (via a thermo-couple) to operate an ear-plug speaker." The design also allowed the use of cow dung, the emanating biogas harnessed to generate heat and power the apparatus. The *Tin Can Radio* emerged from a collective speculative design project for the United Nations Educational, Scientific and Cultural Organization (UNESCO).[80] The prototype, which Papanek designed with student George Seegers for Indigenous Indonesian users, drew on a McLuhanesque vision of a "global village" united through the availability of freely dispersed communication apparatus.

Early in the summer of 1966, a couple of years after Ulm School of Design tutor Bonsiepe had paid an exploratory visit to his product design program in Raleigh, Papanek reciprocated with a visit to Germany.[81] There he presented a lecture in German for the Ulm School professors, the title of which was translated in English as "Industrial Design: Design, Bionics, Man and Environment," with the *Tin Can Radio* as its focal point.

In his slide show on his cutting-edge work for the Global South, Papanek projected two versions of the *Tin Can Radio* for his audience. The first version was the stark "9 cents" low-tech, unornamented upcycling prototype, presented as a democratic solution to expanding globalized power relations of modern media communication, intended for dispersal among

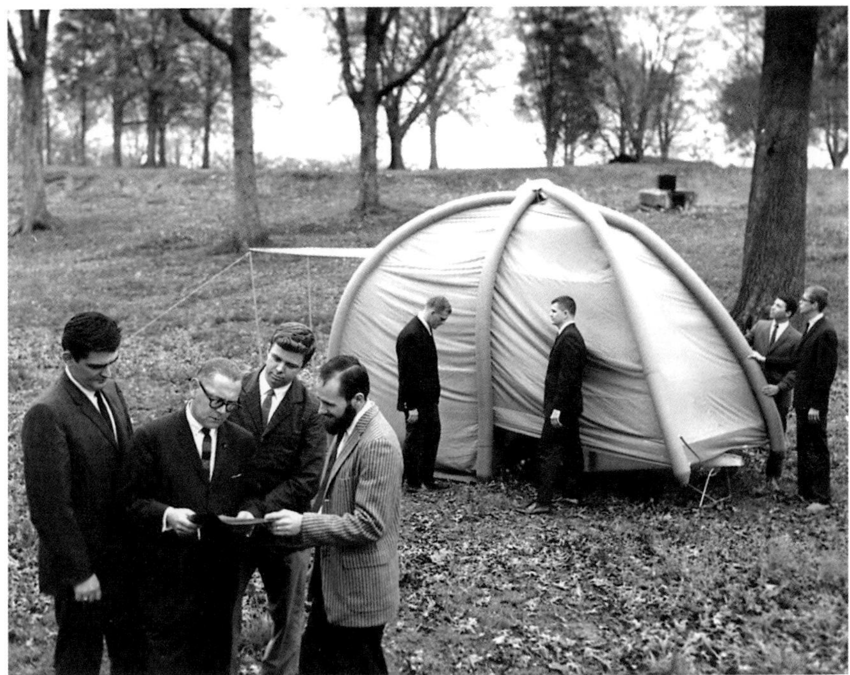

Figure 5.4
Mini-Camp inflatable tent for humanitarian or military application. © Special Collections Research Center, North Carolina State University Libraries, Raleigh

nonliterate Indigenous communities and allowing them to receive broadcasts such as governmental health advice. At a design school specializing in industrial design and visual communication and geared toward an alternative postwar German politics, the pared-down tin can apparatus offered a potentially interesting proposition to the audience. But the Ulm professors looked on with increasing unease as the slide carousel clunked around to reveal the second version of the *Tin Can Radio*, where it was decorated with a highly colored layer of embroidered appliqué, shells, and sequins, illustrating how Indigenous users might aesthetically appropriate the objects into their everyday cultural rituals (see figure 5.6). In Papanek's account, the Ulm faculty condemned the design for its "ugliness" and its lack of "formal" design, the professors in the audience quitting the lecture in open disdain. "Of course, the radio *is* ugly," protested Papanek, "but there is a reason for this ugliness. . . . I feel that I had no right to make aesthetic or 'good taste' decisions that would affect millions of people in Indonesia, who are members of a different culture."[82]

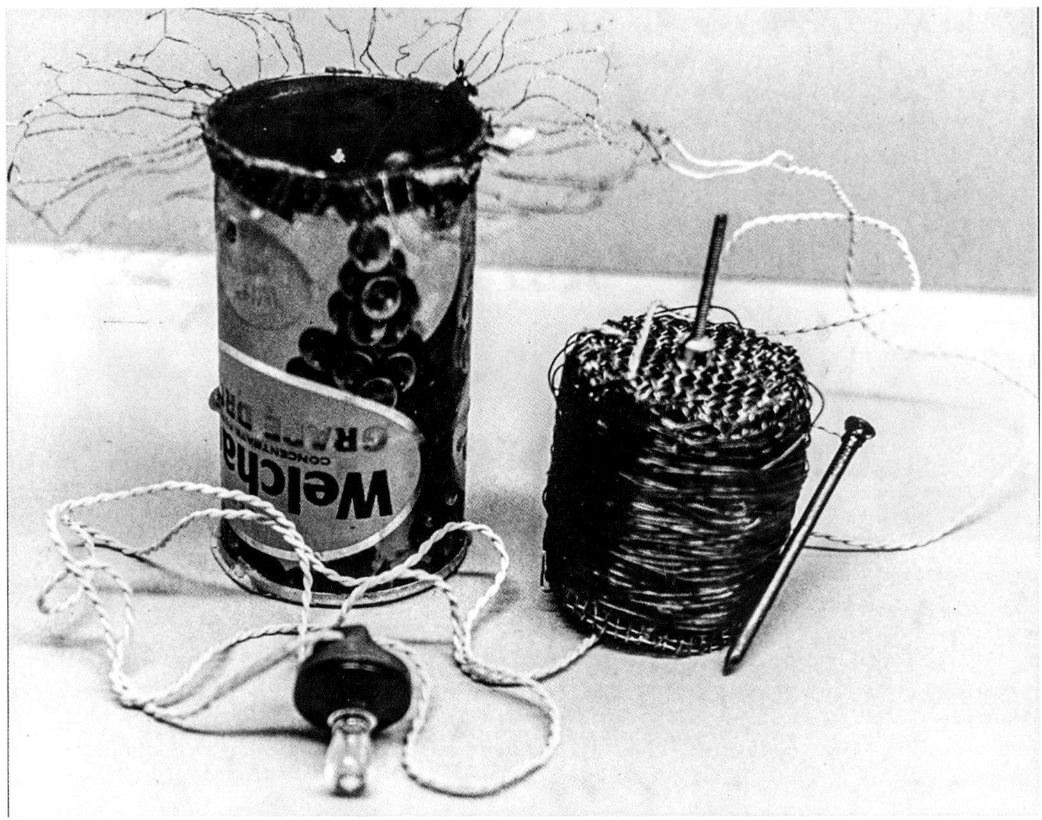

Figure 5.5
Tin Can Radio designed by Victor Papanek with student George Seegers, "Radio Receiver, concept for the Third World." © Victor J. Papanek Foundation, University of Applied Arts Vienna

This doomed Ulm experience would later provide a pivotal and provocative anecdote in the pages of *Design for the Real World*, through which Papanek cast himself as a humanitarian designer thrown to the lions of high modernist elitism.

Retrospectively, the ill-fated visit to Ulm served Papanek well in promoting his brand of humanitarian design as a counteraction to the dilettantism of his rationalist contemporaries, at least in the way that he retold the event. Yet a first-person recollection from at least one participant of the 1966 lecture offers an alternative perspective on the reception of Papanek's work. Dieter Raffler, an idealistic student of industrial design at the Ulm School, attended the "Industrial Design: Design, Bionics, Man and Environment" talk with a keen interest in expanding his knowledge of cutting-edge advances in the United States. Instead, according

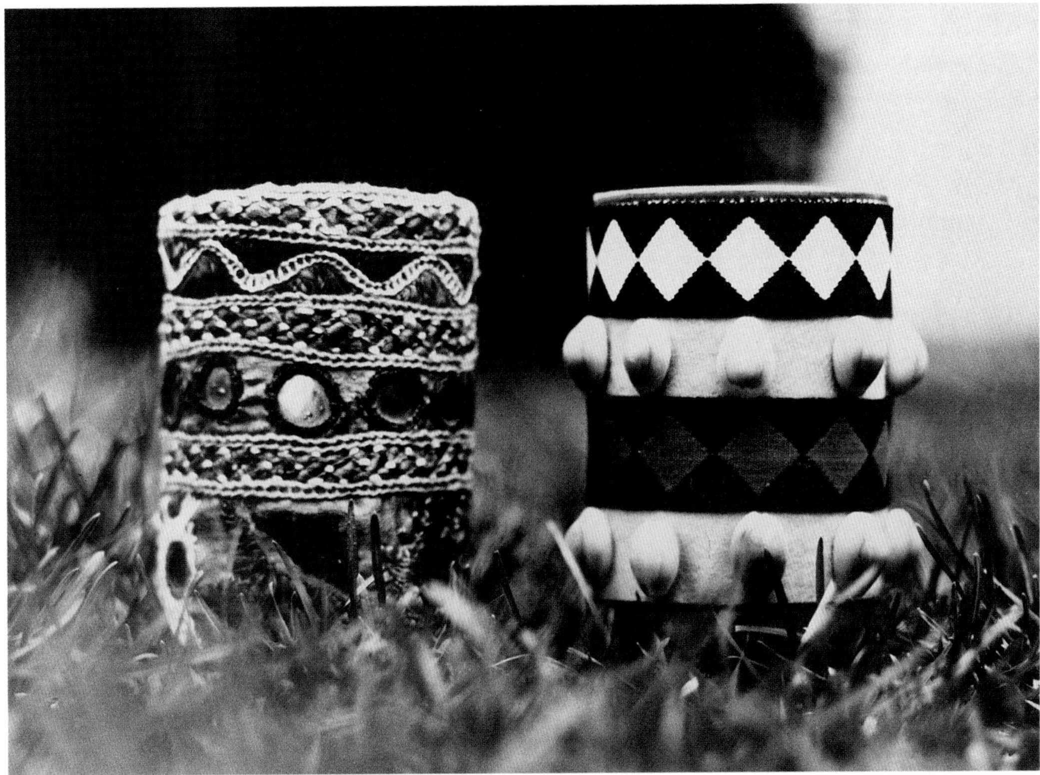

Figure 5.6
Tin Can Radio designed by Victor Papanek with student George Seegers, customized by users. © Victor J. Papanek Foundation, University of Applied Arts Vienna

to his account, he and fellow audience members found themselves horrified by Papanek's succession of design proposals and prototypes for military intervention—including what appeared to be an aerial design mechanism for the dispersal of napalm. Collectively opposed to the US military intervention in Vietnam, the Ulm audience found Papanek's open support of the war and chemical warfare—he was willing to design prototypes for the more efficient distribution of herbicides and devasting chemical weaponary—morally repugnant. At the close of Papanek's lecture, which members of the audience did indeed leave in protest (although not, as the designer suggested, over the tastelessness of the decorated tin can radio), members of the faculty refused to offer the guest speaker a ride to the train station, so reluctant were they to engage with the man or his politics.[83] Designs such as the marauder suit and bionic napalm distributor had not been the impetus behind the Ulm School's reaching

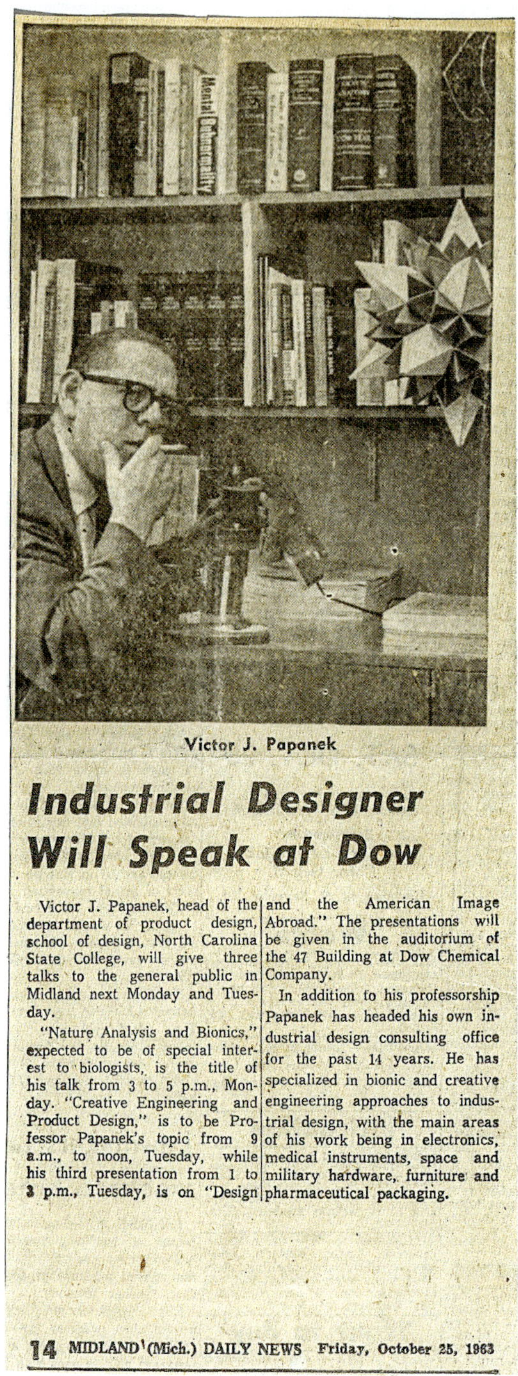

Figure 5.7

"Industrial Designer Will Speak at Dow," *Midland Daily News*. © Midland Daily News, Midland, Michigan, October 25, 1963. Courtesy Victor J. Papanek Foundation, University of Applied Arts Vienna

out to Papanek as a speaker; on the contrary, students and professors had envisioned an appeal to the social and politically radical role of design in the postwar context.

Despite these contradictions, by 1968, the year of the legendary student uprisings, Papanek would find himself addressing radical student activists from the podium of cutting-edge design symposia across Northern Europe, where he condemned the horrors of the Vietnam War and the wanton waste of US consumer corporations. Yet in 1963, he had willingly taken up a spot as guest speaker at the Dow Chemical Company branch in North Carolina, boasting of his decade and a half of professional background in "bionics, electronics, medical instruments, space and military hardware" (see figure 5.7). In the main auditorium there he delivered lectures exploring "Nature Analysis and Bionics," "Creative Engineering and Product Design," and "Design and the American Imagination Abroad."[84] Papanek's seductive polemic, his rage against the corporations and the military, all belies the contradictions inherent to the politics of design and the notion of the industrial designer as an advocate of Western democracy in the Cold War era. Throughout this period, the US military and the consumer and chemical industries were the major sponsors of experimental work in industrial design such as Papanek's bionic marauder suit. Buckminster Fuller's open involvement, with the first full-scale realization of the geodesic dome created for the US military in 1954 to house the DEW line, built in the Arctic region of Canada and stretching across the Faroe Islands, Greenland, and Iceland, provided Papanek a perfect role model, to which he added the spin of socially responsible design. Despite the vitriolic tirade against the establishment that the agent provocateur had honed by the time of his seminal book's publication, in practice Papanek keenly collaborated with the arch players of the military-industrial complex. By the late 1960s, the Dow Chemical Company had become notoriously synonymous with the United States' crimes against humanity: it was the infamous manufacturer of the devastating napalm and Agent Orange compounds used by the nation in chemical warfare during the Vietnam War, prompting the widespread student demonstrations of which Papanek was fully aware.

In July 1964, utilizing the training he had received at the mid-1950s Creative Engineering conferences at MIT, Papanek launched his own Creative Problem Solving Workshop at North Carolina State University at Raleigh.[85] The promotional brochure accompanying the event stated that the workshop leader, Professor Victor Papanek, had conducted seminars for leading organizations such as "Canadian Westinghouse, Ltd; Army Command and Management School, Fort Belvoir, Virginia; Sylvania Electronic Products, Spaulding Fibre Products, Buffalo, N.Y.; Naval Intelligence, Washington, D.C.; Counter Insurgence Branch, Special Warfare Division." The extent to which the details on the brochure were inflated in terms of being a true representation of Papanek's affiliations remains a moot point. What these details offer instead is insight into the makeup of the anticipated audience for the NCSU workshop and the reputation Papanek had accumulated during his role as head of product design.

Leaving North Carolina: The Bionics Professor Gets Fired

By late July, when Papanek was scheduled to deliver his Creative Problem Solving Workshop, he had in fact already been ousted from the university; the brochure was hurriedly altered to reflect this change with the words "Professor Victor J. Papanek, formerly Head of the Department of Product Design." Despite his open collaborations with Dow Chemical and the US military, in a twist of irony, it was his overt and public liberal stance that precipitated sudden fall from grace.

Following the assassination of John F. Kennedy on November 22, 1963 (coinciding with Papanek's fortieth birthday), the designer penned a fiery and disgruntled letter to the widely read regional newspaper the *Raleigh News and Observer* condemning (in the acerbic prose he normally preserved for his design criticism) the outright negligence of the state of North Carolina to enforce respectful shutdown time for commercial and public enterprises in the days following the iconic leader's death.[86] Traveling en route from Buffalo via Washington, DC, Papanek had picked up the regional newspapers at the airport, and had carefully taken note of the various state responses to the nation's mourning and public preparations for the president's funeral:

> To The Editor:
>
> Returning to Raleigh, after having spent the last few days in Buffalo, N.Y. and Washington, is somewhat like landing on Mars or entering a foreign country. Every university, college, public, private, parochial and preparatory school will be closed all day Monday in Virginia, South Carolina, Maryland, Delaware, the District of Columbia, New York, and New Jersey to my knowledge. We must assume that this is true of most, if not all other, 49 states. . . . N.C. State, instead of closing is merely extending its lunch hour by 60 minutes. The papers are crammed with advertisements of pre-Thanksgiving sales, stores will, for the most part, be open. It seems to this writer, that this direct disregard for President Lyndon B. Johnson's proclamation of November 23 is not only incredible, but hardly an act that will give the state a good name. As far as I can determine, our attitude of "business as usual" (with an extra 60 minutes for lunch!) is shared only by Red China and Mr. Castro's Cuba.[87]

The provocative letter was duly published with Papanek's full university address and credentials laid bare, for all to see. Days later he received threatening telephone calls at his home, callers accusing him of being a "Kennedy-loving son of a bitch."[88] Branding him as a radical progressive, the ensuing barrage of threats received by Papanek following publication forced him to reconsider his position in North Carolina. It also fully explained why Papanek kept a revolver stashed in the side pocket of his vehicle; far from being a prop of macho bravado, it revealed the genuine concern Papanek felt for his safety.

Although the design professor stated carefully that his letter was "a personal statement," he fully implicated the NCSU faculty and students by pointedly making mention of his hope that they shared "his sentiments." The board of trustees demanded he write an apology to the newspaper, retracting his criticisms; Papanek refused. The president of the university duly served Papanek his notice, forcing him to resign at the end of the academic year in June 1964.[89] Having secured a prestigious tenure-track professorial position at one of North America's leading postwar design schools, and having built up a unique and award-winning department of innovative industrial design, Papanek had within months of taking up the post alienated himself by refusing to kowtow to the region's conservative politics or his university's authorities.

That same summer of 1964, Papanek's application for membership in the Industrial Designers Society of America (IDSA) was rejected.[90] His waning professional fortune echoed in his personal life, when a week before leading the Creative Problem Solving Workshop as a now-disgraced professor, his fourth wife Diana Jean Kleefeld secured the first legal stage in divorce proceedings against him.[91]

Despite the inherent contradiction, his work at North Carolina would prove to be the most significant of his career, paving the way for the ideas and designs showcased in his seminal *Design for the Real World*, which, in turn, would transform the work of a generation of industrial designers under the rubric of social design. What his later lectures and publications failed to acknowledge was his complicity in military experiments, and the more obscure and ambivalent design projects with which he engaged during his time as Head of Product Design in North Carolina, such as the creation of a low-cost, plastic and steel, self-assembly "revolver 822 single shot gun" designed for "backward areas" and produced "at less than $1.50."[92]

6 Prescription for Rebellion? From Styrofoam Domes to Animatronic Women

In April 1970, the widely circulated national newspaper magazine of the *Chicago Tribune* ran a full-color feature under the byline "Design and the New Environment," introducing the American public to one of the nation's leading critics of design for consumer culture. The *Tribune* reporter told of a professor who, with a "combination of mordant wit and persistent disenchantment," took to task the makers of environmentally hazardous products that "befoul our world."[1] This strident academic, Victor Papanek, embarked on a full-on tirade against the US automobile industry: "First they build the most perfect killing machine yet devised by man. As it runs, it pollutes the visual environment. Its offal pollutes the air. And when it expires at an early date, the result of planned obsolescence, its remains clutter up the landscape and waste space."[2] Sketching out an "education revolution" in the form of "total design," the piece described in detail the social design projects overseen by its interviewee in his capacity as chair and founder of the Department of Industrial and Environmental Design at Purdue University, Indiana; these projects ranged from designs for a cerebral palsy sensory-play environment to adapted bicycles for disabled children, antierosion soil burrs, and low-cost televisions for distribution in remote parts of Africa. Making no mention of the professor's consultancy role with Dow Chemical Company, or his prior US military research sponsorship for bionic weaponry and defense projects, the feature reported how Papanek was in high demand as a consultant with prestigious organizations such as NASA and UNESCO.

The regional press images released to accompany his designs for "underdeveloped countries" during his tenure in North Carolina had shown a dour, straight-suited, bespectacled professor flanked by his clean-cut, all-male design protégées. In marked contrast, the *Chicago Tribune* profiled a tweed-jacketed intellectual who tutored a progressive coeducational cohort of black-turtleneck-wearing industrial designers. One design project for a mathematical metric conversion device executed by a female student, Anne Cooper, warranted particular mention, though the reporter felt compelled to offset her talents as a design engineer by flagging her other role as "a pretty girl reporter for the student newspaper."[3] Located deep in the US Midwest, Papanek's student cohort was nevertheless more diverse than it had ever been. Furthermore, his student projects and institutional affiliations now extended into the upper reaches of northern Europe through his collaborative pedagogic initiatives in Finland, Denmark, Norway, and Sweden as well as guest critic stints at the highly progressive Parsons School of Design in New York.

In the six years since his hasty departure from the College of Design at North Carolina State University following his dismissal over the provocative pro-Kennedy letter he had published in a local newspaper, Papanek had risen from the ashes of a plummeting career to the extent that his design criticism and ideologies now warranted national mainstream reportage. The details surrounding his appointment at Purdue University, shortly after his departure from NCSU in the summer of 1964, remain hazy. But the position certainly put him in closer geographic proximity to Buckminster Fuller, whose base at Southern Illinois University, Carbondale, provided the location, in 1965, for Fuller's techno-utopian *World Game* resources strategy. Perhaps the most vivid expression of Fuller's thesis of technology replacing politics, the military-inspired *World Game* consisted of "a giant computerized electronic data-visualization map of global events, conditions, and resources, [rendering] visible the inexhaustible wealth of spaceship Earth's "four billion, billionaires heir apparent."[4] That same year, Papanek received a gift of one of the maverick systems theorist's lesser-known early works, *Nine-Chains to the Moon: An Adventure Story of Thought* (1938), inscribed with the words: "To Victor Papanek, in admiration and friendship, Buckminster Fuller, 23 March 1965." The gift is testament to their continued intellectual connection in the mid-1960s, hinting that Fuller may even have provided a recommendation for Papanek's position at Purdue. At the very least, Papanek's informal affiliation with Fuller would have considerably bolstered his candidacy for the post.[5]

Purdue University, like NCSU, sat at the hub of Cold War funding and experimentation, and he found there the perfect fertile ground for his continued exploration within the field of bionics. Much to Papanek's wry amusement, his design studio initially was placed within the home economics division. Before long, however, the newly installed professor made his demands known to the university administrators, campaigning for a fully equipped "wet laboratory" that would allow him access not just to flora for his seed experimentation but to fauna through which his team's bionic prototyping experimentation could be furthered. His appeals to the university administration were rejected on formal grounds (animal experimentation took part only in the licensed areas of the university), and wariness on their part regarding an apparent mismatch between the newly appointed professor's ambitions and his provable scientific qualifications. As his close friend and Purdue colleague, graphic designer Al Gowan, later commented, "They did not think we were qualified to work on animals, and they were right. That sort of experiment was already taking place in the agriculture department, where I had seen a cow with a round observatory window in her side."[6] Other facets of Cold War experimentation were, however, fully sanctioned by the university administration. The facilities and expertise at Purdue in areas such as biology, audiovisuals, and filmmaking allowed Papanek to delve deeply into the integrated dimensions of design, alongside the socially responsible projects that had come to define his genre of design.

Bio Graphics: A Transdisciplinary Filmic Design Experiment

Biological experimentation at Purdue formed the basis of a whole gamut of Papanek and his student team's work; the *Chicago Tribune Magazine* featured one such example in the form of plastic burrs that anchored the ground surface, designed by student James Herold, for distribution in underdeveloped countries to prevent soil erosion (see figure 6.1).[7] But in 1966, Papanek initiated a far more ambitious collaboration with a botany professor from Purdue's Department of Biological Sciences, Samuel N. Postlethwait, who stood out in his utilization of radical, haptic multimedia approaches to teaching that included audio tutorials, time-lapse photography, and visual modeling, all aimed at fostering a more immersive approach to bioscience and laboratory work.[8] Postlethwait devised individual learning booths fitted with audio equipment for focused self-learning, coupled with collective laboratories installed with larger shared equipment such as microscopes and spectrophotometers. Students reported to a weekly general class assembly to share individual findings. The interactive, creative atmosphere appealed to Papanek's notion of comprehensive design and systems theory and an expanded version of creative problem solving he had first explored at MIT in the mid-1950s. Crucially, Postlethwait's biotechnical system also plugged seamlessly into a McLuhanist model of media interdependence.

In 1966, the imaginations of Papanek and his colleague Gowan were specifically captured by the design-pedagogic potential of Postlethwait's cell-splicing, exploratory audiovisual experiments. "Amazed by the visual richness" of the vividly colored microscopic botanical elements, they set about exploring the filmic possibilities of a technique "that offered a virtual trip through a three dimensional object" (see figure 6.2).[9] Using the leftover film footage Postlethwait had generated for auto-tutorials conducted in the individualized learning booths, the Purdue Audio Visual Center spliced selected footage together with a jaunty jazz soundtrack under the direction of Papanek and Gowan and with the guidance of cinematography tutor Roy Mills. Program notes accompanying the later film showings in Finland offer insight into the ways *Bio Graphics* was used to promote a broader ideological stance on "Total Design" espoused by Papanek: "The film 'Bio Graphics' was planned and compiled using what appears to be 'found' biological and parasitological material. We did not make it only

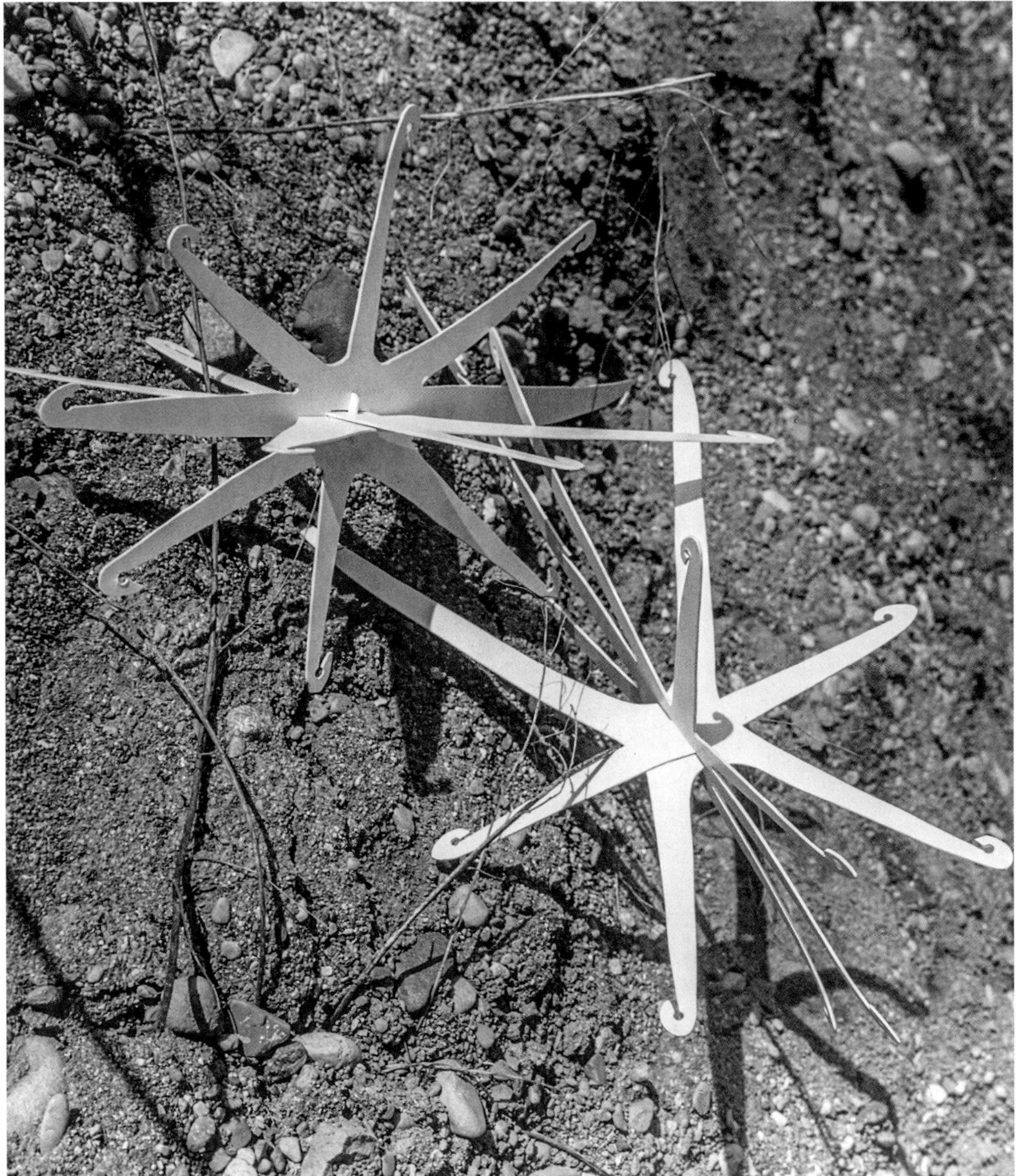

Figure 6.1
"Artifical Burrs," design by James Herold und Jolan Truan, students of Victor Papanek at Purdue University. © Victor J. Papanek Foundation, University of Applied Arts Vienna

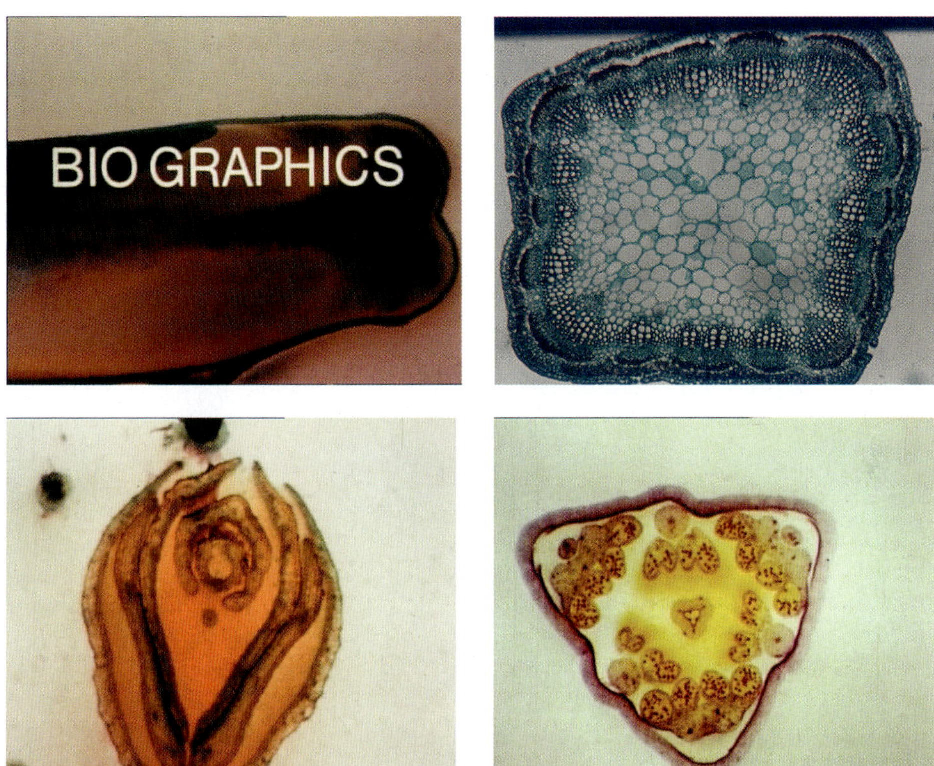

Figure 6.2
Stills of plant cells from *Bio Graphics* film, 1967 by Victor Papanek and Al Gowan, Purdue University.
© Victor J. Papanek Foundation, University of Applied Arts Vienna

because we felt that these visual forms as such are worth seeing, which is also characteristic of our attitude to design. At Purdue University, work in both industrial and visual design is carried out in close cooperation with biologists, engineers, anthropologists, psychiatrists, sociologists and representatives of many other fields."[10]

Filmmaking emerged as an integral component of Papanek's "Minimal Design Team" in the diagrammatic mapping of the politics of design he sketched out as a central tenet of *Design for the Real World*, a diagram first devised in 1969 as a didactic tool for consciousness raising in the field of social design (see figure 6.3). At first glance *Bio Graphics* passes as a purely aesthetic exercise in biocinemacy: a psychedelic, color-saturated, countercultural gesture challenging conservative understandings of design process pitched at invoking the admiration of the youthful student contingent of the 1968 International Design

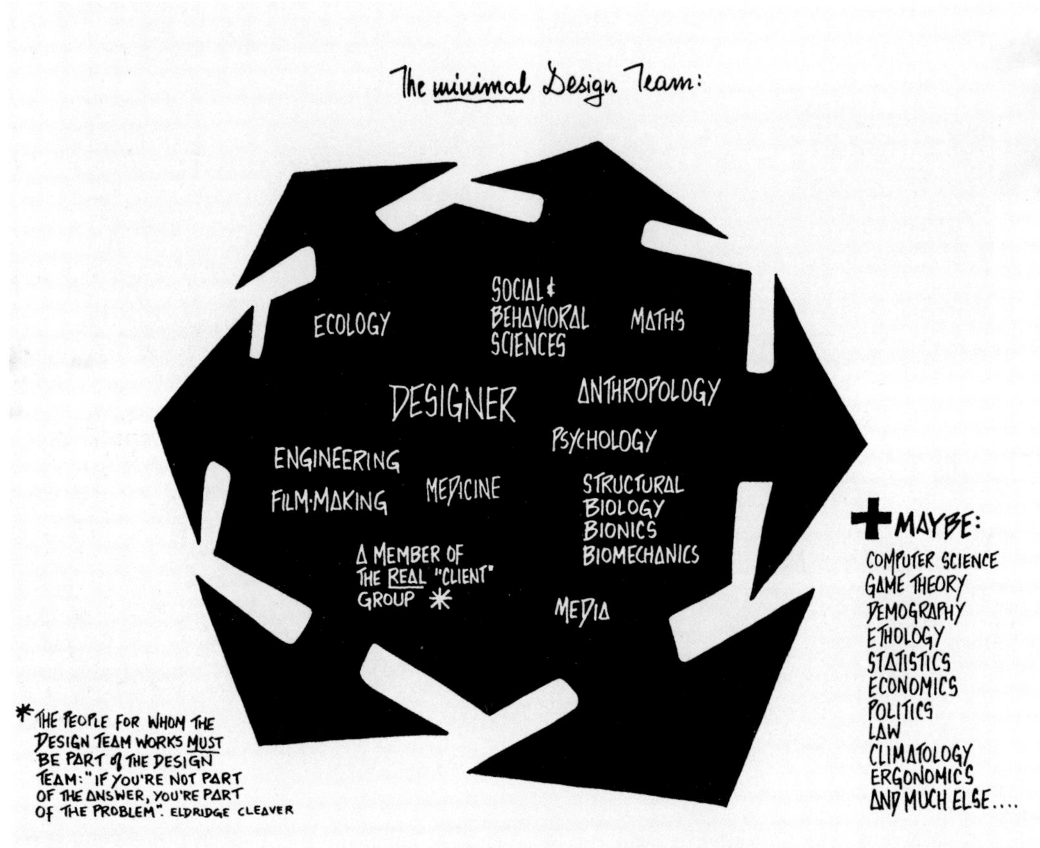

Figure 6.3
"The Minimal Design Team," section of "Big Character Poster No. 1: Work Chart for Designers" [1969] 1973. © Victor J. Papanek Foundation, University of Applied Arts Vienna

Conference in Aspen. Yet that same year Papanek had received a letter at Purdue University from the Chief Scientist of the US Army Research Office addressing him as leader of the "Graduate Industrial Design Group" and outlining the requisite reporting procedure for research conducted "under contract" with the US Army.[11] Although it is unclear whether the communication pertains to a specific project or to a more general research initiative (most likely the extended experiments in bionic design), the letter offers irrefutable evidence that Papanek's transdisciplinary design experimentation served a covertly dichotomous purpose.

On the one hand, this research in the field of bionics fed the burgeoning appetite for a shift in design practice toward societal and ecological issues anticipated by a newly politicized generation of students; on the other hand, it addressed the escalating demand for research innovation promulgated by the behemoth that was the military-industrial complex. In this regard, the accounts of David Hausmann, a former member of Papanek's Graduate Industrial Design Group at Purdue, offer a rare glimpse into the ways in which the design curricula intersected with these two seemingly opposed agendas. In the course of his studies, Hausmann focused predominantly on design for healthcare and medical equipment, through which he devised a color-coded combination-lock childproof pill container that Papanek would later hail, in the pages of *Design for the Real World*, as an exemplary object of socially responsible design. Similarly, the student's work was routinely used as an illustration of the ethos of social-needs-based thinking in Papanek's lecturing. Yet, by his own account, Hausmann coupled this social design with his role as an integral member of the Graduate Industrial Design Group, which received intermittent sponsorship, starting in 1968, from the US Army Weapons Command, Rock Island Arsenal of Illinois. The group, and Papanek as its professor, was involved with the military institution to such an extent that, in 1968, its members were invited to the Biomechanical Engineering conference to immerse themselves in the most advanced thinking in the field. Details of the conference themes and speakers give insight into the interdisciplinary ambitions of the military research agenda more broadly, and its intersection with cutting-edge design and academic research of the period.[12]

Of particular personal interest to Papanek was a fellow Austrian American speaker, University of Illinois Professor Heinz von Foerster, who presented his research on the tracking of snake movement and its potential application to the engineering of vehicle treads in desert environments. Von Foerster offered an extraordinarily apt role model for the ambitious Papanek, having founded and led the celebrated Biological Computer Laboratory at the University of Illinois since 1958, with research centered on the potentialities of biological knowledge and its intersections with contemporary media and technologies. But the Rock Island Arsenal conference offered up more insights, and equally compelling role models, regarding the application of leading academic research to military purpose. Renowned neurophysiologist and cybernetician MIT Professor Warren Sturgis McCulloch (collaborator of Norbert Wiener, the famed exponent of cybernetics) followed von Foerster up to the speakers' platform, addressing the similarities between the neurosynaptic connections of the human brain and those of the computer. As a feted "intellectual showman" and "rebel genius," McCulloch's trailblazing interdisciplinary approach melded brain science, engineering, and philosophy, easily matching and even surpassing the maverick charisma of Papanek's erstwhile hero Buckminster Fuller.[13] These eminent figures, McCulloch and von Foerster, together with fellow

American cybernetician and mathematician Norbert Wiener, among others, were hailed as the leading architects of second-order cybernetics and the exponents of early theories of artificial intelligence, and unquestionably impacted Papanek's vision of "Total Design," in which socioenvironmental, psychological, biological and technological, and analog and virtual elements of design were to be understood as one interconnected entity.[14]

At the most basic level, the conference offered perspectives on the ways in which the intersection of design, mechanical engineering, and biological principles could, by Hausmann's account, "potentially improve peoples' quality of life." On offer was even a model closely resembling the one espoused by Papanek, who professed to use "teaching the research of nature and biology as prototypes for the design of man-made objects and systems" (see figures 6.4, 6.5, and 6.6).[15] The explorations on the interface between biology, industrial design, and bionics—and "the adaptability of certain botanical principles" encompassing methods from model making to photography, film, and feedback testing, according to Hausmann's recollection—primarily consisted of research "toward utilization in new weapons and weapons systems of an offensive and defensive nature" (see figure 6.7).[16]

A designer born of the extreme political and social circumstances of World War II, which he experienced personally as a refugee and professionally in terms of the thrust of his design training from the 1940s onward, Papanek found himself immersed in a design culture geared exclusively to the immediate postwar and then Cold War effort. From his earliest pedagogic experiments in camouflage technologies at the Cooper Union School of Art through to his involvement in the exclusive MIT Creative Engineering initiatives in the mid-1950s, military defense and the bolstering of the consumer democracy as an ideological weapon in countering the Soviet threat lay at the core of the design practice in which he was immersed.

Indeed, the specter of military application left few Cold War research and design departments untouched. Purdue University Audio Visual Center, the unit that had facilitated Gowan and Papanek's experimental film *Bio Graphics*, was also home to the Instructional Media Research Unit (IMRU). The Title VII provision of the National Defense Education Act (NDEA), which supported the effective use of technology for educational purposes as part of the Cold War project, funded IMRU to experiment with the use of film beyond educational training purposes, exploring its potential in military applications including psychometric testing, and focusing in particular on the manipulation of stimuli and memory through analysis of the exposure to moving visuals. The head and assistant head of the research unit, semanticist Warren F. Seibert and education psychologist Richard E. Snow, specifically emphasized in a research paper arising from research in the unit, "Cine-Psychometry" (1965), that "many human decisions are made within contexts that are not static, not principally verbal, and not characterized by print."[17] At this juncture in the Cold War period, this applied, mind-altering film experimentation, according to historian Felicity D. Scott, was recognized as harboring a

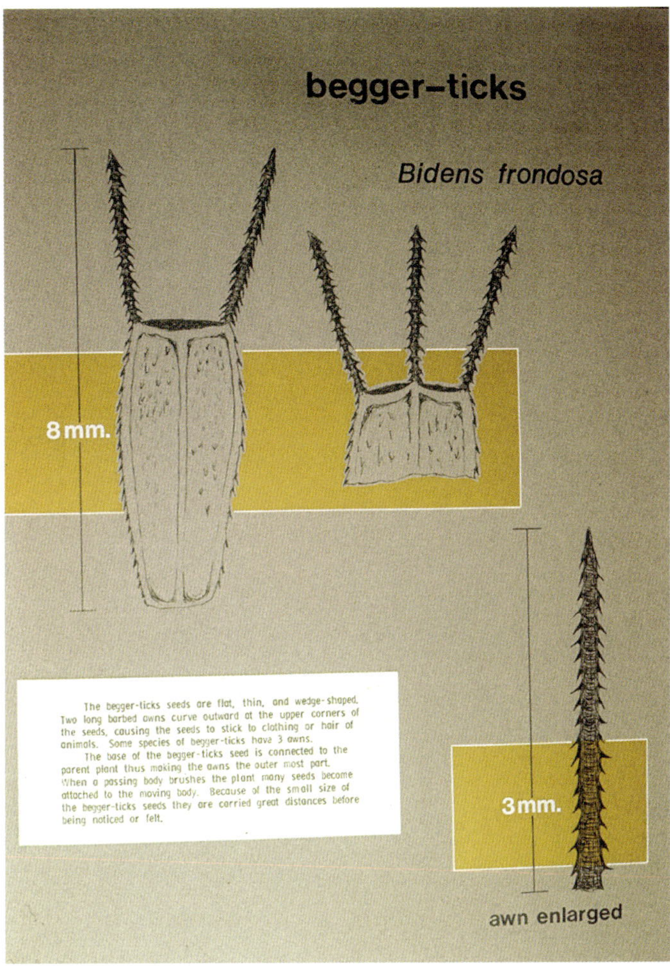

Figure 6.4

"Begger-Ticks [*sic*]," part of Victor Papanek's teaching program of bionic studies used to conceptualize military applications and environmental weaponry. © Victor J. Papanek Foundation, University of Applied Arts Vienna

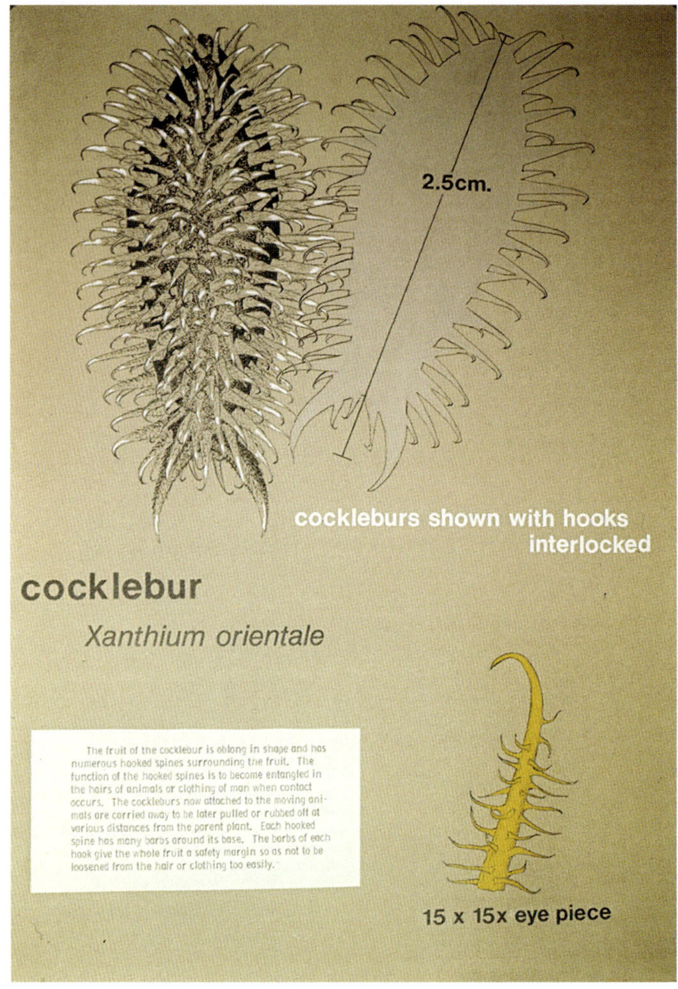

Figure 6.5
"Cocklebur," part of Victor Papanek's teaching program of bionic studies used to conceptualize military applications and environmental weaponry. © Victor J. Papanek Foundation, University of Applied Arts Vienna

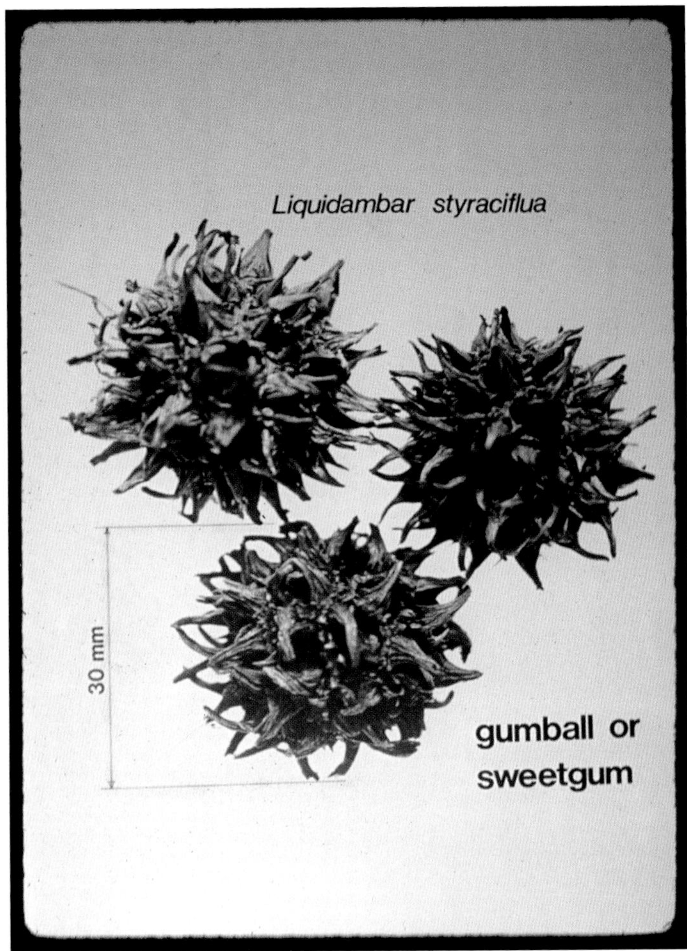

Figure 6.6
"Gumball or sweetgum," part of Victor Papanek's teaching program of bionic studies used to conceptualize military applications and environmental weaponry. © Victor J. Papanek Foundation, University of Applied Arts Vienna

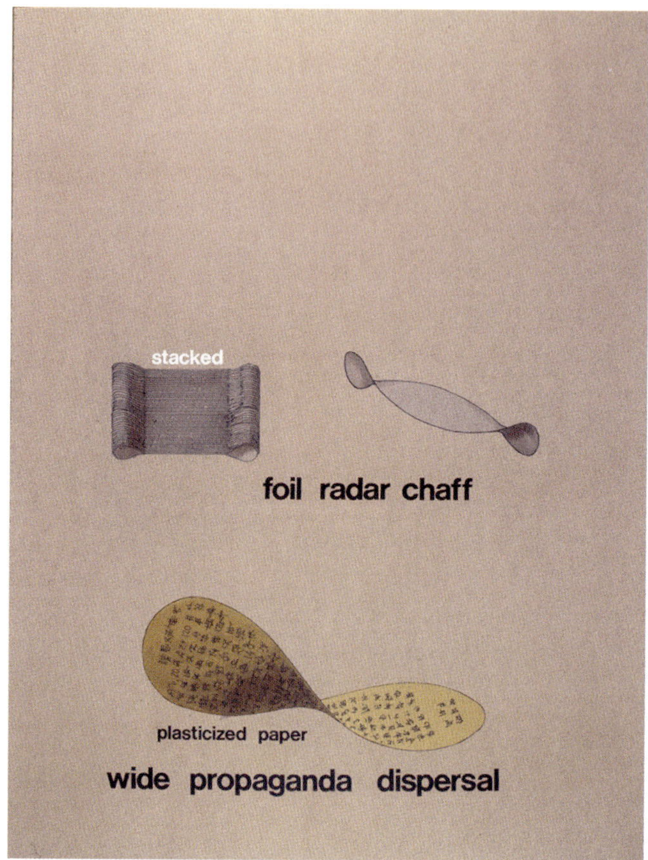

Figure 6.7

"Foil Radar Chaff: Wide Propaganda Dispersal," military application of bionic studies conducted by Victor Papanek at Purdue University. © Victor J. Papanek Foundation, University of Applied Arts Vienna

particular capacity to be "mobilized in remote locations, hence facilitating teaching at a distance."[18] Just as Papanek's *Tin Can Radio* might be harnessed for the propagandizing of nonliterate cultures, or conversely utlitized in the empowerment of communities cut off from the benefits of a burgeoning globalized media phenomenon, research experiments emerging from the Purdue University Audio Visual Center blurred the lines between humanitarian, artistic, and military imperatives under the rubric of interdisciplinarity. In contexualizing the inhererent ambivalence of Papanek's design initiatives during this period, it is useful to invoke historian Pamela M. Lee's concept of the "Think Tank Aesthetic," in which she questions whether the Cold War shift to interdisciplinarity should be understood "less as a kind of radical border crossing" and rather "something closer to a colonizing gesture, in which the specific interests of the humanities are assimilated into the larger programs and methods of the social and hard sciences."[19]

Rewriting Military Design Experiments for a Social Design Agenda

Despite its ubiquitous influence, the dark shadow of military complicity in Papanek's bionic research was completely and disingenuously eradicated in one fell swoop with the appearance of *Design for the Real World,* in 1971. Under the quasi-spiritual chapter heading "The Tree of Knowledge: Bionics, The Use of Biological Prototypes in the Design of Man-Made Systems," the Cold War research projects are rewritten in the tone of an uplifting narrative of ecological and cultural sensitivity, with seed experiments and the observation of biological principles providing the impetus for a multitudinous array of worldwide, socially responsible design initiatives. Papanek provides for his readers a reassuring definition of the term *bionics* ("the use of biological prototypes for the use of man-made synthetic systems") while supplanting the hardcore science of McCulloch and von Foerster with a benign rendering of the phenomenon as little more than finding inspiration in the observation of nature by any member of the general public taking a gentle amble in a local park: "All around us are manifestations in nature of rather primitive structures that have never been properly investigated, exploited, or used by designers, biological schemes that the investigation and accessible to anyone free for a walk on Sunday afternoon."[20]

Recasting himself as a design messiah, and putting to rights otherwise unsuccessful top-down governmental and NGO initiatives in the Global South, Papanek shifted the emphasis on bionics from US Cold War technology to a mode of empathetic design anthropology born of "interdisciplinary teams" whose expertise made a mockery of clumsy Western, rationalist solutions. Written in the style of a modern parable, one such vignette aims to cast derision on the naivete of modernizing initiatives made by the techno-colonializing powers, while boosting Papanek in his self-serving role as an alternative brand of white savior:

> Several years ago a new low-cost plow, designed, built, and distributed in areas of Southeast Asia that until then had used forked sticks weighted down by rock to till the soil. After a few years it was discovered that the plows were not in use and were, in effect, rusting away. According to the religious beliefs of the inhabitants, metal makes the soil sick, it offends the Earth-mother. I was able to recommend that the plows be dipped in a plastic compound similar to Nylon 60. As people were not offended by the technology of plastics, the new plows were accepted and usefully employed.[21]

The compound Nylon 60, by Papanek's account, has no quarrel with the Southeast Asian Earth-mother and is free of detrimental ecological implication. "The point of this anecdote," Papanek reiterates, "is that the use of a cross-disciplinary design team, including anthropologists, engineers, biologists, psychologists, sociologists, etc., would have prevented the original mis-design."[22] This vignette, and many comparable examples invoked throughout *Design for the Real World*, emphasized anthropological sensitivity toward Indigenous cultures as a core aspect of socially responsible design. In the twenty-first century, design anthropology has emerged is one of the key legacies of Papanek's ideas, whereby ethnography and user empathy form an integral aspect of design process.[23] Yet it is prescient that his genre of design anthropology was actually coined in a period of hegemonic Cold War US expansionism during which, under the rubric of "design for development" ideology applied to the so-called developing nations, industrial design and anthropological fieldwork were coupled in the pursuit of "native knowledge."[24] By the late 1960s onward, Papanek and the work of his students built on the rhetoric of anthropological, cultural empathy through codesign initiatives and ethnographic immersion. But as much as their innovative bionic designs remain shrouded in the veil of military experimentation, the specter of the colonizing, instrumentalized model of design for development throws into critical focus the overarching precept of design anthropology as a necessarily progressive, inclusive phenomenon; in the Cold War context it was more about the effective infiltration the local peoples and environments than promoting social inclusion.

Bionic experimentation proves to be the single most significant element in sharpening Papanek's conceptual framework of *Total Design*, and according to his résumés referring to this period of his career he occupied the significant position of president of the American Bionics Society (though the author has found no records to confirm existence of the organization). For it is bionics, biology, and their related fields that he identifies as "the greatest area for creative new insight by the designer," transforming the industrial designer from the mere stylist of tangible things, toward his new role as a "comprehensive synthesist."[25] "Only in industrial and environmental design is education *horizontally cross-disciplinary*," argues Papanek, with the designer forming the bridge between specialist knowledge sets in a context in which "the design of a single product unrelated to its sociological, psychological, cityscape surroundings is no longer possible or desirable."[26]

Addressing the semantics of *Total Design* head-on (incorporating a thinly veiled attack on his nemesis the corporate US stylist Raymond Loewy), Papanek dismissed earlier analog iterations of the concept as an expanded branding exercise—whereby the redesign of a steam iron stretched to the design of a new logo, point-of-sale displayer, and packaging—as having been rendered conceptually redundant. Having shifted from the mechanical era of the industrial revolution to the technological era of the previous sixty years, in *Design for the Real World*, Papanek argues that design now operates in the contemporary *biomorphic era* that requires systems thinking: "'Total Design' in the future will mean seeing the steam iron as well its [manufacturing] plant and promotional gimmicks merely as links in a lengthy biomorphic phylogenetic chain reaching back to heated rocks in stove-irons, and forward to the final extinction of the phylum 'steam iron' by mass introduction of 'perma-pressed' and 'stay press' fabrics."[27]

By the time of his writing *Design for the Real World*, between 1969 and 1970, Papanek's seed experimentation and other bionic research were recast as an oppositional gesture against military-funded and conceived interventions in the field. He comments on the paucity of existing literature in the field of bionics, with the throwaway comment that "virtually nothing has been written in the area of bionics" with the exception of a handful of reports "prepared by the armed services." And while these studies "concern themselves with man-computer-control relationships only and deal with the interface between Cybernetics and neurophysiology," his experimentation pushes design to the forefront in exploring "the possibilities of examining how nature makes things happen, *the interrelation of parts, the existence of systems*."[28]

The seed experiments conducted by the Graduate Industrial Design Group in Papanek's design studio-as-laboratory at Purdue University are represented as benign, applied to socio-environmental problems at one end of the spectrum, and NASA space technology at the other. Maple seeds inspired the design of antierosion devices for the "reforestation of the extreme northern tundra areas of Alaska, Canada, Lapland, and the Soviet Union, as well as re-stocking of these areas with fish" (achieved through water-soluble seeds loaded with fish spores or eggs), while the hook closures of the common cocklebur plant generated designs for an astronaut's null gravity-appropriate footwear.[29] No mention is made of the parachuting napalm herbicide distribution mechanism shown to a horrified audience of professors and students by Papanek at the Ulm School of Design in 1966, inspired by the parachuting "umbrella-like" seeds of the wild onion (see figure 6.8). Papanek did showcase one of his own designs: a "bionically derived" suppository dispenser inspired by the propelling capacity of the peapod.[30]

Double-Think: The Vietnam War and the Styrofoam Domes

In addressing design-activist student crowds prior to the publication of *Design for the Real World*, Papanek pays passing lip service to an antiwar stance. Retrospectively, however, close colleagues during the 1960s, such as graphic designer Gowan, have commented with incredulity at his political contrariness and oftentimes his outright hypocrisy. The same year, 1968, when Papanek had won sponsorship for his Graduate Industrial Design Group from the US military's Rock Island Arsenal in Illinois, there were widespread demonstrations against the Vietnam War on American college campuses, including Purdue, yet Papanek "proudly wore the Rock Island Arsenal security pin on his lapel, and was close-lipped about it."[31]

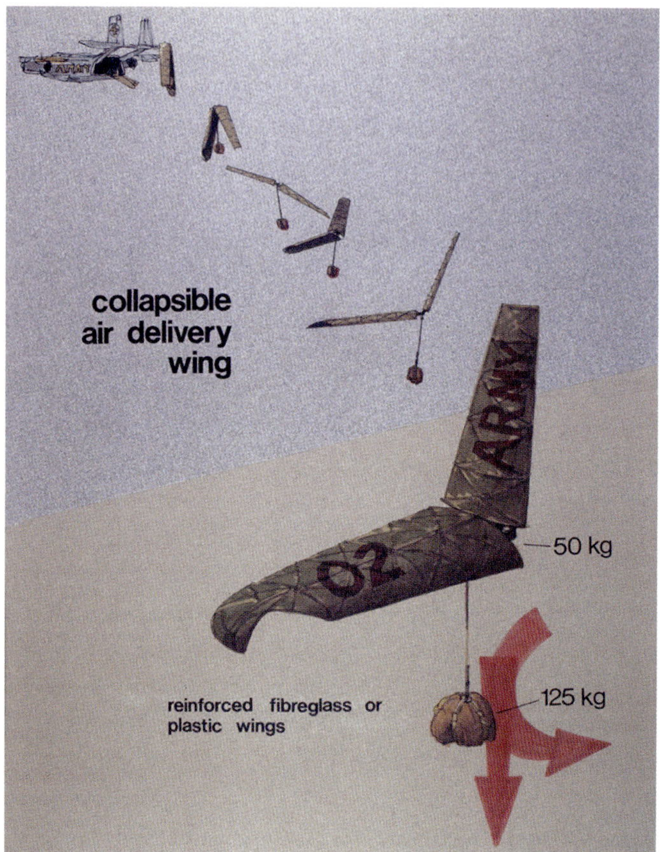

Figure 6.8
"Collapsible Air Delivery Wing," design concept for a US military aerial dispersal system, Victor Papanek color slide, c. 1962–1963. © Victor J. Papanek Foundation, University of Applied Arts Vienna

The highly detailed "Bionics" chapter of *Design for the Real World* gives us both the most succinct account of Papanek's Cold War experimentations, and the most vividly articulated evidence of his reorchestration of them. Intriguingly, though, it also offers up one of the few overt references to the Vietnam War, in which his design practice is so obviously culturally and technologically intertwined. For in "Bionics," Papanek directly equates the moral turpitude of Robert McNamara, former US Secretary of Defense (and his ambivalence "regarding our involvement in Southeast Asia") with his equally suspect role as promulgator of planned obsolescence allowed by his former position as president of the nation's leading automobile manufacturer: "It may be salutary to think that the same Mr. McNamara, while with the Ford

Automotive Company, shared the then prevailing automotive infatuation with tailfins, hood ornaments, and other small and neo-Freudian ephemera," sneered Papanek.[32] In contrast to the liberal press and his contemporaries, who widely criticized McNamara for his escalation of the Vietnam War, Papanek spins a line of contorted logic to champion his own brand of social design by accusing the former Defense Secretary of celebrating obsolescence in favor of fostering design's capacity for social transformation: "Had the American automotive Industry brought its production know-how to something like the self-generating Styrofoam domes developed by Dow International, some 250 million shelters might have 'grown' in South-east Asia by now, and the sociological pressures leading to civil wars and American involvement might never have happened."[33]

The reference to Dow Chemical Company as a socially responsible corporation, whose involvement as a manufacturer might have resolved the Vietnam War, must have struck contemporary readers of *Design for the Real World* as jarringly incongruous. Those same readers may have taken part in the widespread protests against the Dow Chemical across the United States in 1969, or at the very least been familiar with Dow's explicit involvement in chemical warfare through its manufacture and supply to the US military. Dow's infamy as the major provider for chemicals and psychotropic drugs utilized in the Vietnam War extended way beyond America. In 1967, for example, acclaimed Swedish graphic artist Sture Johannesson designed a provocative anticorporate, anti-American poster titled "Dow Shalt Not Kill! U.S.A.—Union of Stoned Anarchists" showing then-president Lyndon B. Johnson surrounded by a temple-shaped collage of American corporate branding logos, and flanked by star-spangled banners. Johnson is poised as if to do God's bidding, with the handwritten scrawl "Dow Shalt Not Kill!" laid before him on his Oval Office desk. American Christian morality is mocked here, with the rhetoric of the commercial corporation and its visual propaganda acting as God, its ideology taking precedence over any greater humanitarian concern. As Papanek's first version of *Design for the Real World* was published in the Swedish language, by a Swedish publisher, his building on the anticorporate and anti-American political activism of young Swedes, many of whom (like Johannesson) were aware of Dow's suspected complicity in biological and pharmaceutical experiments in Vietnam, seems all the more contradictory.

Since his first involvement with Dow in his capacity as head of product design at North Carolina State University, Papanek had continued working as a consultant for the company, and claimed to have initiated one of its more offbeat design collaborations during his tenure at Purdue University. In referring in exaggerated terms to the self-growing Styrofoam domes that Defense Secretary McNamara might have utilized to prevent the emergence of the Vietnam War, Papanek actually alluded to a real-life architectural project for a women's clinic constructed close to Purdue University in the late 1960s that was made almost entirely from a form of Styrofoam. Dubbed, dubiously, "Titty City" by certain members of the local community and Purdue faculty (apparently on account of the structure consisting of one large and six smaller domes, each with a Plexiglas "nipple" as a skylight at its pinnacle), by Papanek's account the initiative arose from his professional connections with Dow Chemical Company, which built the structure from premolded Styrofoam over the course of three days in 1966.[34] There is no evidence to support the veracity of Papanek's claim to have been involved in the women's clinic initiative (though his professional connection to Dow Chemical may have been an ancillary factor). But the recounting of this Styrofoam anecdote—and the sniggers its telling surely garnered at Papanek's collegial dinner parties, where guests sat on his prototypes for the wooden Shibumi chair, and ate from the finest Finnish tableware collected from his Nordic travels—certainly matches accounts of the design critic's propensity for sardonic humor and an attitude toward gender politics that increasingly fell out of kilter with that of the young students he instructed.

The Rising Star of Bionics: Parsons School of Design, 1966–1968

In July 1966, having divorced from his fourth wife, Diana Jean, forty-two-year-old Papanek married his fifth wife, Harlanne Herdman, who was one of his former undergraduate students, twenty years his junior. A textile designer hailing from Delaware, Herdman had taken his classes as a minor subject, but was greatly inspired by the social agenda of his teaching. Following their marriage she remained with Papanek as a key ally of his design agenda, and an artist in her own right, until she initiated their divorce in 1989.[35] Along with his new partner, Papanek remained loyally supported by his elderly Austrian mother, Helene, who also resided in Lafayette until her death, at age eighty-three, in October 1967.[36] As she was the last remaining tangible link to his Viennese and émigré past, his mother's passing exerted an enormously disruptive impact on Papanek. In an uncharacteristically emotional letter to a distant colleague, he had already admitted to having a serious nervous breakdown the previous year, shortly before his marriage.[37] Despite struggling with his mental health, and the demands of a burgeoning new domestic life, Papanek launched himself full-throttle into an international career as a visiting lecturer in his newly circumscribed fields of bionics and social design—marking his debut as a keynote speaker for the Industrial Designers Society of America (IDSA) annual conference at Parsons School of Design in 1966.

At the prestigious gathering of design professionals, he found himself billed alongside design luminary Arthur Pulos, professor of industrial design at Syracuse University.[38] Delivering his keynote address immediately prior to Pulos—an eminently more established design speaker with a prominent profile as a former president of the Industrial Designers Education Association (IDEA), Papanek may have initially considered himself as the less significant, warm-up act. As it panned out, though, the reality proved quite the contrary. In presenting his vision of future design pedagogy, Pulos came across as the unwitting foil to Papanek's vanguardism—with the Syracuse design professor merely trotting out a lackluster list of mundane design curriculum requirements (color and light theory, physics, mathematics, drawing, the history of civilization, etc.). Going against the contemporary shift to transdisciplinary practice, Pulos warned design students against the seduction of disciplines outside their field (as "practising them was not [a design student's] *forte*"). In a final gesture of perceived conservativism and technophobia, Pulos reminded the audience to take heed because "scientism can become an end in itself if we are not careful."[39]

By way of contrast, Papanek's lecture, under the headline "Enter Bionics," was singled out for comment by the on-the-ground design journalist from the prestigious British *Design Journal*, who used Papanek's breakthrough keynote as a means of illustrating the shifting, transdisciplinary terrain of design in America.[40] "Victor J. Papanek, professor of industrial design at Purdue University," read the commentary, "entered a plea for the examination of biological prototypes—bionics—during design training. Why bionics? Because bionics is the nature of change, and we must learn today to design for non-static, self-changing, self-compensation constitutions."[41]

The *Design Journal* paraphrased Papanek's stirring keynote address, offering its readers a succinct summary of the bionic design pedagogic approach: "There is a need to develop the most flexible syllabus for students of design [and] 12 hours of kinematics, 12 hours of drawing, etc. is not the way to do it. Students need to receive a creative background in art and science, and a study of bionics gives us this." The report stressed that by Papanek's account, engineering courses had been dropped from the Purdue curriculum, with students instead assigned "wicked problems" (as they would be coined by 1967) "which will force them to take the courses needed to solve them."[42]

After his nearly two decades away from New York City, the Parsons School of Design proved the perfect setting for Papanek to relaunch his career under the auspices of bionic creativity, most importantly bringing him back into engagement with an innovative East Coast design fraternity. Parsons School of Design was considered one of the most progressive design schools in the States, and the two-day inaugural IDSA conference coincided with Parsons' seismic shift toward an open agenda of social accountability and citizenship engagement, which, by the end of the decade, was further cemented by its merger with the New School for Social Research, the renowned institution of progressive thinking.

In this respect, the pitch of Papanek's keynote acted as a highly opportune calling card, his design polemic segueing fortuitously with Parsons' own recent conceptual developments in industrial design. In the three years following his initial conference speech, the social design and bionics professor was invited to join the Parsons industrial design faculty as a regular critic and lecturer, involved with an "environmental research probes" initiative. Strikingly similar to Papanek's own pedagogical projects, the "probes" were offered (as an additional fourth year of study) only to appropriately qualified students who wished "to develop projects in the areas of human needs, including medical-life equipment, shelter and omni-transportation products."[43]

But this was not his only connection to New York during this period. There were two significant figures from Papanek's early émigré life in New York City that reinforced the connections to his past there: one an elderly art history scholar, the other a distant Viennese relative of international acclaim.

During his education in architectural design at Cooper Union School of Art in the 1940s, one of the most influential figures on Papanek's early ideas around the intersection of design and humanities thinking had been the Jewish Berlin émigré, art historian, and critic Paul Zucker. Originally part of the New School faculty following his escape from Nazi Germany in 1938, Zucker eventually moved to Cooper Union and delivered a substantial part of the humanities element of the architectural and design curriculum. By the late 1960s, at the age of eighty-three, Zucker had joined the Parsons faculty, delivering to the design students an extensive art history and interpretation course.[44] Meanwhile, during the same period, Ernst Papanek, the renowned Viennese childhood psychologist and distant relative of Victor, was a key figure in the groundbreaking Human Relations Center at the New School, which just a couple of years prior to its formal merger with the Parsons School of Design, maintained close affiliations.[45] Whether or not Papanek met in person with either of these formative figures from his earlier life during his visits to New York, he undoubtedly was aware of their close proximity. Furthermore, following his mother's death in 1967, the connections of Zucker and Ernst Papanek made Parsons an even more significant destination for him, these visits summoning up memories of his early life as a young struggling émigré designer, with only his vulnerable widowed Viennese mother by his side.

Besides rekindling personal ties and provoking poignant memories, at Parsons Papanek rediscovered the intellectual intensity of a design school set amid the vibrant, ethnically mixed and culturally liberal metropolis that had shaped his first design practice, the Design Clinic, in the 1940s. It surely threw into sharp relief the ingrained conservatism of the environs of the Midwest and Deep South in which he had since carved out his career and design ethos. Parsons, and its affiliated institution the New School, welcomed America's top critical thinkers, political figures, and radical intelligentsia into their fold, and at the time of Papanek's new role there as design critic it had become a hotbed for the development and debate of gender issues. The design students were exposed to a strongly progressive curriculum informed by critical social sciences, and unlike NCSU and Purdue, women formed an integral part of the industrial design faculty.

The best-documented aspect of Parsons' 1960s shift to an overtly social agenda is the transformation of the longstanding and renowned interior design department, which, following its adoption of an experimental and social participatory agenda, emerged as the newly created Environmental Design Department. The turning point had come in 1965, when in response to questions regarding the social value of design, the Interior Design graduates mounted the exhibition *A Place to Live*, addressing the issue of substandard inner city housing rather than conventional, aesthetic interior design and consumer furnishings.[46] Viewed through the broader prism of the 1960s emergence of social and environmental counterculture, Papanek's social and bionic design polemic proffered a perfect pedagogical addition to the school—so long as he omitted reference to his Cold War military research agenda.

Design Course Journal: A Radical Prescription for Rebellion?

Back at Purdue University, the student design body was also avidly engaged in addressing issues of social injustice, gender, and race equality. One key focus and evidential record of these activities is the journal *Design Course*, first initiated in 1968 by Al Gowan, Papanek's close ally and advocate of socially responsible design (see figure 6.9). *Design Course* drew on debates raging at the 1967 and 1968 International Design Conference in Aspen (IDCA), themed "Order and Disorder" and "Dialogues: Europe/America," respectively.[47] The first edition appeared in the spring of 1969 as an international quarterly, with an opening editorial by Gowan titled "The State of Design," celebrating that finally "more than 20 years after World War II, the multitude of designers are vocal and verbally articulate." Along with a reprint of the lyrics to Buckminster Fuller's song "Roam Home to a Dome" came a weighty article by British design critic Richard Carr exploring the "Industrial Design Establishment in Britain" and pieces dealing with the visualization of environmental catastrophe.[48] Funded directly by Purdue University, the design magazine was distributed freely among design faculty and students, as well as being handed out to the audience at the IDCA to promote the Midwest university as a forward-thinking institution.

In this respect, the final issue, published in the fall of 1969, proved the most striking of all, for it featured evidence, rarely found in conventional historical accounts of this period, of direct intellectual debate around US race politics and industrial design. Under the unapologetically provocative title "Me, Design and the Niggertown Slums"[49] and lodged between a conventional overview of Japanese corporate design and a discussion of future of design education sat this compelling call to action by an African American designer, Stephen Frazier. One of very few high-profile Black designers of the period, Frazier worked for the Center for Advanced Research in Design, Container Corporation of America (CCA), Chicago, at a point when the company occupied the center stage of twentieth-century public US design

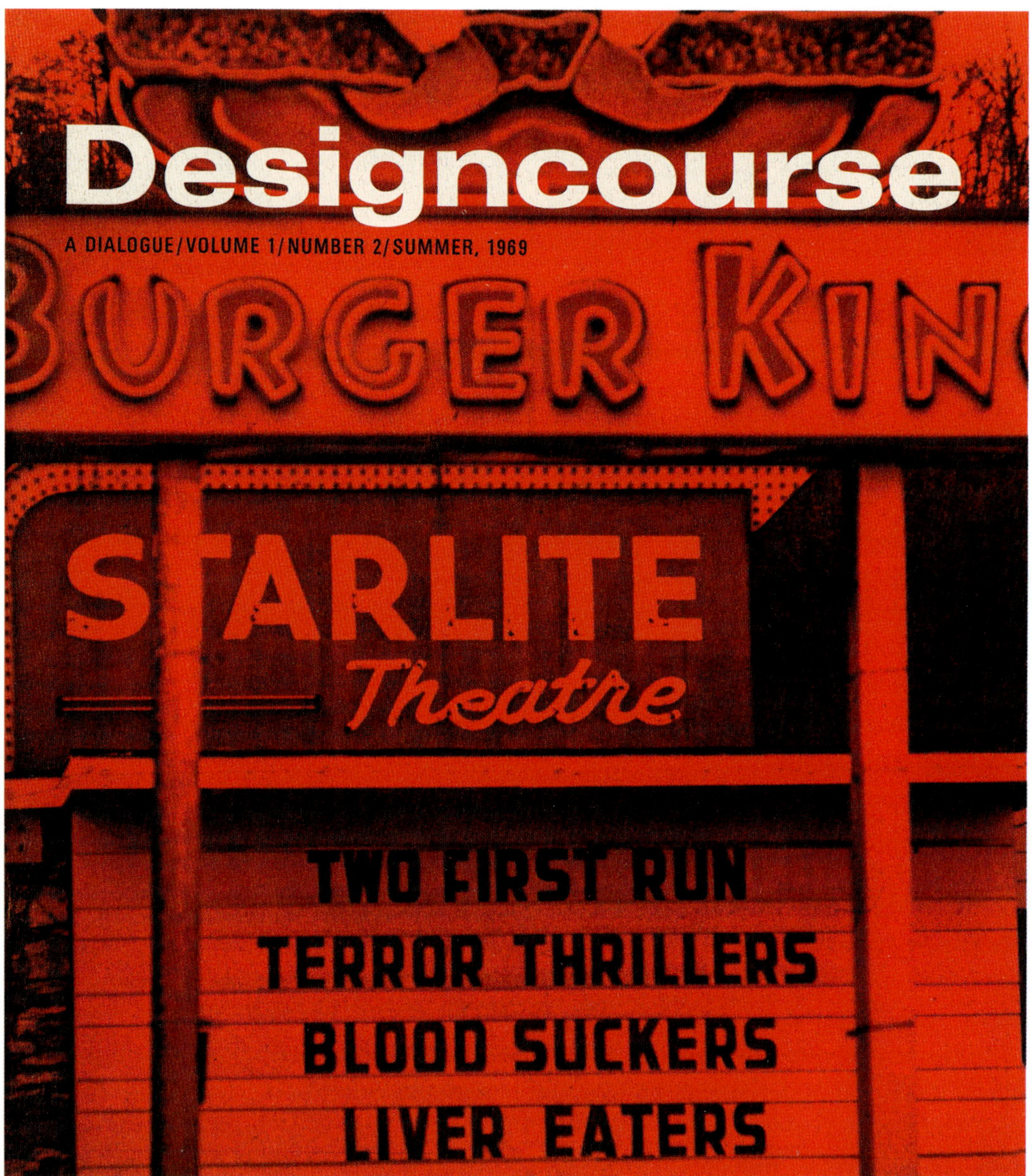

Figure 6.9
Design Course 1 (Spring 1969), edited by Al Gowan. Courtesy Victor J. Papanek Foundation, University of Applied Arts Vienna. © Purdue University, Indiana

discourse. CCA's chairman, Walter Paepcke (a fervent advocate of progressive design under the rubric of European modernism), was the founder of the Aspen Institute, a key supporter of the Chicago Institute of Design, and an instigator of the celebrated IDCA that Purdue design students and design faculty (including Gowan and Papanek) avidly attended each year. While the Center for Advanced Research dealt ostensibly with corporate identity design for major external clients, Frazier's own words stretched well beyond the boundaries of mainstream corporate design to focus on a section of society invisible to, entirely erased by, or alienated from design.

Addressing the complacency of the white liberal student readership of *Design Course* from the get-go, Frazier incorporated a clarifying footnote explaining the intention behind his provocative article title: "I use the term 'Niggertown Slums' simply because the word ghetto is too respectable a word to be used in reference to the appalling conditions under which some human beings live."[50] He continued, "If you find the above title offensive, if the words 'niggertown' and 'slum' jab harder at your sensitivity and moral consciousness than does the word 'ghetto,' then read on."[51]

Outlining the transformative power of youthful creativity, Frazier argued that "from man's origin in East Africa at Olduvai Gorge and El Bedari down through practically every age, be it Bronze or Atomic, a truly creative artist has sought to design a better world."[52] He carefully provided a further footnote as a caveat intended to distance mainstream commercial design from that driven by a social agenda: "Somehow I cannot believe that the fins and other superfluous ornamentation on some of the cars of the fifties were the products of creative minds."[53] Embracing the anticorporate rhetoric so familiar to Papanek's polemical outbursts, Frazier continued by singling out companies, and their complacent leaders, as the cause of much disaffection among the American youth and their desire for societal change: "The revolution of which they sing is full of hope. It says, 'Take Care, Mr. Businessman,' and 'Listen to what I'm Saying,' 'Open Up the Door An' I'll Get it Myself,' it admonishes industry. It demands that big business become more conscious of human needs."[54]

Despite making clear the necessity of designers to take up similar strategies as those of the Black Power movement, Frazier's overarching message focused on the corporation as the seat of the revolution, stating that "big business plays a major role in the educational, governmental, and indeed even the social systems of America. It follows then that the place to begin reform is at big business level."[55] Adopting a pro bono model later advocated (or borrowed) by Papanek as a key element in forging a socially responsible design practice, Frazier divided his corporate design position with that of a motivational lecturer at Black American high schools.[56] As initiator in 1968 of the CCA American Youth Motivation Program, he incorporated the jazz music from his band, Steve Frazier and the Soul Experience, into consciousness-raising and empowerment events for young African Americans. "I felt," wrote Frazier of the initiative, "that the time was right to involve industry in the true scope of Black Power."[57]

Critical of complacent white liberals and "conservative black bourgeois 'Uncle Toms,'" whom he accused of being "as instrumental in perpetuating racism in America as the "card-carrying segregationists," under the subtitle "What Can Designers Do?," Frazier called out the necessity to "redevelop the role designers play in the revolution" through resetting the design course in its entirety so that design meets "its ever present obligation to society."[58] The first reset would be "white America" facing up to its "decade upon decade" of "oblivious complacency and outright bigotry" being the root cause of the creation of the "ghetto," and the second reset being a challenge to the cultural hegemony of design itself: "A white architectural designer who's lived his whole life in virtually lily-white Deerfield, Illinois, cannot possibly design housing for a black family on A.D.C. [Aid to Dependent Children] if the only contact he's ever had with the ghetto is the view from the [socially] 'elevated' or the black football player he has in his rhetoric 102 class," concluded the designer.[59]

Crucially, Frazier's contribution to *Design Course* (most likely commissioned through Gowan and Papanek after first encountering his ideas at an Aspen conference) is clear evidence that Papanek, and his student cohort, had firsthand access to discourse on the complex intertwined politics of race and design in the United States. For Frazier's polemic pushes way beyond the Cold War framework of white patronage for "undeveloped peoples" that shadows much of Papanek's design for the Global South, in spite of Papanek's claims of inclusion through codesign, user participation, and transdisciplinarity. Instead this polemic offers a stark, unmediated insight into the lived reality of top-down power relations in America, and the US industrial design industry. "If the question comes to mind 'can white designers design for the blacks in the ghetto?,'" writes Frazier, "The answer is as obvious as if the question reversed ('can blacks design for whites'?). Of course blacks can. They can because blacks understand the needs of white Americans; that is not so of white Americans regarding the needs of blacks. It follows, then, that black Americans must now become educators of themselves, but more importantly re-educators of white Americans about themselves."[60]

It is no coincidence that around this period, Papanek's lectures, particularly those presented outside the United States, began to include an increasing number of references to "ghettoes," and Black communities whose design needs contrasted greatly with those of the consumer group of white American suburbanites that mainstream product design was aimed at. A case in point is a Swedish television interview of 1968, in which Papanek illustrated with a crudely misconceived example why design students needed to address head-on the principle of cultural appropriateness and user perspective: small-capacity refrigerators designed for nuclear, middle-class families failed to meet the practical and aesthetic needs of high-density living in "African American ghettoes," Papanek explained, where the refrigerator was shared communally and acted as a collective, rather than a private, design object.[61]

The Volita Project: A Reactionary Among the Progressives?

The *Design Course* journal's declared agenda had been to engage a newly politicized student cohort, as evidenced by Frazier's uncompromising contribution. That article threw into sharp relief an anachronistic intervention Papanek had made in the first edition of *Design Course* (of which he was coeditor). Papanek's "Volita Project" article (named after Nabokov's infamous pedophilic novel, *Lolita*, made into a Stanley Kubrick film in 1962) described the logistics of designing and manufacturing a plastic, animatronic woman. Written in the guise of a leaked memo exchange between the "CEO of POW Chemicals" (a barely disguised reference to Dow Chemical Company for which Papanek had, unbeknownst to his students, worked as a consultant) and its "Synectics Team Chief," it appeared as a "letter to the editor" posing as an insider's corporate exposé. The spoof exchange attempted to mock the moral vapidity and social ineptitude of corporate research by drawing on the popular currency of pseudo-science fiction like *Barbarella* (1968), a film that famously featured 1960s icon Jane Fonda as its highly sexualized rubber-bondage-wear-clad central figure, and which later garnered widespread criticism for its regressive gender politics dressed up as sexual liberation.[62]

"If the first pilot model should turn out to be about 48-16-40 and sort of Euroasian-looking [*sic*] around the eyes," pleaded Papanek in the guise of a corporate executive, "then send her over."[63] If the intention had been to capture the anticorporate sentiment of the activist student body, the licentious tone of the piece, the details of which were written with undisguised relish, seemed distinctly off-color for a middle-aged male professor who wielded power on a daily basis over female students: "Since women are increasingly thought of as expensive and capricious sexual playthings, the Volita Project should satisfy what has by now become a powerful drive. Both single and married men will find it possible to install one of our units in some small room closet, and activate her only when and if needed. Cross-cultural experimentation with women of varied racial backgrounds (i.e. Chinese, Hottentot, Australoid, Bushman, Pygmy, Eskimo, etc.) would now become feasible without fast expenses for travel, hotels, etc., or without courting social censure by one's peers."[64] "Consumers who show marked preferences for unusually young or elderly Volita sub-types can also be accommodated through our Special Products Division," continued Papanek in his assumed role of a sociologist of the POW Chemicals interdisciplinary design team.[65]

Rather than making a radical intervention into the politics of 1960s design, the faux exchange offers up an insight ino the subconscious of Papanek and a potential insight into the psyche of a generation of male design professors vexed by their relation to an increasingly liberated cohort of women (viewed here through an "exoticized" intersection with racial stereotyping). Seen as a tangible threat, women are "problem-solved" and silenced by the process of design, rendering them controlled and standardized for wholesale commodification or "made to order" according to the buyer's preference.

Perhaps this uncomfortable, anachronistic attempt at humor betrays the growing gulf that lay between the progressive thinking of the late 1960s student cohort (the rhetoric of which Papanek paid lip service to at youth design conferences across Europe and America) and the generationally entrenched and subconscious racism and misogyny that lingered. Fighting the specter of his own growing irrelevance and fossilization within a rapidly shifting terrain of sexual and gender politics, Papanek took on the mantle (in the guise of a whimsical literary charade) of the generation of patriarchal design bigots he professed to be rebelling against when he took the platform of design activist conferences and events across the United States and Europe. Dressed up as a wry critique of the plastics consumer commodity industry's boundless attempts at innovation at any cost, the fictional writing also offers an unparalleled insight into Papanek's continued fascination with fetishistic pornography, tracing back to his first paid design commission as an illustrator interpreting the photographs by notorious "glamour pin-up" purveyor Irving Klaw, in 1940s New York City.

Papanek was not, as such, disingenuous in his clarion call for a shift toward social design. Yet, ever chameleon-like, he consciously sought to maintain currency through his association with the progressivism of the students and radical figures that surrounded him.

Design Course (despite Papanek's off-kilter contribution) was a strikingly innovative initiative and a testament to the spreading internationalism of design discourse during Papanek's directorship of the Department of Industrial and Environmental Design at Purdue. Its editorial content also revealed Papanek's growing allegiance with the ideas of Nordic and Scandinavian designers and activists.

Under the title "Design and Youth Revolution," Per Johansson, a young Swedish designer from the Industrial Design department of Konstfacksolan, Stockholm, contributed a compelling article to the second, and penultimate edition of the journal.[66] As general secretary of the Scandinavian Design Organization, a student organization "that united the students of Sweden, Norway, Denmark, Iceland and Finland," Johansson offered insights into the radical design and anticonsumerist activities of a distinctly more radical set of students than those studying design with Papanek at Purdue University. As co-organizer of a succession of student conferences and cutting-edge design workshops that had taken place in seven institutions across the Nordic and Scandinavian countries, Johansson staked a claim for a new politics of design that fervently rejected the vision of their predecessors. "What do I mean by revolution?" Johansson asked rhetorically before continuing, "it is not very hard to see that today's youth is much more highly qualified than most adults to form an opinion of what tomorrow's world will look like if today's trends and systems continue. Youth all of the world by now knows that our system holds values with no ideals whatsoever . . . a totally blind generation is ruling the world today."[67]

All evidence suggests that it was a group of obscure student design activists emanating from Finland, and later the broader Scandinavian and Nordic countries, that bolstered Papanek's ideas and professional profile by providing him with the ideal bandwagon on which he could hitch a ride, and ultimately the impetus to produce the first version of *Design for the Real World* in Swedish in 1970 (which, controversially included the incongruous "Volita Project"—later removed from the English language edition after feminist protests to the publisher).

Through his association with this network of activists, Papanek finally recast himself as the Marshall McLuhan of industrial design—an iconic figure worthy of sharing the speakers' platform with design and architecture's highest-profile maverick, Buckminster Fuller, as they made their way across Northern Europe together, spreading the word of innovative American design in a region perilously close to the Soviet threat, and ideally placed as a stage upon which to act out the techno-geopolitics of the late Cold War.

Victor Papanek's career took a profound turn toward international recognition from 1966 through the early 1970s. Drawing enormous audiences through his appearances at prominent cultural events across the Nordic region, the design critic found himself entangled in the "ideological battlefield" of late Cold War Finnish design.[1] It was in 1966 that Papanek first attended Finland's celebrated Jyväskylä Summer Arts Festival (Kulttuuripäivät Jyväsky-län Kesä) as a speaker co-leading a seminar with Indian graphic designer Abhijit Barua.[2] Originating predominantly as a musical festival overseen by the Finnish composer Seppo Nummi, who had been instrumental in launching the Jyväskylä Arts Festival in 1955, by the 1960s the event had begun to focus on broader cultural issues, including visual arts and design. Home to Finland's premier architect and designer, Alvar Aalto, the festival built on the city's preexisting international reputation in the design sphere. Through the incorporation of a Visual Arts Conference into the festival program, led by architect Tapio Periäinen, design emerged at Jyväskylä as a cultural practice deemed just as worthy of serious critical attention as classical and contemporary music.

Victor Papanek Meets Real-World Geopolitics

The tenth Jyväskylä Arts Festival embraced the theme of "problems common to East and West," a topic evolving from UNESCO's involvement in the cultural event from 1964 onward.[3] In representing the annual event to the public, in 1966, Nummi took pains to emphasize its broader agenda "to create a harmonious comprehensive image of our civilization," pointing to that year's program (which included a seminar titled "Western Cultural Imperialism and the Cultures of the East") as exemplary of that ambition. Set in the heart of Finland 150 miles north of Helsinki, amid a landscape of endlessly unfolding hectares of birch forests interspersed with lakes, the 1966 festival, even by contemporary standards, attracted extraordinarily diverse participants. These included the cultural adviser to the Shah of Persia, the rector of the Islamic University of Cairo, the leading Japanese composer Toshiro Mayuzumi, famed Indian film director Satyajit Ray, Austrian American architect Richard Neutra, Japanese designer Sori Yanagi, Indian cultural journalist Krishan Sondhi, Polish literary expert Jan Kott, and numerous Finnish cultural experts, among them contemporary composer Joonas Kokkonen, film director Jörn Donner, and designer Kaj Franck.[4] The varied format consisted of film screenings, music recitals, open lectures, dialogues between selected cultural experts, seminars, and even an interfaith worship gathering (including Buddhist, Lutheran, Hindu, Jewish, Christian, and Marxist speakers).

The sheer scale of the event—as well as its chosen theme of bringing together Finnish and Western European critics, with a range of speakers drawn predominantly from the nations of Japan, Poland, and India—was a clear indication of Jyväskylä Arts Festival's renewed geopolitical relevance. Affiliated with UNESCO (and undoubtedly backed by a range of other Western Cold War stakeholders), and staged just hours away from the Russian border, the festival played a vital and performative role in forging cultural ties and global allegiances. The "East and West" theme specifically aimed at facilitating cultural dialogue between Eastern and Western democracies, soft communist countries (those more open to exchange with the capitalist West), liberal Muslim nations, and the most powerful developing nation: India.

The festival's potent geopolitical agenda did concede to national interests too, with the local press release dutifully informing the "festivalists" that the homegrown, international icon of mid-century Finnish architectural design, the "academician Alvar Aalto" would "participate and present his work."[5] Yet despite Aalto's stature, a younger generation of Finnish designers had begun to rebel against the aesthetic and conceptual contraints of Nordic design that the septuagenarian and his contemporaries perpetuated. By the mid-1960s, according to the renowned Finnish textile designer Vuokko Eskolin-Nurmesniemi (a member of the post-Aalto generation), the formerly revered national hero was met with a growing sense of alienation within the contemporary Finnish design community—a patriarch of design betrayed by the very generation he considered to have been the direct beneficiaries of his international renown.[6]

All the more pertinent, then, that as Aalto's star waned an unknown designer stepped in to become the design darling of the 1966 proceedings. "Design is also represented by Professor Victor Papanek, one of the designers involved in the US space programme," stated the relatively anodyne press release. This appeared in the same paragraph where Aalto's appearance was mentioned, written without foresight of the impact the American speaker's fervent anticonsumer, antidesign rhetorical interventions would have on the audience.[7]

Prior to his debut at Jyväskylä, where the night of his arrival he had inauspiciously ended up sleeping in a communal men's hostel at the bus station (the organizers having forgotten to book his accommodation), Papanek had been entirely unknown to the Finnish design community. It was only with persistent and dogged determination that the designer had secured an invitation to the festival, and in so doing, an entrée into a thriving cultural scene boosted by funding from Western Cold War infrastructures keen to protect the region from Soviet influence. Obsessed with accessing the European, and more specifically, Scandinavian and Nordic design scene, he began his quest months previously, by penning a cold-call letter, simultaneously self-deprecating and self-promoting, to the director of a major art and design school at a Swedish University in the Finnish industrial city of Turku. "Please forgive me if I address this letter to you, a complete stranger, and if the letter is tiresomely long and tedious. . . . Forgive me if the next few sentences sound somewhat egotistical and self-praising: Unfortunately I know of no other way of saying it," began Papanek, before launching into a lengthy and vaguely sardonic list of his accomplishments. "I am an informed, dedicated and (occasionally) amusing speaker," he continued, then emphasizing his close relationship with Frank Lloyd Wright, with whom he described having "spent several years studying and working."[8] Papanek framed his proposed trip to Europe as a research exercise in understanding "the delicate balance between handcrafts and prototypal mass-consumer design, especially in Scandinavia."[9] Aware of the value and pertinence of US technological knowledge in the Cold War context, the ambitious designer exaggerated his expertise in "design for non-Terran [*sic*] environments," emphasizing that "the National Aeronautics and Space Administration" constituted one of his clients.[10] Fortuitously for Papanek, just a few weeks after forwarding his speculative pitch, it was passed on to the desk of Periäinen, director of the Visual Arts Section of the Jyväskylä Arts Festival. Largely on the pretext of his engagement with the "U.S. space programme" (as stated in the pre-event PR program outline) the organizers were persuaded to invite the American designer, ignorant of the fact that this claim was the single most spurious detail of his résumé.[11]

Extraterrestrial design, it transpired, did not feature at all in Papanek's talks at the 1966 festival. Rather, under the title "Design and the Good of Society," he stuck to his familiarly strident dismissal of the follies of degenerate and wasteful commodity culture. After a flurry of hyperbolic press reports focusing on Papanek's provocative lectures, the design critic embarked on a campaign to persuade the organizers of his renewed relevance as an expert speaker for following year's gathering, themed "Man and Science": "I am more eager than ever to participate, especially as the general subject matter is even more closely in my own area of competence than last year's. Also I have greatly added to my 35mm slide collection specifically with Finland and Jyväskylä in view," he wrote, pleadingly.[12]

Buckminster Fuller and Victor Papanek: The Cold War Design Dream Team

At the 1966 Jyväskylä event, Papanek had shared the speakers' roster with the internationally acclaimed fellow Austrian American émigré Richard Neutra, whose work and writings Papanek had keenly followed throughout his career to the extent of borrowing the architect's best-known written works, *Survival Through Design*, for the title of one of his own 1950s *Design Dimensions* TV broadcasts. Yet Neutra's presence elicited no expression of excitement or interest on Papanek's part; quite the contrary. In later comments to the organizers regarding potential improvements to future festival programs, Papanek conjured up the basic trappings of the US counterculture, suggesting the parking lot be made into a dancing space for nightly parties along with "arrangements for a 'tent city' or sleeping bags."[13] But his greatest pearl of wisdom centered on ousting the old guard, personified by Neutra, and the celebration of forward-thinking designer-architects such as Fuller (and himself): "Your speakers should be <u>youthful!</u> By 'youthful' I am not referring to the chronological age (I'm 43 myself) but to <u>outlook.</u> For instance: Bucky Fuller (at 73) is a lot more youthful than Richard Neutra (at 69). There never was any excuse for someone like Neutra to speak."[14]

When Papanek discovered Fuller would be speaking at the following year's Jyväskylä Arts Festival, his determination to secure his own spot beside his ally-cum-rival intensified. After some arduous to-ing and fro-ing of letters and telegrams, Papanek finally received notification of his acceptance to participate, and for the first time in his career he would feature, with equal billing, alongside Fuller on the program proceedings of a prestigious international forum. In his final letter confirming travel arrangements to Helsinki, he wrote gleefully, "I am looking forward to meeting with Bucky Fuller, whom I have not seen for a couple of years but is a wonderful person" (see figure 7.1).[15]

The Jyväskylä Arts Festival 1967, and his meeting there with Fuller, would prove pivotal in forging Papanek's association as one of a triad of contemporary theorists: Fuller representing architecture and technology; McLuhan, media; and Papanek, design. That same year, a collection of essays edited by McLuhan, under the title *Explorations: Verbi-Voco-Visual*, explored the transition to new modes of orality, featuring a tautologous attempt at contemporary art-historical theory by one "V. J. Papanek."[16] First published as the eighth and last edition of Marshall McLuhan and Edmund Snow Carpenter's cutting-edge journal *Explorations: Studies in Culture and Communication* in 1957, its reprinting saw Papanek formally and publicly linked to the world-renowned media theorist. This intellectual tie to McLuhan in turn reinforced his connection with Fuller. For McLuhan and Fuller had maintained a mutually respectful relationship, with regular correspondence, since their first personal encounter in July 1963 aboard a floating think tank, the seafaring vessel *New Hellas*. The yacht set sail from Athens for an eight-day tour of the Greek Islands billed as a transdisciplinary

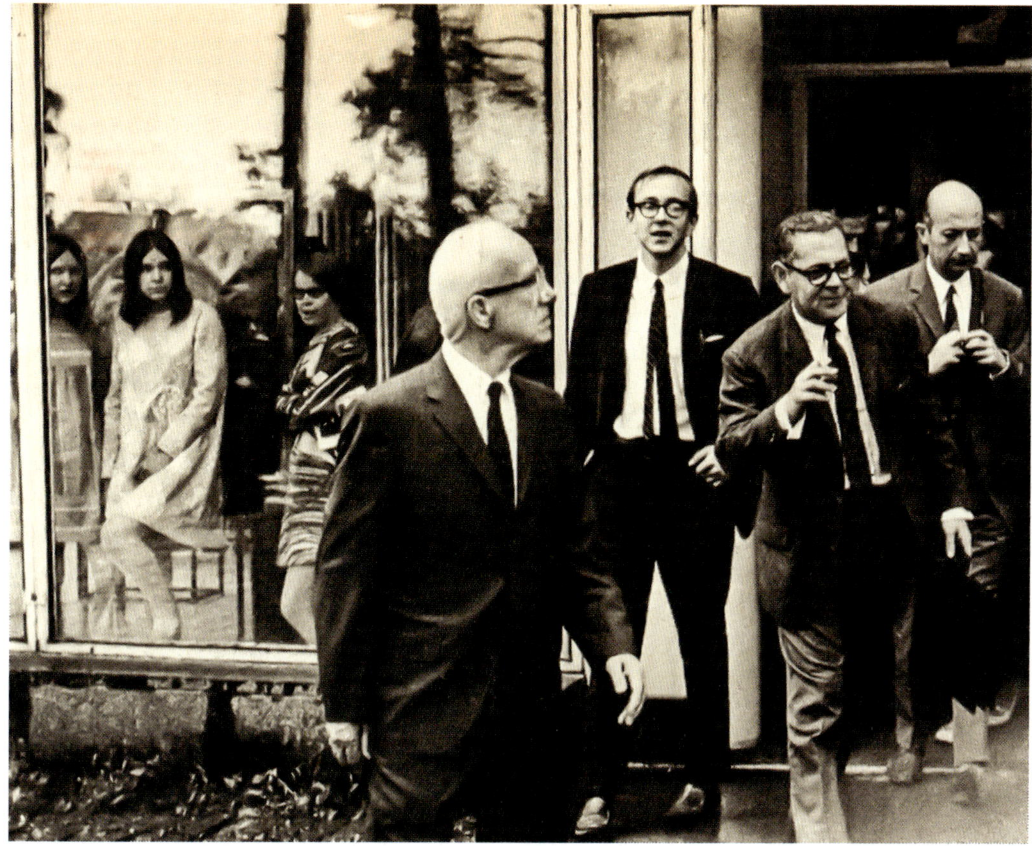

Figure 7.1
Victor Papanek and Buckminster Fuller (front center) and colleagues at the 11th Jyväskylä Arts Festival,
Finland. Photograph featured in the Finnish magazine *Me Naiset*, September 27, 1967. © Me Naiset

"symposion" (the original Ancient Greek for "symposium" that invoked the social networking aspects of the academic gathering), with thirty-two intellectuals, including leading
anthropologist Margaret Mead, onboard debating the future of architecture and city planning.[17] In the course of the think-tank cruise Fuller (as with Papanek two years later) gifted
the media and communications theorist a copy of his recently reprinted and little-known
early work *Nine Chains to the Moon* as a means of reiterating to McLuhan that he had "been
describing technology as an extension of the body" ever since the book's first publication in
1938, well before McLuhan's theories of networked electronics, and before the term *global
village* had even been coined.[18] Despite Fuller's sensitivity toward the potential interloper,

according to architectural historian Mark Wigley, the two men recognized the overlap in the respective theories and "defended each other's work, seeking out opportunity to be together and pursuing the global implications of prosthetics and networks to the limit."[19] McLuhan certainly sought to flatter Fuller, as his intellectual senior, seeking his advice on pushing the boundaries of his own theories and readily sharing his ideas, as an excerpt from a letter written just a year following their first meeting reveals: "Dear Bucky, have had a good deal of luck in analysing various problems lately. I enclose a note on one of these. If one says that any technology creates a new environment, that is better than saying the medium is the message. . . . Supersonic flight will create a new environment which makes our present cities somewhat useless. . . . Would appreciate your suggestions about readings in the matter of technology as creative environment. Today the environment itself becomes the artefact."[20]

In turn, Fuller, made it publicly clear that some of his core ideas formed the basis of the development of McLuhan's own, and that this was a mutually acknowledged fact. In response to one academic inquirer, keen to inculcate Fuller in a hostile critique of the communications theorist's off-the-wall theories and potential plagiarism of Fuller's own work, the equally eccentric technology maverick came to the defense of his intellectual ally: "Regarding McLuhan he acknowledges use of my concept and phrasing of the 'Mechanical' and other 'Extensions of Man' which was first published in the 'predictions' in my preface to *Nine Chains to the Moon* . . . I speak about such phenomena as a scientist. McLuhan speaks as a Professor of Literature. I don't always agree with his viewpoint, but I greatly enjoy his foot and rapier work. I like him, personally, respect him and appreciate the respectful friendliness he shows toward my own work."[21]

As a design theorist, Papanek could never claim to have matched the stellar international profiles of Fuller or McLuhan, the scale and extent of their theoretical dissemination, or the intimacy of their close intellectual relationship. But his impact on industrial design practice publicly and as a discipline certainly stood up to comparison with the impact the media theorist and architectural maverick made on their respective disciplines. Both of these eminent figures shared not only a familiarity with the design critic's work—McLuhan having published Papanek's writing and Fuller openly endorsing it with a preface to *Design for the Real World* in 1971—but they both had also known the designer personally as far back as the mid-1950s. Papanek, then, was supremely successful in forging an identity as a designer, theorist, and critic occupying a space perfectly equidistant between McLuhan and Fuller. Each meeting, appearance, or association with either man's work reinforced his growing reputation as part of triadic tour de force.

Reunited with Fuller in Jyväskylä, July 1967, Papanek delivered an open lecture titled "The Demands on Design in an Increasingly Mechanized Society," a couple of days following

Fuller's own *studia generalia* (open university) lecture "Man's Relation to Nature and in Different Stages of his Culture."[22] Equipped with bespoke slide-carrying tote bags, used to haul his four-thousand-piece slide collection on international travel, and a reel of his 16 mm experimental film *Bio Graphics*, Papanek received a rapturous reception, at least by the standards of the festival committee. "I am delighted," wrote the secretary-general of the organizing committee, "to see that your lectures have had an influence of Finnish cultural life, especially on our young designers who have been discussing your ideas a lot. There are many people like me who have been thinking that your lectures are a stimulating injection for finnish [*sic*] design."[23]

Despite the fact that the Jyväskylä Arts Festival was notable for its lack of female speakers, this theme struck a particular chord with the mainstream women's press in both Finland and Sweden, which enthusiastically embraced his call for ethical consumerism. In September 1967, Finland's largest-circulation women's weekly magazine, *Me Naiset*, featured a lead article entitled "The Third Way" with a large-scale close-up portrait of the visiting American designer, topped off with a quote attributed to Papanek: "[Arthur] Koestler has said that he suffers for chronic acute social rage. I have the same disease."[24] The interview piece incorporated stirring quotes from the witty design critic, adopting a consciousness-raising tone for its readers with its lead paragraph: "Have you ever drunk from a glass that is difficult to drink from? Ever wondered why some saucers are more expensive than others? Have you ever taken up a brush handle and noticed that it is too short? Many people have! Victor Papanek feels that things are designed that are either beautiful and expensive, or cheap and of poor quality . . . he speaks on behalf of those who have little money and many problems."[25]

By the time the magazines hit the shelves, Papanek, safely removed by a five-thousand-mile distance from his peers in the United States, took the opportunity to further puff up his résumé with the addition of his role as "adviser to the U.S. government in housing issues."[26] Any connections to the US military or Dow Chemical Company were again noticeably erased from his past career: "Papanek has turned down offers from the two sectors where American designers can make money: the automobile and weapons industries," the reporter unwittingly observed.[27] The face-to-face interview for an obscure (to Papanek) Finnish women's magazine provided an irresistible opportunity to further embellish his unproven credentials as a former sidekick of America's greatest living architect. On this occasion, he brazenly inculcated the tale of his rebellion against Wright as justification for his turn to industrial design as a superior socially responsible practice: "I worked at one time with Frank Lloyd Wright, who was a charming person, but I gradually began to be bothered by the fact that we designed only for the select few and the rich. And nonetheless there is an immense need for housing in the world, some six hundred million people without homes. I came to the conclusion that architecture can never be the answer. We have to find a system that in some way can ensure housing for the masses. At that stage I decided to become an industrial designer."[28]

Despite the widespread, generally enthusiastic coverage of Papanek's interventions, in both the design and popular national press, voices of dissent toward his wholesale cultural appropriation of Finnish design began to emerge from diverse quarters. Following Papanek's attendance of the Jyväskylä Arts Festival in 1967, Lauri Perkki, a student activist and journalist, condemned the patronage of well-meaning outsiders in the daily central Finnish newspaper *Keskisuomalaine*: "We cannot afford to bring specialists from behind the oceans just to tell jokes to the plebs."[29] The basis of Perkki's critique was that the Institute of Special Education at Jyväskylä University, with which Perkki was involved, had not been engaged in the design project for the special equipment developed by Papanek during his visit for the rehabilitation of the physically impaired. Papanek had expanded his normal social design and bionic pedagogic repertoire (without consulting or bringing on board local specialists such as Perkki) by addressing the subject of "Aids for Children with Disabilities." Certainly Papanek interacted mostly with the general public during his initial visit to Finland, rather than designers and architects. This is made evident by the prominent articles regarding his design reform manifesto appearing first in Finnish and Swedish women's magazines rather than the design press—due in part to the fact that Papanek had been a late addition to the Jyväskylä Arts Festival 1967 program, again actively seeking the invitation himself rather than being independently approached as a key speaker. Consequently, his lectures were delivered for the *studia generalia* and attended by large audiences, including students, the press, and residents of Jyväskylä. In the context of a preexisting national design culture centered so fervently on famed and fêted individual design personalities, Papanek's ambiguous status (was he a real designer, or just a critical commentator?) also lent to the disquiet in some quarters of the design profession.

Nevertheless, he and Fuller (who also received widespread press coverage) evolved as a kind of design double act between 1966 and 1968, on the road together attending symposia, proselytizing in an oratory style startlingly distinctive in its cocksureness from that of their modest, understated Finnish hosts. In the midst of the Jyväskylä Arts Festival, Papanek and Fuller took time out to travel south together to attend a Nordic Education Seminar in Otaniemi, in the district of Espoo bordering the nation's capital. The invitation to attend this, the first in a series of pan-Scandinavian Design Students' Organization (SDO) events, had been extended by a twenty-five-year-old graduate student activist: Yrjö Sotamaa, president of the students' union of the Institute of Industrial Arts, Helsinki. Intellectually ambitious, Sotamaa would emerge as one of the most prominent figures in forging Papanek's relations with Finland, and the broader Nordic and Scandinavian countries. In 1969, Sotamaa followed the design professor to Purdue University as an assistant professor, relishing the proximity to the world's leading socially responsible designer and Fuller with his *World Game* project in Carbondale, Illinois, in equal measure.[30] As a friend to Papanek and acquaintance of Fuller, Sotamaa provided a further bridge between the two men, also reinforcing their respective connections to Finland.

Northern Lights: Finland as an Alternative Economy of Design

Throughout the late 1960s, Papanek used his trips and engagements with Finnish designers and cultural figures to bolster his version of ethical design in the mainstream US design press. With its relatively recent urbanization, socialist infrastructure, and "good design" pedigree born of an economy of need rather than the corrupted economy of consumer desire sketched by Papanek's US contemporaries Vance Packard and Ralph Nader, Finland proffered an ideal backdrop for his theories of socially responsible design.

Following one of his earliest visits in 1966, the design critic penned an article poetically (and somewhat clichéd) titled "Northern Lights," in which he explained to the American design profession how "almost all excellence of Finnish design has grown out of honest need."[31] He detailed how the lineage from handicraft to mass production had evolved naturally, thus eluding the ceaseless and purposelessness of a megalithic US industrial model geared toward generating design obsolescence. Finnish design, by Papanek's account, belonged to a sustainable ecology exemplified by the influence of environmentally determining factors: long winters, small living spaces, and, predictably, the traditions of the sauna. The fully illustrated "Northern Lights" piece appeared in *Industrial Design*, a journal loosely affiliated with the Industrial Designers Society of America (IDSA), the organization that Papanek would later claim had "blackballed" him following publication of the English version of *Design for the Real World* in 1971. Juxtaposing Finnish design, defined as "good form" born of a democratic society, with the endless flotsam of gadgetry and "toys for adults" that defined late twentieth-century North American design, the feature was intended as a provocative, open critique of the design industry that the journal itself sought to represent. Readers of *I.D.* would have been fully aware of Finland's sensitive proximity to the USSR and the Cold War politics of design.[32] Within the context of the conservative strictures of the US design industry, Papanek's expression of an overt sympathy for a socialist economic system that prioritized social relations over individual aspiration, and state-supported amenities over privatized wealth, stood as a form of barely covert activism. His contempt for an American culture defined by commerce, in which even the most traditional communities were beholden to "the market," was made explicit: where Finland boasted a surviving arts and crafts tradition, in the United States, lamented Papanek, there were "five-and-ten cent stores, supermarkets and discount houses in every town and hamlet with the worst excesses of trash product culture filtering down into even 'the tiniest village.'"[33]

In contrast to the superficiality and alienation of US commodity culture, Papanek's feature outlined an idyllic vision of Finnish design centered on the home and its familial and social relations: "The northern climate is partially responsible for the fact that the dwelling is regarded not only as a place where one eats and sleeps, but as the true frame around family life. In the South one meets friends in bars and inns; in Finland you invite them to your home, hence the house and its furnishings are of social interest."[34]

In his romanticized, pseudo-anthropological analysis of late 1960s Finnish life, dreary winters resulted in the bright, kaleidoscopic colors of Marimekko fabrics and Vuokko dresses; small farmhouses inspired Finns to develop closely packed furnishings and stackable, multi-function homewares; and wood technology derived from the making of stave boats "greatly influenced the Finnish designer's approach to wood, as the early furniture of Alvar Aalto exemplifies."[35] Furthermore, Papanek observed, "unlike England, France and the United States fashion is designed by people who *don't* hate women."[36]

The iconic Marimekko dress in particular was lionized for the ways in which it expressed nuanced cultural sensitivity and fitness for purpose wholly in tune with the versatility of the emancipated 1960s Finnish woman (Papanek's notion of liberation being confined to the ability to wear a minidress suitable for domestic and social multi-tasking): "If we look at the circumstances under which a Marimekko dress is worn (housework in the morning, shopping in the afternoon, cocktails and dinner out, and then a film with only the sandals changed to heels and stockings, and then a final plunge in the Sauna), it is easy to see how the simple and comfortable lines reflect these many activities."[37]

Papanek's singling out the Marimekko dress for celebration was by no means arbitrary. Jacqueline Kennedy's taste for easy-to-wear, brightly colored, "liberal" European Marimekko dresses was well known; she was famously pictured wearing a Marimekko sundress on the front cover of *Sports Illustrated* in 1960, the launch year of John F. Kennedy's presidential campaign, and was widely photographed wearing the brand throughout the 1960s. Finnish design, even for a general US audience, read as shorthand for a brand of utopian liberalism.

The leading Finnish design companies Artek, Iittala, Marimekko, and Nuutajarvi distributed their wares during the 1960s to a predominantly liberal, bourgeois elite through the Design Research Store, which had a dozen branches throughout the States, including New York City, Boston, and San Francisco. The items these companies produced, ranging from textiles and bentwood furniture to tableware and glassware, presented themselves to Papanek with a type of anthropological integrity that far exceeded modernist functionality. "These products," he commented, "have made real inroads in the everyday culture of Finland and play an important part in the average man's way of life."[38]

Papanek's vision of a liberal, homogenized, and culturally authentic Finnish life devoid of the vulgar trappings of popular culture rehearsed the search for a preindustrial style advocated by the turn-of-the-twentieth-century Arts and Crafts reform movements. Yet, as design historian Korvenmaa has argued, Finland during this period was the most Americanized of all the Nordic countries, and it was not "until radical and . . . left-wing political currents swept over the intellectual landscape of Finland in the late 1960s that the culture of design became ideologically politicized."[39]

The essentialist tone of Papanek's homage to Finnish design and culture, his setting up of its authenticity in opposition to the degeneracy of American commodity capitalism, is undoubtedly flawed as a serious piece of anthropologically researched critique. The suggestion that Finns did not embrace popular commodity culture, or the influences of contemporary media, in the 1960s is clearly fallacious. Furthermore, Papanek's article acted in part as a type of advertorial for the Finnish design industry, as awards, including the Finnish Cultural Affairs Travel Grant, supported a number of Papanek's trips to Finland. However, in arguing for a *societal* understanding of the practice of design and the necessity to frame aesthetics in the broader contexts of local traditions of material culture, Papanek's essay drew attention to the inseparability of design and the politics of everyday life and sketched out a socially embedded, anthropologized approach to design that would become the hallmark of his approach. As such, it represented the beginning of a genre of design activism, shaped directly by Papanek's experience of Finnish design and material culture, that would feed into the broader design activism of Europe and North America throughout the 1970s.

Prior to the 1960s and 1970s, Finnish design culture had been dominated by the impact of governmental policy and the broader representation of Scandinavian design and mid-century modernism typified by the *Design in Scandinavia* exhibition that toured the United States and Canada from 1954 through 1957, the effect of which extended well into the mid-1960s.[40] The government policies that boosted Finnish design had made national heroes of designers such as Tapio Wirkkala, Kaj Franck, and Timo Sarpaneva. The dramatic urbanization of Finland in the 1960s had seen the country transformed from an agrarian to an industrialized nation that prompted an urgency for "national strategies to steer this new and more complex union of design, industries and growing foreign trade."[41]

While Papanek's consumer critique found favor in sectors of the Finnish popular media, and his depiction of Finland proved compelling to the US design audience, intriguingly it was a fellow North American who emerged as one of Papanek's fiercest critics, accusing the designer of wholesale cultural appropriation of the Finnish culture. Donald Willcox, a self-taught expert on Finnish material culture, keenly followed Papanek's involvement with

Finnish national design throughout the late 1960s and early 1970s. This culminated, in 1973, with the publication of an impressively nuanced account of Finnish design as a form of extended critique on the misappropriation of Indigenous material culture. In *Finnish Design: Facts and Fancy,* Willcox titled an entire section of the postscript "Victor Papanek." Taking Papanek to task for his false representation of Finland, he described him as a "design parasite" with a superficial understanding of the nation's culture: "[Papanek's] opinions ring with the same sounds as that of an American tourist visiting Helsinki on a 24-hour stopover . . . he has been guided from place to place without the opportunity of going it alone. His opinions about Finland have come from surface glances, rather than digging below the surface to see what makes Finland tick."[42] Willcox goes on to tell how a Finnish friend compared "Mr Papanek's visits to Finland to those of the circuit-riding gospel preacher who travels from town to town delivering the same Sunday sermon, and who sometimes forgets where he is, and delivers the same message a second time."[43] Willcox certainly had a valid point regarding Papanek's cursory and often stereotypical depiction of Finland, its objects, and people, and his extensive reliance on the ideas of its innovative student activists.[44]

Finland formed the hub of what by 1968 had become Papanek's broader self-consciously engineered Nordic and Scandinavian enterprise, which radiated out to design pedagogical explorations of Denmark, Sweden, and Norway over the following five years. This expansion originated in his repeated involvement with the Jyväskylä Arts Festival that, in the summer of 1968, he combined with the attendance of a watershed design activist event held on a fortress island off Helsinki, where he was once more accompanied by Fuller.

Decolonizing Design: *Project Ujamah,* a TV for Africa

"During the international design festival at Jyväskylä, Finland, in 1968," reflected Papanek in his book *Design for the Real World,* written shortly thereafter, "I participated as a part of a UNESCO team of international design experts to develop new ideas for Black Africa . . . that can be built in Africa by Africans . . . that will by-pass private profit, corporate structures, exploitation, and neo-colonialism."[45]

Under the theme of the "New African Society," the festival had brought together Nordic and African expert speakers to address issues ranging from development cooperation strategies to urban planning and African traditions. Papanek delivered a lecture under the title "The Need for Design in a Tradition-Bound Society," before initiating a new design project for an "inexpensive educational TV set" having been informed by "black men from seven nations" at the event that this was one of the "greatest needs" of Africa (see figures 7.2 and 7.3).[46] A practical application of McLuhan's "global village" paradigm was merging with a Cold War exercise in the effective dissemination of education or propaganda.

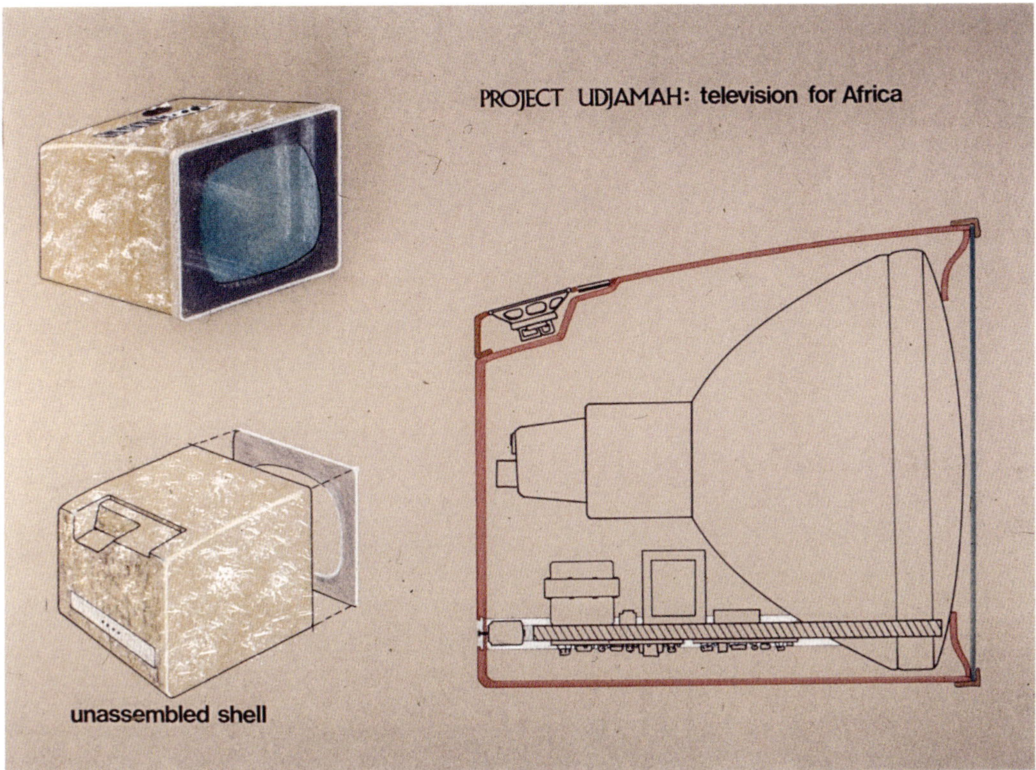

Figure 7.2
Project Ujamah: television for Africa, unassembled shell, cross section of design. © Victor J. Papanek Foundation, University of Applied Arts Vienna

Papanek and his Purdue University students set about conducting a multidisciplinary feasibility study: "Our research had to consider climatology, anthropology, electrics and electronics, population densities, prevalence of African languages in various areas, terrain (for transmission reasons), social attitudes, and many other guidelines of design."[47] A nine-cent device fit for the tropical conditions of the African continent, far from being a conduit of Americanized popular culture with thirty-six-channels and fancy styling, Papanek reassured his audience of the product's ominously prescriptive use: "We plan to broadcast educational material only, and the set will always be 'on.'"[48] Part of his growing repertoire for Third World design, when completed, the apparatus would be "given to UNESCO . . . to join our non-electric, thermocoupled, cow-dung radio (designed for Indonesia)," the social designer explained.[49]

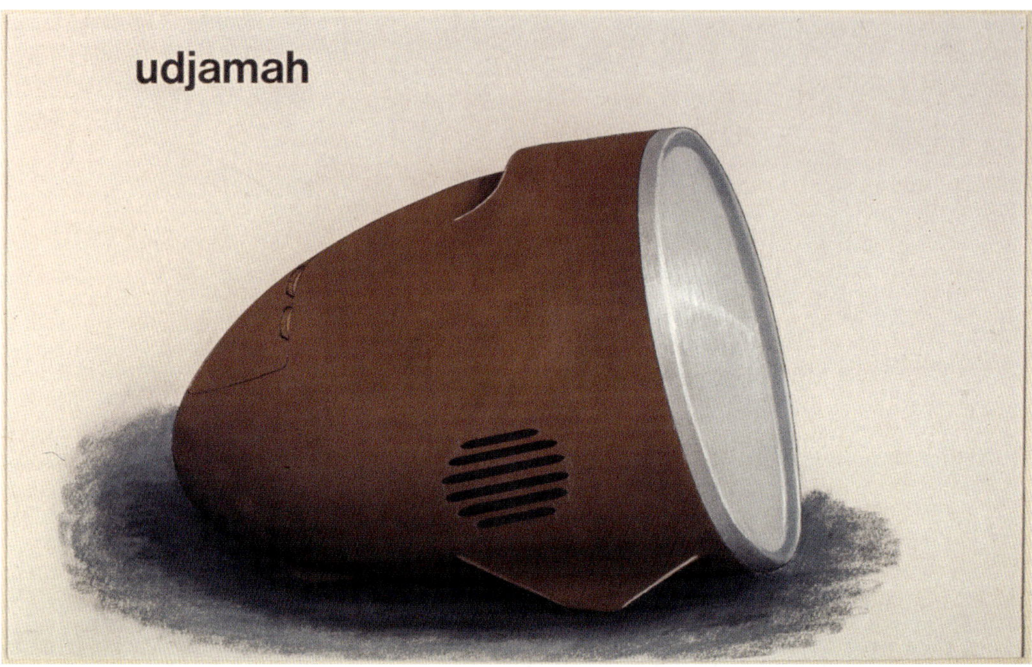

Figure 7.3
Project Ujamah, television for Africa, prototype. © Victor J. Papanek Foundation, University of Applied Arts Vienna

The discourse of the "New African Society"–themed festival took place against the backdrop of raging political debates and demonstrations around the Biafra War, and the horror of its ensuing famine. That being the case, many of its discussions and lectures were touched by pan-Africanist ideas, in the search for a self-determined, nonhegemonic development politics. Although the television for Africa design appeared firmly embedded in the Cold War military weaponry and defense schemes that Papanek and his students experimented with at Purdue University, the adopted nomenclature spoke to a vastly different politics. Dubbed *Project Ujamah*, the "low-cost educational TV set to be built by Africans in Africa" borrowed from the Swahili word *ujamaa* (broadly meaning "family hood"), a term used by Tanzania's first president, Julius K. Nyerere, to define a form of African socialism centered on self-reliance and a communitarian approach to African societies. The theme of the 1968 Jyväskylä Arts Festival (affiliated, as it was, with UNESCO) was almost certainly formulated in response both to the controversy over the handling of the Biafra famine, and the Arusha Declaration of January 1967 that marked independent Tanzania's commitment to the "creation of an egalitarian socialist society."[50]

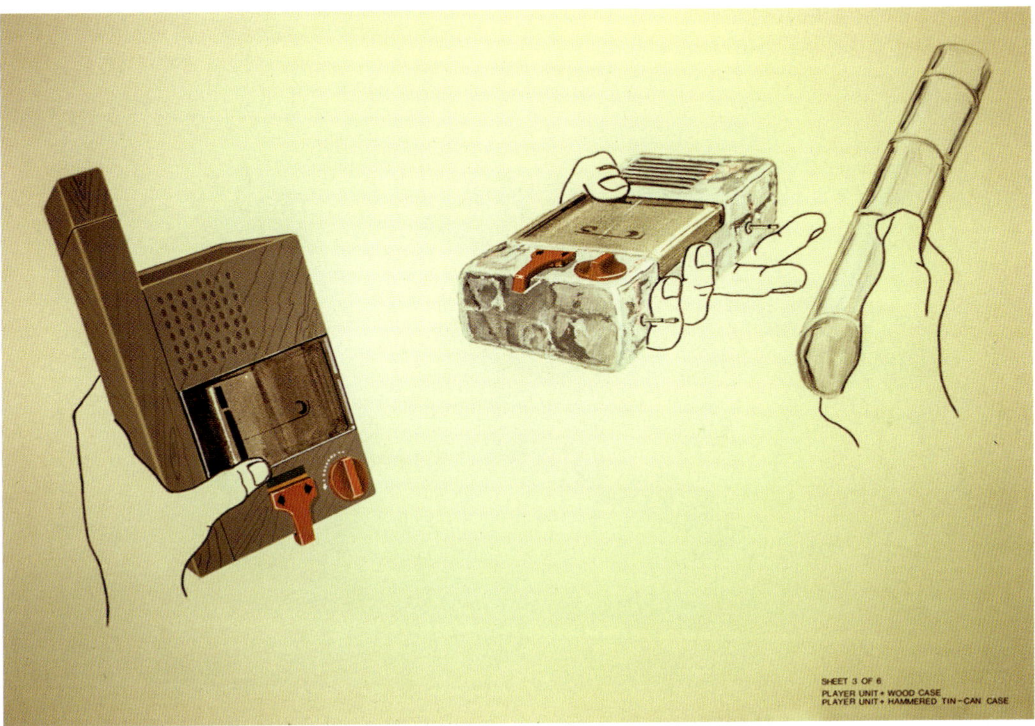

Figure 7.4
Project Batta-Kōya, designed by Victor Papanek and Mohammed Azali Bin Abdul Rahim, 1973. © Victor J. Papanek Foundation, University of Applied Arts Vienna

Considered an exemplary model of transdisciplinary codesign, *Project Ujamah* led to later explorations in the potentialities of hybridized modes of industrial design that sought to combine tradition with contemporary media technologies. In 1973, *Project Batta-Kōya* (translated as "Talking Teacher" from the Chadic language of Hausa), codesigned between Papanek and his student Mohammed Azali Bin Abdul Rahim, incorporated observations of oral storytelling traditions into the design of a customized cassette player, part of which could be housed in a calabash or bamboo casing (see figures 7.4 and 7.5). Sponsored by the Tanzanian and Nigerian governments, under the auspices of UNESCO, the project was underpinned by ethnographic research identifying the need to communicate orally on issues such as nutrition and public health with a disparate group of communities who spoke over two hundred dialects. Importantly, drawing on the anti-neocolonialist politics of development that Papanek first encountered at the 1968 Jyväskylä Arts Festival, *Project Batta-Kōya* made explicit the anthropological and participatory framework of its design process, which was duly credited

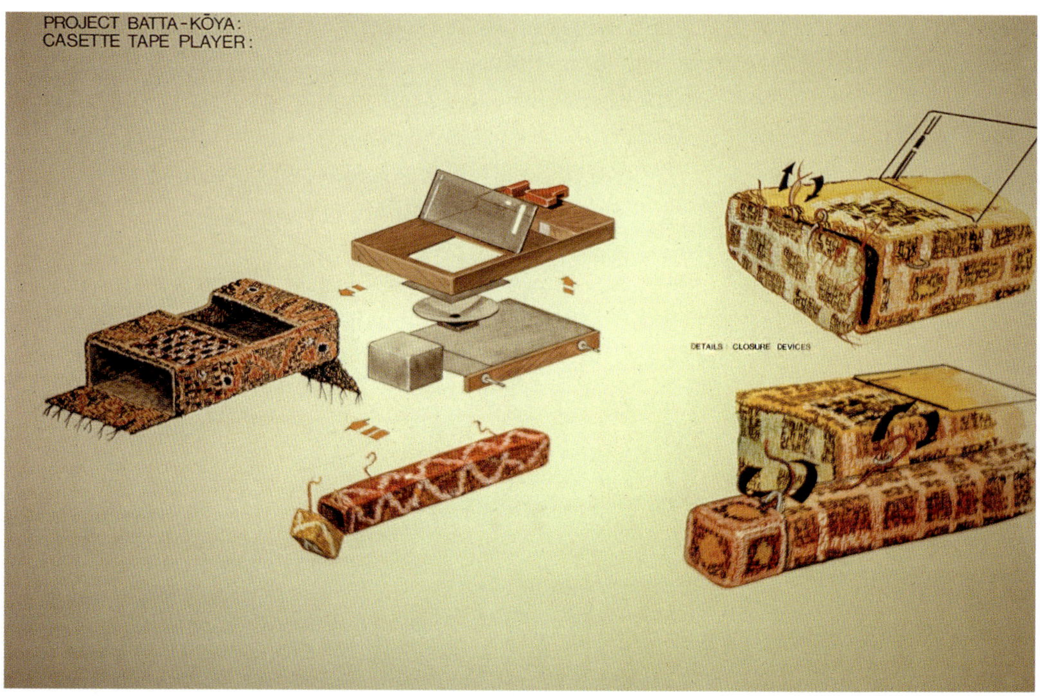

Figure 7.5
Project Batta-Kōya, rendering showing camouflage covers, designed by Victor Papanek and Mohammed Azali Bin Abdul Rahim, 1973. © Victor J. Papanek Foundation, University of Applied Arts Vienna

to "Victor Papanek and Mohammed Azali Bin Abdul Rahim of Malaysia in consultation with peoples from West Africa, Central Equatorial Africa and East Africa."[51]

Predicting the expanded application of *Project Ujamah* in individual community settings, Papanek had forecast in *Design for the Real World* that "there can no doubt that, especially in the areas of education video cartridges will completely revolutionize the development of the Third World."[52] Similarly, the calabash cassette player (a simplified version of a Phillips 3302 battery-powered recorder) was envisioned as a means of decentralizing the hegemony of government information through its use of more than two hundred dialects, bridging the gap between preliterate and postliterate information systems and societies, and lending itself to localized consumption and the furtherance of community empowerment.

As hybrid modes of material culture melding local and globalized concerns, Papanek's low-tech media technologies, including the customizable cow dung *Tin Can Radio, Project Ujamah,* and *Project Batta-Kōya* offered up a genre of design attuned to the contemporary critical discourse over "aid as imperialism," and the understanding of Western interventions into Third World problems as a symptom of neocolonialism.[53] A year prior to the publication of *Design for the Real World*, which (as the title suggests) pivoted on the advocacy of a reimagined humanitarian role for design, Papanek mused that it might in fact be altogether preferable if designers refrained from practicing at all: "The truth is simple: the environmental quality of life on earth and the gulf dividing the haves and have-nots is what war and survival are about. Through design man can positively participate in society. Most designers love design, not people. Just stopping designing completely might help."[54]

As a solution to the quandary of design's indisputably negative impact on the depletion of planetary resources and its stoking of material and social inequality, Papanek's vernacularized, design-anthropological projects stood as moral objects, pitched to a newly enlightened youth and pitted against the debauchery of a Western consumer capitalism generated by a previous generation. Cold War transdisciplinary experimentation is reapplied to an awakened social sensitivity: "Working in interdisciplinary teams away from the 'Disneyland' of marketing, with needs <u>real</u> to people, design can become a meaningful moral act."[55]

The projects, however, left themselves vulnerable to accusations of neocolonial interventionism, and the question of how such apparently benign media technologies might be just as easily appropriated as stealth objects for US government and military propaganda. Following publication of his *Batta-Kōya* project, Papanek complained that some social scientists "in the West" would argue against the device not fitting into a concept of Indigenous authenticity rendering it neocolonial in intent. But he countered, "to withhold it because of theoretical sociological minority reasons smells of professional imperialism, or bourgeois romanticism camouflaged behind revolutionary rhetoric."[56]

In 1974, following the release of the Italian edition of *Design for the Real World* (*Progettare per il mondo reale*,1973), Papanek found himself publicly condemned for his perceived neocolonial endeavors, in the widely read and revered Italian design magazine *Casabella*. The review article was penned by none other than Papanek's Ulm alumnus friend the German designer Gui Bonsiepe, who had visited the author's North Carolina design studio in 1964. Despite his earlier admiration for Papanek's work, Bonsiepe ripped into the self-proclaimed socially responsible designer's manifesto as being nothing more than pallid pseudopolitical gesture: "Even when the activity of design groups could be drastically shifted . . . towards projects of social interest, the results of the work would still be part of the networks of capital. [Papanek's] pale crusade of the petit bourgeois is manifested in the fact that no mention is ever made of the organization of relationships of production and the role of productive forces, especially that of the working class, whose participation constitutes an indispensable premise for any change of design praxis."[57] More damagingly, Bonsiepe's convincing and highly sophisticated takedown of *Design for the Real World* specifically identified the cow dung–powered *Tin Can Radio* design as evidence of Papanek's wanton collaboration with the

US military: "The radio constitutes a tool of ideological penetration and control, and what drove the development of the project has now been transformed into nothing less than a tool of UNESCO pageantry."[58] "Maybe the author," continued Bonsiepe, hammering the final nail into the coffin of Papanek's reputation as a social designer, "during his stay in the United States and his attacks against the design 'establishment' thought he could find allies in military circles, a tactical, strategically disastrous error."[59]

The *Casabella* review proved particularly pertinent, the journal acting as the principal mouthpiece of the avant-garde Italian design radicals, many of whom shared with Papanek a direct interest in exploring alternative models of design, beyond capitalist commodity culture.[60] The most striking similarities were to be found with the Italian Global Tools design collective initiative (1973–1975) and their critical reappraisal of design process, through the celebration of holistic vernacular and anthropological modes of design. Global Tools cofounder, the designer Franco Raggi, who as editor of *Casabella* between 1971 and 1975 had commissioned Bonsiepe's uncompromising review, later reflected that "Papanek was less ideological and more practical, more pragmatic and didactic with respect to the theoretical (abstract) assumptions of Global Tools. Among other things, the goal of Papanek was almost that of a missionary, trying to spread a 'decolonized' design idea."[61]

Fortunately for Papanek, the detailed condemnation of his attempt at decolonizing design, published exclusively in Italian, remained largely confined to small, elite audience of designers. Nevertheless, humiliated and keen to defend his reputation against this genre of attack, Papanek routinely included anonymized accounts of this encounter into explanations of his decolonizing artefacts, arguing, "Tools of this sort are by their nature trans-national, culture-preserving and meta-political."[62]

Under its "New African Society" theme, the Jyväskylä Arts Festival of 1968 helped consolidate the very "tools" Papanek would choose to pursue his "real world" design convictions. Controversy over the efficacy of "an autochthonous low-technology" like the *Tin Can Radio* as a decolonizing design solution, predated the *Casabella* review and continued through Papanek's career and into the present day.[63] But it is no coincidence, considering the extent to which Papanek's ideas relied on his interactions with the country, that it was a Finnish designer and former student, Barbro Kulvik-Siltavuori, who would offer the most public defense of his pragmatic, decolonizing approach: "Today there is much controversy about design responsibility. And inevitably in discussions and articles Victor Papanek is mentioned. Some say he's too political, others that he is not political enough; some that he encourages neo-colonial exploitation, others that he is selling out the white race. But let us remember that he was the first to tell us that as designers we had any responsibility at all. . . . And we, his students, have gone to Tanzania and Kenya, into hospitals and clinics. We have been too busy to defend him, too busy designing."[64]

Actions Speak Louder: Pan-Scandinavian Design Activism

Finnish design activists, such as Kulvik-Siltavuori, had first encountered Papanek along with Fuller in 1967, on the occasion of the first in a series of pan-Scandinavian Design Students' Organization (SDO) events, this one held in Otaniemi, exploring Nordic design education and the potential for progressive design pedagogies. Following on from the Otaniemi meeting, in the summer of 1968, picturesque, fortressed islands in the port of Helsinki acted as the dramatic backdrop for the second larger-scale SDO event, which also cast Fuller and Papanek as its lead figures (see figure 7.6). Here students such as Kulvik-Siltavuori and Pirkko "Tintti" Sotamaa (partner of Yrjö) had a further opportunity to consider how their regional

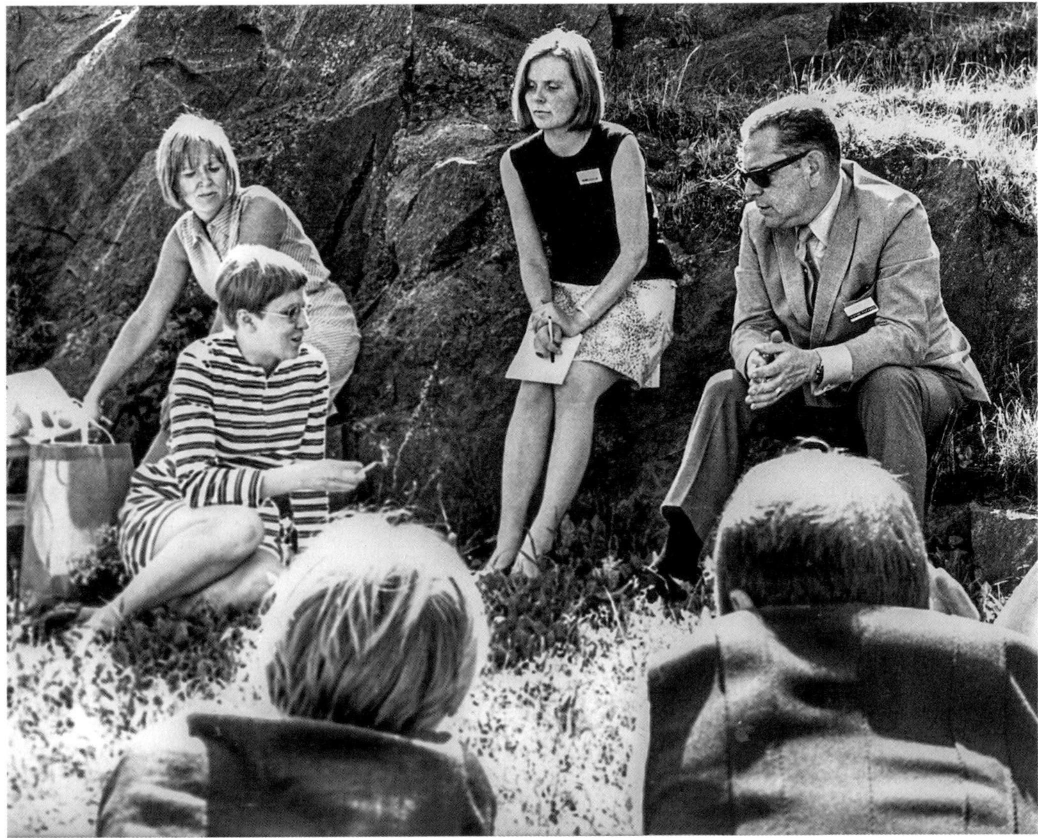

Figure 7.6
Victor Papanek in discussion with students, Scandinavian Design Students' Organization symposium, Suomenlinna, Finland 1968. © Kristian Runeberg. Courtesy Yrjö Sotamaa

design politics might progress within a broader international framework, as well as gender and social class issues (Pirkko Sotamaa, for example, designed accessible and inclusive packaging for contraceptive pills devised principally for use by nonliterate women).

Fuller's participation at the pan-Scandinavian event had been nothing short of a coup for the Finnish students.[65] By 1968, his international profile had magnified, and according to architectural historian Felicity D. Scott, his status had shifted from "obscure and eccentric inventor to a household name in American ingenuity."[66] The iconic biosphere devised for the American Pavilion at the Montreal Universal Exposition of 1967 (the theme of which was "Man and His World") captured the imagination of a mass popular audience. The futurism proffered by Fuller's construction jarred with the quaintness of the Finnish contribution to the Scandinavian Pavilion overseen by Finland's feted sculptor Timo Sarpaneva. Under the slogan "Creative Finland," the display caused consternation for a younger generation of designers keen to extricate themselves from anemic politics and clichés of Scandinavian design and decorative arts. The Finnish exhibit featured abstracted and craft-as-art pieces that were deemed to have miserably failed to represent the exposition's central theme of globalization and mass electronic communication, thus relegating Finland to the status of a backward, handcraft-producing nation. The broad-scale condemnation of Sarpaneva's failed curatorial approach signaled a greater crisis and complacency within a Finnish design culture that, like the rest of postwar Nordic and Scandinavian design, relied almost solely on the promotion of selected, individual designers rather than an engagement with contemporary societal issues and technologies. United in their vision of pan-Scandinavian design reform that challenged the dominance of big-name designers and US models of corporate design, the SDO had on its hands the challenge of unraveling decades of economic and cultural diplomacy.

Finnish design culture had been dominated in the immediate postwar period by a governmental policy constructed around the representation of Scandinavian design under the rubric of mid-century modernism. *The Design in Scandinavia* exhibition that toured the United States and Canada from 1954 to 1957 exemplified this phenomenon, the legacy of which extended well into the mid-1960s.[67] These same government policies geared toward boosting Finnish design had also made national heroes of designers of the applied arts: Tapio Wirkkala, Kaj Franck, and Sarpaneva, for example. The dramatic urbanization of Finland in the 1960s had seen the country transformed from a rural nation, with an agrarian economy, to one of sudden industrialization that, according to one historian, demanded "national strategies to steer this new and more complex union of design, industries and growing foreign trade."[68]

Prior to this widespread urbanization, and through to the late 1960s, Finland occupied a crucial Cold War position with government organizations such as the United States

Information Agency (USIA), working tirelessly on using soft diplomacy aimed at penetrating the nation's cultural politics. In 1953, for example, the Finnish-American Society devised the *American Home* exhibition at the Taidehalli (Art Hall) in Helsinki, a performative installation that would be a precursor to the renowned USIA Moscow 1959 *American National Exhibition*, whose model US home generated the so-called Nixon-Khrushchev Kitchen Debate.[69]

By the 1960s, technological revolution and global communication stood at the forefront of Cold War discourse. Despite the United States' victory over the USSR in the Space Race, international focus on the student protests against the Vietnam War and US racial segregation intensified. So did the need for soft diplomacy aimed at a liberal cultural elite occupying the crucial geopolitical site of Europe's nearest Soviet neighbor.

Exploring the intersections of industry, environment, and product design, the principal aim of the second, extended SDO event was to create a brainstorming forum to reimagine a postobject, postcraft Nordic and Scandinavian design landscape, and a future politics of critical design practice for a new generation of designers. "Nordic collaboration was revived by the young generation now centred on critical thinking and activism," to quote one Finnish design historian, "instead of promotional strategies linked to export industries, manifested in international exhibitions and their object-focused displays."[70]

The summer of 1968, then, saw Finnish architecture and design students along with the recently formed group of pan-Scandinavian design activists immersed in an eight-day-long, think-tank-style event amid the ruins of the historic sea fortress Suomenlinna. Better understood as a happening than a conventional design conference, the event incorporated an impressive lineup of contemporary international architectural and design figures, following the extensive lobbying of national politicians, educators, and policy makers by core Finnish members of the SDO.[71] The location was a strategic choice, not just for its scenic backdrop but also its relation to the dramatic and bloody upheaval of Finnish political history. A former prison camp for the Soviet Red Guard following the Finnish civil war of spring 1918, Suomenlinna was notorious as the site of firing squad executions and mass deaths from hunger and disease. As a gesture to its complex and contested history, parts of the fortification were dubbed with allegorical or political affiliation. So it was that the activists came to hold their progressive design discussions in the shadow of the aptly named Bastion of Good Conscience (*Hyvän omantunnon linnake*), flanked by a monument of unsettling historical resonance.

If Suomenlinna stood as an apposite stage for a dramatic Cold War student intervention, so did the nation itself: a northern European country poised precariously close to the Soviet Union, Finland had defiantly rejected aid (in the form of the Marshall Plan) from the United States after World War II and ran a successful state welfare system whose collectivist ideologies stood in stark contrast to those of the free-market United States. The vested interest the

States had in reconstructing war-torn Finland in the 1940s, during its reconstruction phase, and the role architect and designer Alvar Aalto had in bridging the US-Finnish relationship have been well documented.[72] But the overtly radical nature of this new Finnish-American relationship, in the form of Papanek and Fuller's participation, made a significant detour from an established trajectory.

The SDO had no qualms in spelling out its "Alternative Mission Statement" to the broadest mainstream design audience. "We oppose capitalism with every means at our disposal to attain a dynamic socialist system that will make total social justice possible," they declared in the Swedish design magazine *FORM*.[73] A published manifesto further clarified their position: "We want to put an end to a system in which invented needs are satisfied at the expense of genuine needs. We want to carry out surveys of the genuine needs of people in different areas in the world. We want to analyse the results of these surveys and use them as a basis for participation in satisfying these genuine needs. We want to put an end to a system that misuses our shared resources on the globe. We want to support national liberation movements actively."[74]

The symposium audience and speakers consisted of a varied and interdisciplinary crowd, drawn from corollary design disciplines such as engineering and pedagogy, with the intention of carving out the possibility of a design practice that looked beyond the paradigm of the profit-driven market. With around twenty-five local and international speakers, lectures shifted from presentations to informal open-ended discussions and heated debates, some continuing into the early hours of the morning and others consciously inverting student-expert power relations through their informality. Divisions among design, research, technology, architecture, and pedagogy were suspended. The UK-based Design Research Unit (a user-oriented design consultancy) spoke, for example, alongside architectural theorist Christopher Alexander, who addressed the audience with his lecture titled "The Organization of Design Pattern." Engineer Graham Whitehead considered the issue of "System Ergonomics: The Human Factor in Complex System Design," alongside world-renowned Finnish designers such as Kaj Franck and Antti Nurmesniemi, keen to support a transdisciplinary agenda.

The choice of Fuller and Papanek as keynote speakers for the SDO international event was undoubtedly inspired by intellectual curiosity, but it was also a calculatedly provocative gesture aimed at the Cold War sensitivities of "the establishment." Fuller was already established as a visionary thinker and inventor of the biosphere, while Papanek was familiar among the organizers as an advocate of social design due to his attendance of the earlier Finnish cultural events in 1960s. Most significantly, as a duo, Papanek and Fuller offered the dual advantage of representing the United States (in the formal sense of their citizenship) while simultaneously undermining their nation's standing as the great capitalist power through their critical oratory styles and social, humanitarian, and ecological agendas. They had also proved themselves as strong allies to the cause through their attendance of the first SDO event in 1967. Fuller used his second appearance to hold forth for three hours on his World Resources Inventory project, while Papanek addressed issues ranging from "Human Needs and [the] Designer" to "The Needs of the Underdeveloped and Backward Areas," railing against the crimes of corporate design and weaving into his discussions condemnation of the Vietnam War.[75] Papanek's inflammatory tone chimed with the youthful, feisty audience: "I would like some day to have a knife, a *puukko* [traditional Finnish belt knife], and meet the gentleman who decided in some advertising agency twenty-five years ago that soap and detergent should make a lot of foam. Because that man is responsible for a great deal of pollution in our rivers and streams and even our oceans by now, the fact of the matter is how much foam you get has nothing to do with how well something washes. So when you make major design mistakes, you make them on a global scale."[76]

Following his main lecture, Papanek received what might be considered the ultimate Finnish accolade of the day—an invitation for a late night picnic at the famed Experimental House at Bökars, Porvoo—home to the founder of Marimekko, Armi Ratia and her husband Viljo Ratia. Resplendent in his brightly striped red and pink Marimekko shirt and dark shades, Papanek boarded a speedboat to one of the most coveted design destinations in the Nordic countries, to hang out with some of the most fashionable, and liberally progressive design icons of the late 1960s.

Buckminster Fuller's Reindeer Abattoir and the *CP-1 Cube*: The Design Lab Revolution

The creation of a portable reindeer abattoir encapsulated the audaciousness of the organizers' ambitions (see figure 7.7). The result of an onsite transdisciplinary workshop led by Finnish student Esko Miettinen and overseen by Fuller, the prototype addressed decolonizing and socioeconomic issues head-on: adopting design as an overtly political tool. The structural principles of Fuller's Expo 1967 geodesic dome were crudely applied to a scaled-down model of a mobile slaughterhouse with the effect of resembling a Joseph Beuys-style radical ecosculpture rather than a socially useful design. Recent top-down government policy devised in line with the increasing industrialization of meat production in Finland forced the Sámi (whose semi-Nomadic reindeer herding underpinned their livelihood and traditional way of life) to adhere to unfeasibly strict regulations based on standardized, static methods of cattle slaughter. The design, intended more as a political gesture than a practical solution, responded to the perceived erosion of the rights of the Indigenous peoples of Sápmi. The

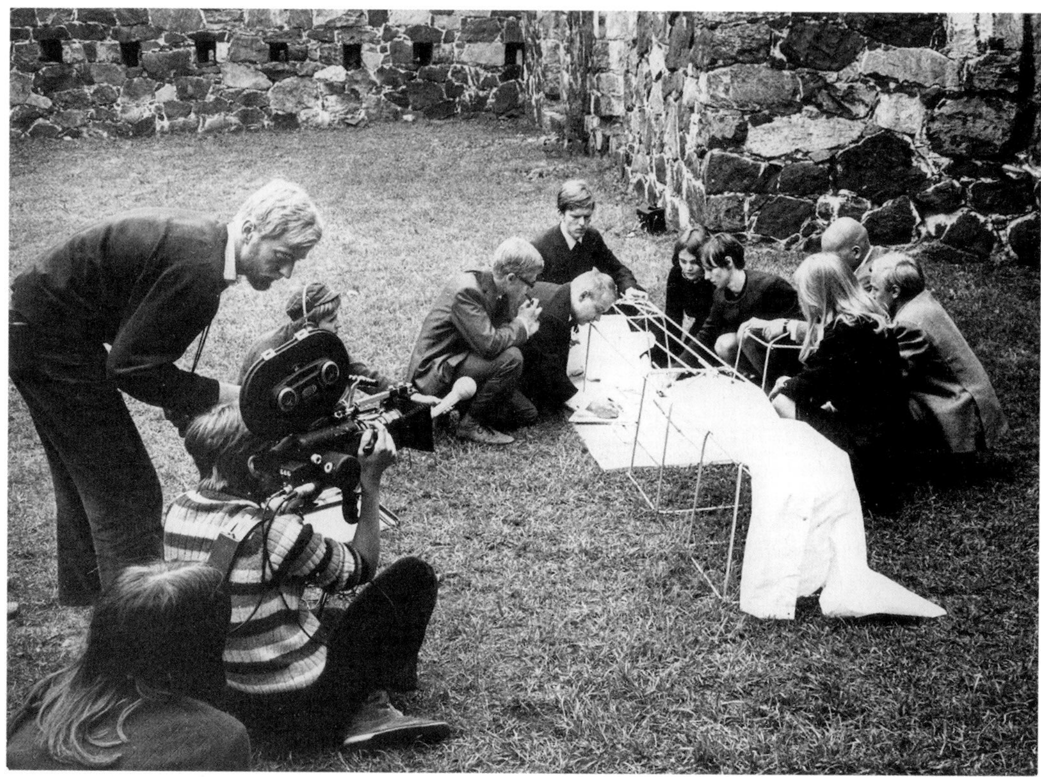

Figure 7.7
Design for Portable Reindeer Abattoir, Scandinavian Design Students' Organization symposium, Suomenlinna, Finland 1968. © Kristian Runeberg. Courtesy Yrjö Sotamaa

prototype exemplified the activists' endeavor to escape the strictures of formalist design in favor a model of experimental, transdisciplinary, and socially purposeful design practice.[77] Coupled with Papanek's group-designed and ethnographically researched rehabilitation installation for children with cerebral palsy (replete with iconic Finnish Marimekko textiles), these hands-on projects came to define the politics of the pan-Scandinavian activism. They concretized the ambition to apply theoretical design conjecture to "real-world" issues and humanist agendas (see figure 7.8).

For it was "action" rather than words that defined the Suomenlinna design seminars, and it was in this experimental setting that Papanek generated one of his most public interventions into the role of design in the area of disability politics. The *CP-1 Prototype*, an environment for children with cerebral palsy devised alongside Fuller's portable reindeer abattoir as

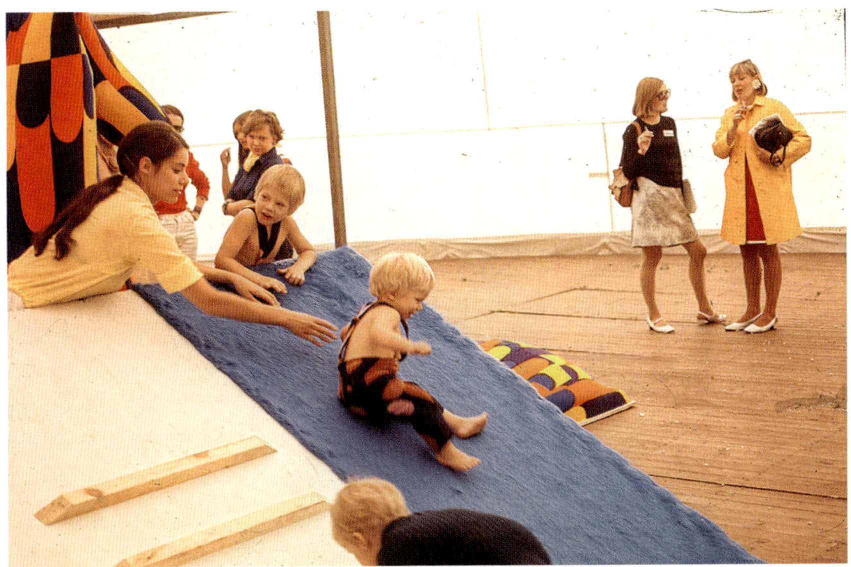

Figure 7.8

CP-1 Prototype portable environment for disabled children, Scandinavian Design Students' Organization symposium, Suomenlinna, Finland 1968. © Victor J. Papanek Foundation, University of Applied Arts Vienna

a sociopolitical intervention, formed the basis of Papanek's contribution to the event. Nine design students hailing from Finland, Sweden, and Hungary carried out the basic research regarding the needs of children with cerebral palsy from an interdisciplinary perspective, conducted face-to-face interviews with clinicians, and engaged in "participant-user" play sessions with children affected by the condition. Starting from the premise that not a single toy had ever been designed specifically for children with cerebral palsy, the team included student designers Immi Tiivola, Marianne Andersson, sculptor Zoltan Popovits, Jorma Vennola, and Yrjö Sotamaa. Facilitated by Papanek, they set about designing a therapeutic and stimulating environment based on playfulness. The action-based, "hands-on" approach involved four meetings of three hours each with the nine-person team, with each research meeting deciding the direction of the next. Gradually, the idea for an environmental, two-meter cube that could travel in knocked-down form from clinic to clinic emerged. Dubbed "CP-1" on the understanding that this prototype would spawn a series (with "CP-2," "CP-3," and so on), thus creating a portable product genre for mass self-build dissemination, each iteration was designed, researched, and customized according to the specific, or special, needs of the user. The project made headlines in popular Finnish magazines: one article, titled "A Place Where

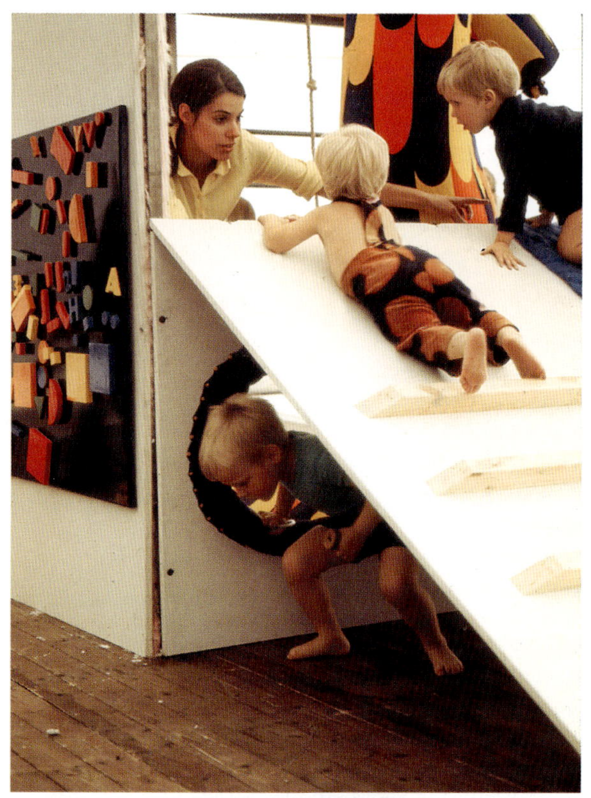

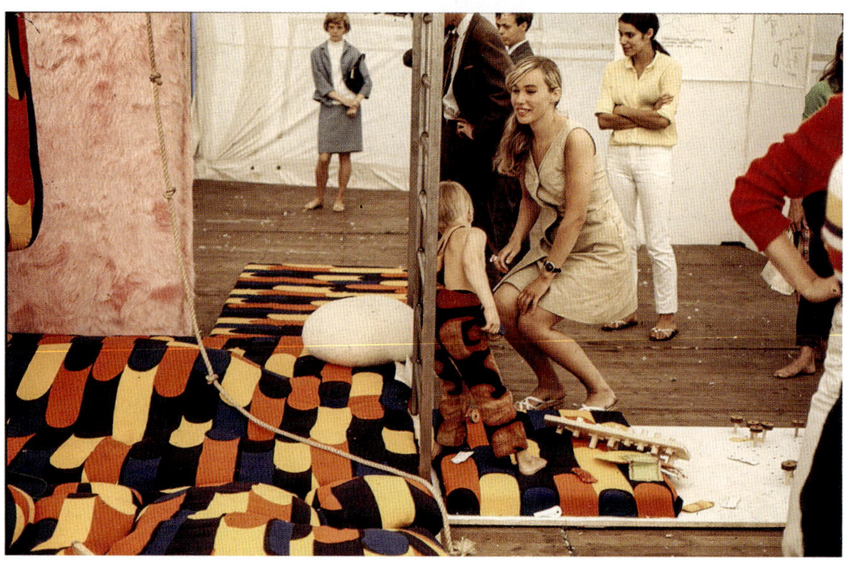

Figure 7.8
Continued

Everyone Can Play," praised the *CP-1* environment as an initiative to include "minorities, those who are easily forgotten because they are socially, racially, psychologically or physically different."[78]

This design-lab approach demonstrated, according to Papanek, "that a design conference need not be a succession of papers presented by the pedantic to the apathetic."[79] The brightly colored cube was essentially an activist installation that the team hoped would draw attention to "a vast area of design neglect" and "make sure that each of the students who had participated afterward would forever feel a little ashamed when designing a 'sexy' coffee percolator with tail fins, a grenade launcher or a transistorized back-scratcher."[80]

When Papanek chose the slogan "Actions Speak Louder" as the title of a feature for the leading US industry journal *I.D.* in 1968, he was clearly alluding to a broader backdrop of radical politics, social upheaval, and student activism occurring in Europe. Neither was it coincidental that he chose to decry the emptiness of conventional design conferences in a journal whose audience was predominantly corporate designers; he announced instead the radical new participatory design processes he had evolved through working with Finnish designers in Suomenlinna.[81]

Contrasting the tedium and lack of creativity of conventional design conferences with the enlightenment of student-originated, informal think tanks, Papanek continued thus: "Too many martinis, slight morning hangovers, overheated hotel rooms . . . boring dry-as-dust-speeches . . . with one or two topics, [such as] 'Is Industrial Design Moving Towards Greater Professionalism?' or 'Do We Have an Identity?'; all of these combined with a degree of back-slapping bonhomie, spell 'design conference.'"[82]

In a thinly disguised critique of the world-famous International Design Conference in Aspen, Colorado, which was dominated by the major US corporate power players and adhered to a conventional speech-led format, Papanek declared, "We made the decision . . . that at the 1968 Suomenlinna conference we would take the heretical step of having designers *design* rather than discuss methodology."[83]

In his exploration of Cold War 1960s political youth activism, historian Nicholas Rutter has described the importance of understanding how the "internationalist sensibilities" of youth movements were mediated and made complex by national specificities.[84] "The youth" of 1968, argues Rutter, are too often understood as a "collective noun, oriented against the state, the parents, the Cold War but not against itself."[85] Comparative and transnational histories have more recently begun to address the complexities of leftist and progressive counterculture movements of the 1960s, commonly described by historians as a dissonant decade. The rise in youth activism and youth movements across Europe shared common threads of discontent purportedly born of intergenerational conflicts, but were nevertheless deeply entrenched in the specifics of localized and broader macropolitics.

It is perhaps no coincidence, then, that the Suomenlinna activists arranged their design symposium four years after Helsinki had served as a "neutral" venue for the 1962 Soviet-sponsored World Festival of Youth and Students. The festival had rendered the city and its youth the focus of intense Cold War international relations. In 1962, along with the KGB and CIA, Helsinki witnessed the strongest presence of US citizens and representatives of developing nations in its history; it is also this period that is broadly recognized as defining the beginning of youth radicalism in Finland.[86]

In an attempt to appeal to the Finnish youth and intellectual vanguard, and to banish critiques regarding racial segregation in the United States, the CIA covertly sponsored artists and musicians of African American and Hispanic origin to attend the World Festival of Youth and Students in Helsinki. One such artist was Howard Smith, an African American sculptor from Philadelphia, who traveled to Finland as a member of the Young America Presents Group and remained there, initially supported by his work at an art-oriented advertising agency.[87] By 1963 Smith, who was unaware of the origins of the grant that had supported his original travel to Finland until the post–Cold War period, presented a private exhibition of his designs in Helsinki and went on to design for major Finnish textile companies. Smith later became an ex-pat friend to Papanek during his visits to Finland in the 1960s, Papanek's trips ostensibly were supported by the US American-Scandinavian Society as part of a broader late Cold War policy of using figures such as Fuller and Papanek to ward off communism. As a counterculture icon, Fuller himself had been the subject of intensive FBI interest, prompted by the sensitive work he carried out in conjunction with the US military and other government agencies. Their main concern, made all the more significant by his regular appearances in Finland, was the possibility that he could have been in contact with Soviet intelligence agents.[88] FBI interest in Fuller, who traveled extensively, including to the Soviet Union, peaked between the mid-1960s and 1968, before ending abruptly with his last visit to Finland. So crucial were Fuller's activities to the nuanced world of Cold War politics that an FBI agent even traveled to the design professor's home at Carbondale, Illinois, to conduct a formal interview with him. The report notes reveal Fuller's filibustering, using the opportunity to provide the agent with an extensive rendition of his philosophy: "The remainder of the interview with Dr. Fuller was taken up," explained the FBI agent's notes, "by the latter's explanation of his 'geosocial revolution,' which he stated is a revolution by design and invention where the world's vast resources are used for the betterment of man."[89]

Framed by these wider national issues and politics, Finnish activists also drew on a groundswell of dissent across Europe that bolstered the ambitions of the pan-Scandinavian progressive designers. Just two months prior to the Suomenlinna symposium, protesters at the May 1968 Milan Triennale had ransacked the annual exposition, railing against its amoral celebration of overt consumerism and the hypocrisy of its choice of theme: "The Greater Number."

With police intervention and the abandonment of the Triennale, crisis befell the design profession.[90] In light of the Milan protests, the organizers of the Industry, Environment, Product Design symposium had striven to avoid the power dynamics and nepotism associated with corporate design events, such as the Aspen conference overseen by prominent US design figures Eliot Noyes and Charles Eames. Instead, they redefined design as a participatory practice whose role was critical, provocative, and distinctly pan-Scandinavian.[91]

Prophet of Our Time: From Finland to the World Stage

In April 1969, Finland's leading public radio station featured a half-hour program focused on Papanek under the title *Prophets of Our Time*, led by Olli Alho, one of Finland's most prominent left-wing cultural personalities and a key figure in the Finnish Broadcasting Company.[92] The discussion expanded the concerns of those first flagged up by the SDO, with the host introducing the three discussants as "experts who have on many occasions opposed the so-called cult of personality when it concerns famous architects or designers."[93] These included Sotamaa (cofounder of the SDO and lead organizer of the Suomenlinna event), Harri Moilanen (leading architect-designer and cultural radical), and Juhani Pallasmaa (Finnish architect and educator).[94] Sotamaa proceeded to dispel the narrow-minded nationalism that had defined postwar Finnish design: "I haven't come across many students of design who think in national terms," he declared. "There may still be some who do but, in general, values now are completely different from those that existed when the present leading names of Finnish design—Sarpaneva, Wirkkala—were students, and now it is felt that the designer should participate in other social activity rather than designing alone."[95]

Papanek (as opposed to Fuller) was singled out as a prophet precisely because, in his Finnish appearances and related media, he had openly and vehemently opposed the cult of personality that dominated the existing Scandinavian design and architecture scene. Papanek, they countered, while charismatic in his presentations, did not adhere to a reactionary "cult of personality," but rather was considered "some kind of prophet," as Moilanen put it, "because product design and design theory and the social aspects of design have lagged behind and Papanek had truly presented a new critique of design."[96]

In keeping with the "internationalist" agenda of the period, Papanek's advocates sought to overturn a culturally essentialist and vernacular notion of Finnish design (promulgated by the nation's most famous design export, Alvar Aalto) in favor of a socially, politically, and technologically progressive internationalist vision. In a strikingly prophetic statement, Moilanen preempted the move away from the author-designer and the shift to a dispersed design culture beyond the object: "I would say that younger people, students of product design, and design students in general, do not see solutions through objects," argued

Moilanen on the radio. "They see powerlessness, the powerlessness of the designer, and they understand that they must also participate in political activity. To be part of groups seeking to correct problems and failings that cannot be done with the aid of design."[97] The great Scandinavian icons of design, according to these critics, had entirely overlooked the pertinent themes of social and ecological concerns, Third World poverty, and social exclusion, as well as the power of a collective, politicized youth in transforming the purpose of design.

The *Prophets of Our Time* panel discussion offered nuanced insight into the emergence of a critical counterculture of design that challenged the authority of the designer and design as a practice of form-making and aesthetic discernment. Through its exchanges, the emergence of design as a nonartifactual, cross-disciplinary mode of political engagement forged in a decade of dissent is clearly traced—an engagement whose trajectory is evident in the critical, speculative, and design anthropology of the twenty-first century.[98] But how did these discrete acts of design activism, forged from a Cold War and pan-Scandinavian agenda, come to impact broader 1970s alternative design strategies, and the struggle for a utopian vision of design for the masses, against design for profit?

By 1969, now returned to Purdue University and inspired by his interactions with the pan-Scandinavian activists and his new young Finnish colleague Yrjö Sotamaa, Papanek began outlining a treatise on the politics of design and expanding his work for social needs as well as promoting the ethical and aesthetic dimensions of Finnish and Scandinavian design more generally. Sotamaa and his wife Pirkko exerted massive influence on the transdisciplinary direction of the Department of Industrial and Environmental Design at Purdue University, helping instigate, for example, a shift toward progressive pedagogic design for children with autism and cerebral palsy (see figure 7.9).

Per Johansson, the young Swedish designer and general secretary of the SDO who had contributed the article "Design and Youth Revolution" to the *Design Course* journal, egged Papanek on to produce a manuscript charting his alternative vision of design for the masses for a politically aware Swedish and Finnish audience. The basis for the first draft of what would become his groundbreaking book *Design for the Real World* began with a series of hand-drawn doodles, plotting a radical design pedagogy of the future. By the early 1970s, the key ideas behind these doodles propelled design departments across Europe into a state of quasi-revolution, as Papanek's manifesto forged from Nordic and Scandinavian design activism took hold of a newly politicized generation of designers.

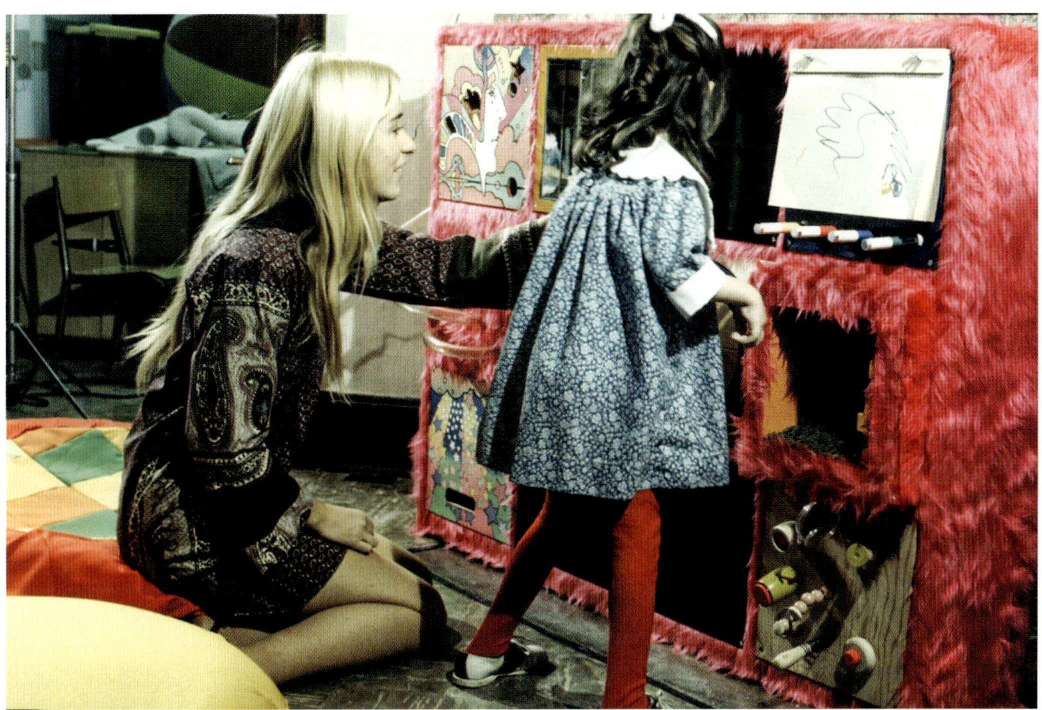

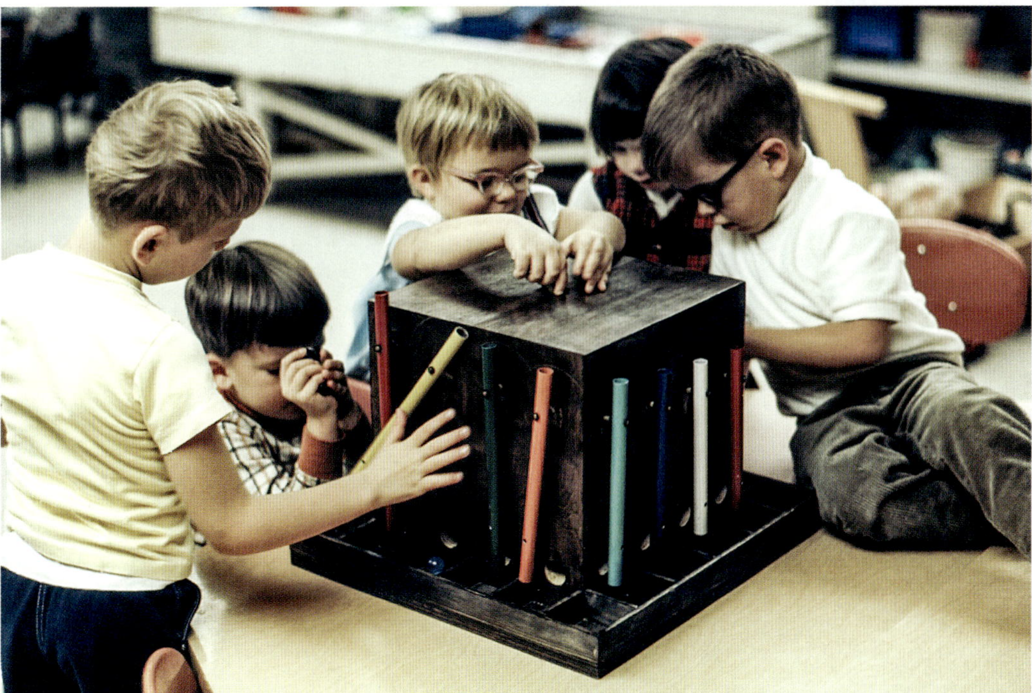

Figure 7.9
Pedagogic play projects designed for disabled children overseen by Yrjö Sotamaa, assistant professor, Purdue University, 1969. © Yrjö Sotamaa

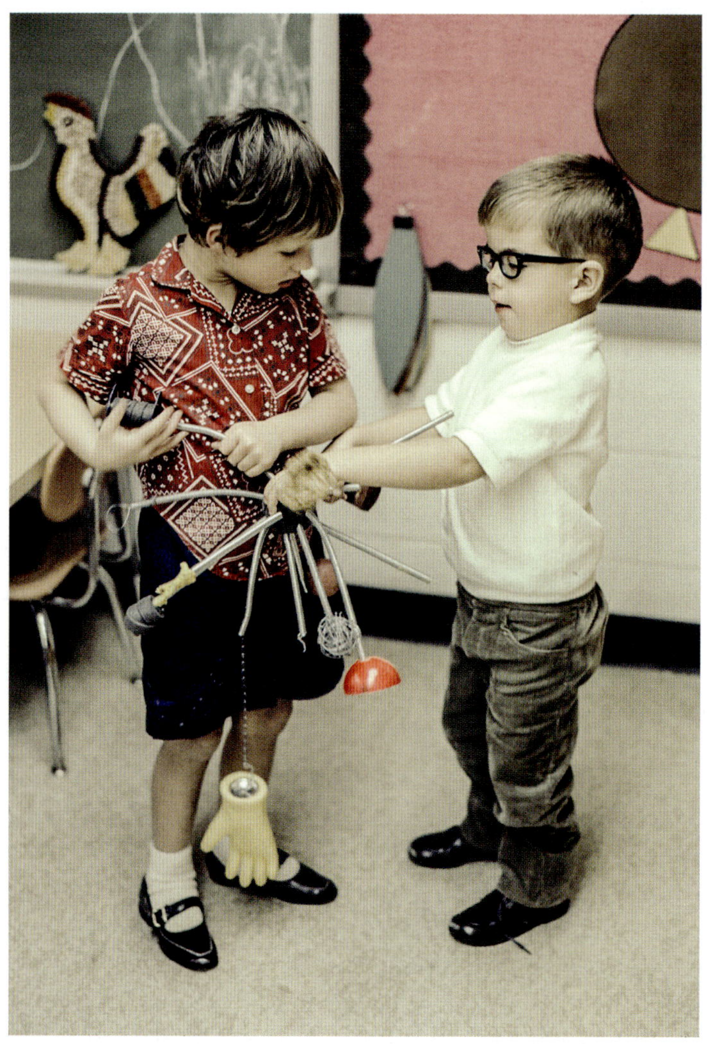

Figure 7.9
Continued

8   Design for the Real World: A Call to Action

In the summer of 1969, Papanek was once again invited—by a group of Danish design students loosely tied to the Royal Danish Academy of Fine Arts—to be a key speaker at an SDO (Scandinavian Students' Design Organization) seminar. Building on his growing following in Northern Europe, the designer led a workshop in Copenhagen exploring the politics of design and the future of design pedagogy, with a specific emphasis on issues of disability and design. An intellectually vibrant and fecund event, the seminar addressed five working groups on the themes of international solidarity, urban environments, health care environments, working environments, and future environments. Engaging with the growing prominence of disability rights, design student Susanne Koefed sketched out a first draft of the wheelchair icon that would go on to become the international symbol for access.[1] The intense gathering of like-minded designers also gave rise to another iconic piece of graphic design, one that came to embody a new socially led politics of design.

The Copenhagen Flowchart: Visualizing the Politics of Design

In the atmospheric eighteenth-century academy building of Charlottenburg Palace, situated in the center of a historical capital city awash with 1960s alternative subcultural and community projects, Papanek sketched out the first draft of a design that would be dubbed the "Copenhagen Flowchart": a visualization of the indices and intersections of power relations of design. Mapped out as a hand-illustrated diagram, the flowchart would be the central tenet of *Design for the Real World* and reproduced in 1973 as the purchasable "Big Character Poster No. 1: Work Chart for Designers" (see figure 8.1).[2]

What began as little more than a codesigned mind-map, sketched out in an interactive design setting, went on to form the basis of a publication that would transform the idea of design practice in the late twentieth century: a map for a new politics of design that argued for the prioritization of ecological and social issues—and the application of design to the

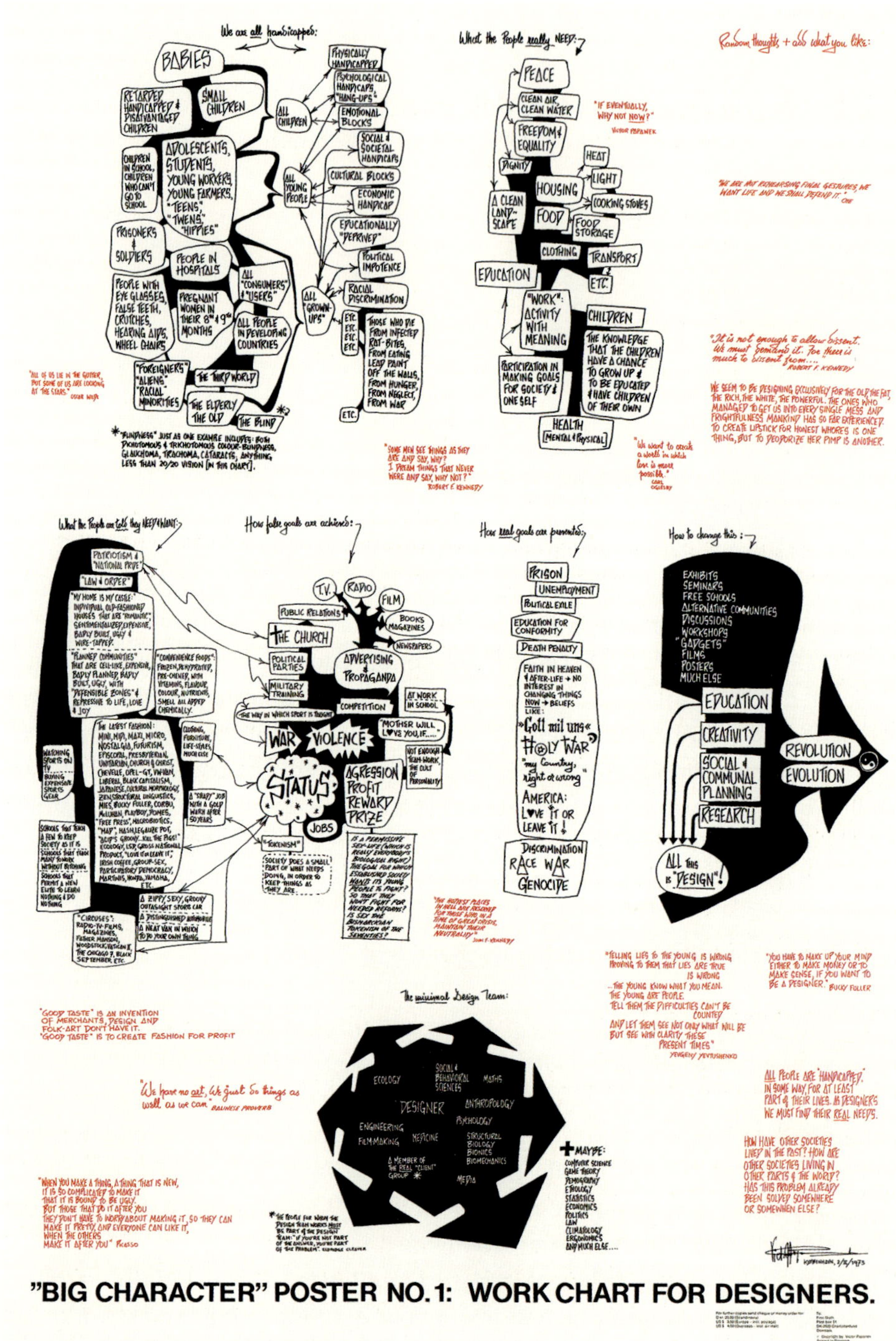

Figure 8.1

"Big Character Poster No. 1: Work Chart for Designers," Victor Papanek [1969] 1973. © Victor J. Papanek Foundation, University of Applied Arts Vienna

needs of the many, against the desires of the elite. "While participating in a design conference held by the Scandinavian Student[s'] Design Organization at Copenhagen in the summer of 1969," Papanek later recounted in the pages of *Design for the Real World*, "it was my job to construct a 'general case' portion of a flow chart concerned with the social and moral responsibility of the designer and his [*sic*] position in a profit-oriented society."[3]

Going on to describe how the flowchart had been deliberately left open as an ongoing entity to be completed as a collective project, the designer reiterated the way in which the chart held the power to "form a social and political blueprint for tomorrow—for society as well as for design."[4] "This is a large job indeed," he continued, before surmising that in fact "this entire book [*Design for the Real World*] attempts to address itself to precisely that question."[5]

Depicting a constellation of agents, disciplines, and methods involved in the design process, the diagram began as an infographic showing the interconnections between the groups excluded from the normal matrix of design concerns. These included babies, infants, and children, students and adolescents, "people in emerging countries," "'Foreigners' and racial minorities," "pregnant women in their 8th and 9th month," "prisoners and soldiers," "people with eye-glasses, false teeth, crutches, hearing aids, wheel chairs"—with a qualifying asterisk defining *blindness* (not included in the "physically handicapped" diagrammatic bubble) as a term including all eye disease, ranging from glaucoma, cataracts, colorblindness, myopia, and "almost 200 more diseases."[6] Though terms such as *retarded* were still commonly accepted and used in Papanek's work, a pointed attempt to complexify and problematize terminology related to modes of disability, and the avoidance of terms such as *Third World*, clearly indicate that Papanek's interaction with a new generation of student activists was impacting his thinking.[7]

The Social Designer Gets Hip

As this same point in his career, Papanek's demeanor and sartorial sensibility gradually began to shift away from an outmoded identity of the clean-cut, dour-faced Cold War industrial designer to that of a loosened up, politically enlightened social polemicist. This transformation did not escape the attention of his followers, some of whom had first encountered Papanek, along with Buckminster Fuller in Finland in 1966, as a stiffly dressed figure of male authority. "There was something suspicious," Gunilla Lundahl, Swedish design journalist and editor of *FORM* journal in the late 1960s later recounted, "about a man who first addresses us wearing a sharp-cut suit, and then a couple of years later leather trousers and a belt with an over-size buckle. Let's just say, he was a man who wanted to make an impression—later he even wore shoes with heels."[8]

According to sociologist Sam Binkley, the process of "getting loose," born of the counter-cultural politics of the late 1960s, went on to define the expressive lifestyle and consumption politics exemplified by alternative, nonmarket-oriented initiatives such as the *Whole Earth Catalog* (1968–1972). The impetus to be "hip" and to free oneself from the constraints of inauthentic commodity culture gave rise to a sensibility in which a "posture of rebellion," borrowed from the politics of the Left, led to the search for purity as an ontological comfort for a middle class faced with shifting power relations.[9] Crucially, this process, according to Binkley, manifested itself through "a pattern of appropriation, emulation, and cooptation of the expressive styles practiced by other groups and other identity movements principally African Americans and Black Power groups, but also women, Asian, Native American, and other ethnic groups."[10]

Regaling students with tales of his time spent embedded in communities of the Inuit, Balinese villages, poor white Appalachian towns, and African American ghettos, and routinely posing for photographs against a backdrop of his study collection of ethnographic objects (which included masks and ritual artifacts arranged across his domestic interior), Papanek sought to imbue himself with the authenticity of the minority groups he professed to serve. During this 1969 visit to Denmark, and later in his capacity as a guest professor at the Royal Academy, a number of Danish students were informed that Harlanne, the young wife who accompanied Papanek, was of "Native American" origin.[11] One of the standard biographical summaries he used during this period of his career suggested, fallaciously, that he had carried out bona fide anthropological fieldwork: "He has lived with an Eskimo tribe for nearly a year, as well with Hopi Indians."[12] The need to appropriate an identity, and a network of social relations outside those belonging to the standard white, male, middle-class American proved essential to a figure set on mastering the discourse of a new politics of ethical consumption.

Intertwined with Papanek's personal reinvention as a right-on social design professor, the Copenhagen Flowchart perfectly illustrates the sociohistorical phenomenon of "getting loose." Despite its progressive tone, *Design for the Real World* also reads as a potential treatise in "appropriation, emulation, and cooptation," coming perilously close to fetishizing socially disadvantaged groups as entities occupying a state of premodern authenticity, their needs perpetually couched in terms of "real" as opposed to "false."[13] The politics of design sketched out by the flowchart and further elaborated in *Design for the Real World* blended the historical shift from Cold War transdisciplinary design experiment through to the antiestablishment counterculture movement, finally culminating in a call for design's role in pursuing a purer, decommodified lifestyle. "We seem to be designing exclusively for the old, the fat, the white, the powerful: the ones who have managed to get us into every single mess and frightfulness mankind has so far experienced," raged Papanek's final words at the end of the flowchart. "To create lipstick for honest whores is one thing, but to create deodorant for her pimp is another," he exhorted in a typically Papanek-like swing at the denigrating state of contemporary society, and the artifacts designed to bolster that denigration.[14]

In the late 1960s, "caught between the proverbial rock of technocratic progress and a hard place of impending social disaster," according to historian Andrew Blauvelt, technology was steered away from corporations and the military-industrial complex toward a utopic vision of progressiveness as social good.[15] As a consequence, a form of "hippie modernism," epitomized by Papanek's genre of Third World and disability designs, generated alternative forms of consumption (experiential, networked, and collectivized) and inspired a turn away from the perceived ills of conventional commodity culture.[16]

In many respects, the Copenhagen Flowchart was "hippie modernism" writ large. Following its two rhetorical sections—"what the people <u>really</u> need" and "what the people are <u>told</u> they need & want"—the infographic offers an explanation of "how false goals are achieved," including the hand-drawn bubbles of "war," "violence," "competition," "money, profit, rewards," "advertising & propaganda," "military training," "the church," "game shows," "TV," "Jobs" and, perhaps interestingly given Papanek's own biography, "Mother will love you, if. . . ." "Status" stands out as the single most graphically eye-catching component of the "False Goals" section, with an arrow pointing to a bubble-box for qualifying comments: "a zippy sexy, groovy, outasight sports car, a distinguished automobile." Written by a cultural design critic who had indeed flaunted his own "zippy, sexy, groovy, out a sight" scarlet-red sports convertible (complete with a handgun in its side pocket) just a few years prior to executing his critical map of design politics, was this a deliberate act of self-mockery or an accidental display of his subconscious?[17]

At its center the "Minimal Design Team"—with "Designer" at its core, surrounded by adjunct disciplines including social and behavioral sciences, ecology, anthropology, psychology, medicine, filmmaking, and engineering—drew on earlier manifestations of the interdisciplinary design team, here presented as a means of generating a more empathetic design process. When the Copenhagen Flowchart was later reproduced as a poster, the quotations of selected voices, from Buckminster Fuller to Pablo Picasso and John F. Kennedy, made it a perfect complement to the ubiquitous 1970s Cuban revolutionary poster that according to Susan Sontag became "one more item in the cultural smorgasbord provided in affluent bourgeois society . . . [the posters'] final resting place not the streets, town squares or factories, but the living room wall."[18] The Copenhagen Flowchart, as the first iteration of his book, and as a pull-out insert and poster, was arguably the single most significant design Papanek ever created.

The Environment and the Millions: Design for Service or Profit?

By late 1969, shortly after the Copenhagen seminar, the SDO began to fizzle out, torn apart by internal politics. In the Scandinavian design press its critics accused it of being little more than a "Scandinavian discussion club for a small number of uninformed students."[19] Nevertheless, following the introduction of Papanek to the Stockholm publishing house Bonniers and brokered by the SDO's general secretary Per Johansson, the legacy of the activist initiative lived on in the form of *The Environment and the Millions: Design for Service or Profit?* (*Miljön och miljonerna: design som tjänst eller förtjänst?*), a radical new treatise of design's transformative social potential published in Swedish in 1970 (see figure 8.2).

While Papanek had been comfortable in his role as a critical foil to the US design industry, ever poised to drop a polemical one-liner, he could never have envisaged the enormous impact his encounter with pan-Scandinavian radicals and his publication of *The Environment and the Millions* would have on his American and worldwide profile. Drawing extensively on Papanek's engagement with the SDO and its activities, this Swedish book was released in English as *Design for the Real World: Human Ecology and Social Change* (1971), selling thousands and popularizing an anticonsumption, alternative design movement. The polemic quickly gained status as a counterculture classic, sandwiched between Rachel Carson's *Silent Spring* (1962) and E. F. Schumacher's *Small Is Beautiful: Economics as If People Mattered* (1973).

A version of "Do-It-Yourself Murder: Social and Moral Responsibilities of Design," a pivotal chapter in *Design for the Real World,* had first been published in the SDO magazine in 1968, offering a searing critique of design and architecture's role in the making of harmful and negligent products. In it Papanek taunted mainstream US design professionals with a series of provocations: "There are professions more harmful than industrial design, but only a very few of them. Never before in history have grown men sat down and seriously designed electric tooth brushes, rhinestone covered file boxes, and mink carpeting for bathrooms, and then drawn up plans to make and sell these gadgets to millions of people."[20]

By way of contrast, drawing on social projects carried out with his own students at Purdue University (including design and research for the TV set for Africa), and those of the Nordic and Scandinavian countries, *Design for the Real World* catalogued tangible examples of a new, socially responsible mode of design. One compelling example drew on the politics of urban renewal, action on child poverty, and the expanded notion of design as a social process, rather than a finished aesthetic outcome. In January 1969, Papanek combined a visit to Oslo State School of Design to deliver a lecture titled "Revolution—Social Change and Design," with a two-week project to create, with students from the design school, a playground environment in a run-down urban neighborhood whose "backyards were given over to garbage

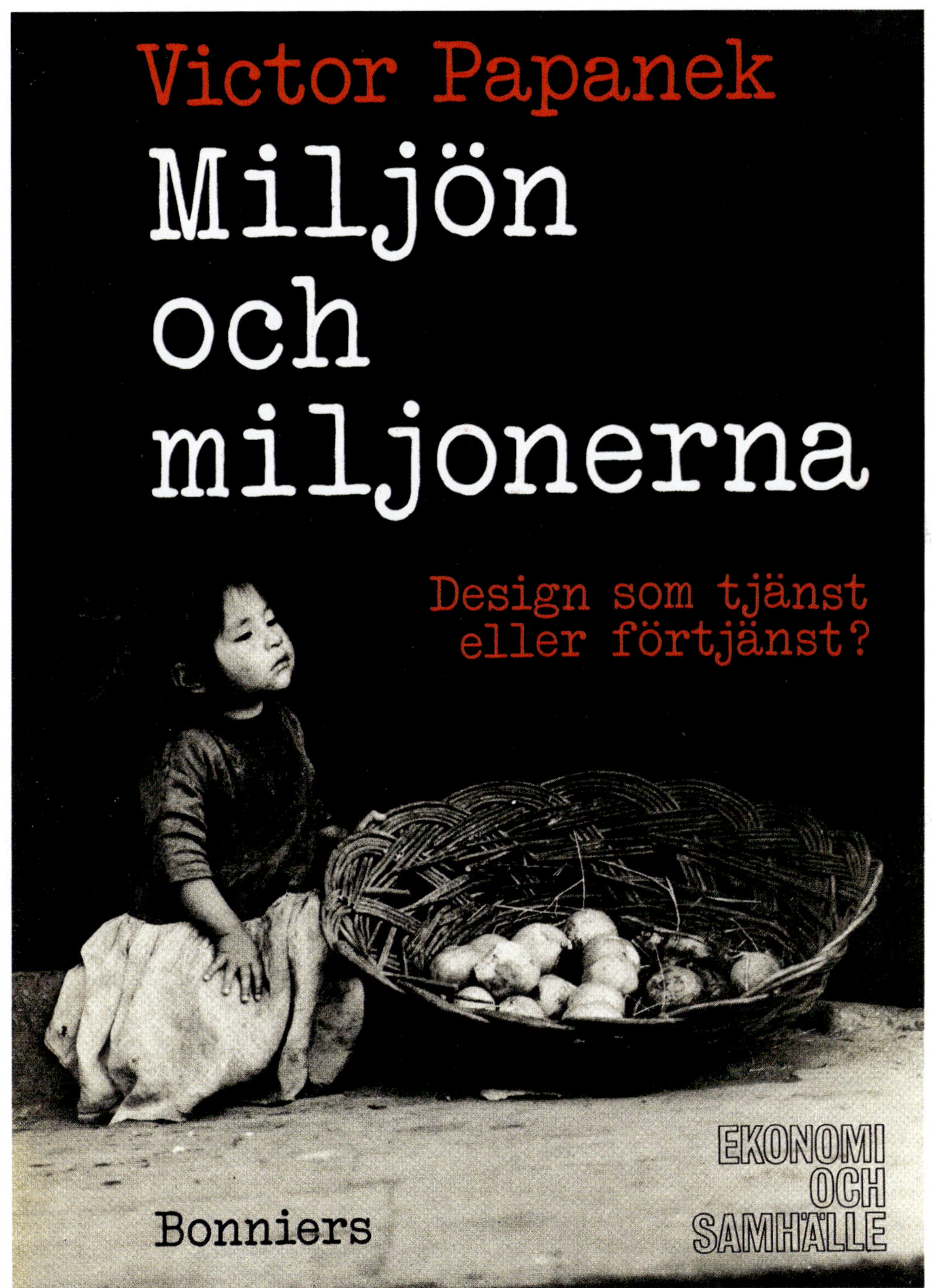

Figure 8.2

Victor Papanek, *Miljön och miljonerna: design som tjänst eller förtjänst?* (*The Environment and the Millions: Design for Service or Profit?*) (Stockholm: Bonniers Förlag, 1970), first Swedish edition of *Design for the Real World*. © Bonniers. Courtesy Victor J. Papanek Foundation, University of Applied Arts Vienna

pails, high metal fences and laundry."[21] The environmental design project, aimed at improving the lives of poorer children, applied immersive ethnographic research techniques to engage genuinely with the local community and their needs. The project therefore took on a life of its own, attracting students from the Schools of Architecture and Landscape, along with Industrial Design, even opening out to the University of Oslo, with students relishing the opportunity to put into practice progressive new theories around community engagement in their respective fields. Interviewing residents from the young to the elderly, parents and children, to explore the "intersecting needs and problems" of the area to be redesigned, the demanding but imaginative research initially captivated the Norwegian students—before some came to the realization that it offered no quick-fix solution.

"I must admit," reflected Papanek, "that at first many of the students became interested because of the novelty of the problem. Later they found that being involved in this type of social design is much more difficult than creating still another teapot or perfect salt cellar. Many were discouraged, and some dropped out."[22] The landscaping, design, and development of play equipment, and the three-dimensional models were, however, all delivered on time by dedicated team members. As part of the social design process, Papanek deemed it particularly important that the students understand the legacy of their intervention in any given community environment—in this instance, by installing an informal program of activities that (from a contemporary perspective) read as a paean to counterculture clichés but had a serious intent: "It was up to the students that outdoor movie showings, guerilla theatre, poetry readings, and 'sing-ins' be brought to the backyard on long summer evenings," reiterated Papanek, because "through engaging in these activities, the students came to a closer and 'operative' understanding of the people's problems; the people, in turn, assumed a more active role in shaping their own future and gained to pride and identity."[23]

Design for Disability: A Social Imperative

Design for disability formed a central theme of Papanek's work, yet unlike his predecessors he viewed the practice as a Trojan horse, according to historian Bess Williamson, aimed at exposing mainstream design's "limited view of target users as affluent, able-bodied, Western consumers."[24] As such this area of research, building on the groundswell of disability rights' activism, was successfully brought into the broader design curriculum of Papanek's students as a central rather than peripheral and medicalized aspect of specialist design.[25] Nowhere was this more evident than in the studio teaching of Yrjö Sotamaa, an assistant professor with Papanek at Purdue University between 1969 and 1970 who team-taught a further development of the Finnish SDO cerebral palsy play-cube project titled "CP-2" as part of the Industrial and Environmental Design program. Reaching out to local clinics, communities, and hospitals, Sotamaa worked with students and users on designs that were featured in a widely circulated article "Design and the New Environment" that appeared in the *Chicago Tribune Sunday Magazine* and the pages of both the Swedish and English-language versions of *Design for the Real World*.[26] Decrying the "Stone Age level" of design applied to "prosthetic devices, wheel-chairs, and other invalid gear," Papanek declared design for disability as a category that fell squarely within the remit of "Areas of Attack for Responsible Design."[27] The subsection "Design of Teaching and Training Devices for the Retarded, the Handicapped, and Disabled, and the Disadvantaged," documented projects for muscle-building bicycles, toys for the visually impaired, hydrotherapy water vehicles for disabled children, and sensory objects for children affected by cerebral palsy such as Finnish designer Jorma Vennola's *Fingerma-jig*, a brightly colored collection that Papanek used as props in press interviews to illustrate what an artifact of social design might look like.[28] Arising from the research conducted in 1968 on the *CP-1 Cube*, Vennola's design consisted of a ball-like configuration with flexible, protruding dowels made in eight bright colors. Described by Papanek as "an ideal design development" due to the trajectory of success in taking a well-researched prototype to the mass market as an inexpensive social intervention, it provided a "superb exercise of hand muscles for all children, as well as those with cerebral palsy, some types of paraplegia, and myasthenia gravis."[29]

Although many of these designs, such as a "perch or reclining structure" design by Purdue design student Steven Lynch for handicapped or "restless" children in the classroom, would not pass muster by present-day standards of ergonomics or user-generated solutions to disability, they reveal a genuine, persistent engagement with an area otherwise neglected in conventional design programs of the period (and indeed today).[30]

The attention to design for disability, and the broader area of user design and ergonomics, was heavily influenced by Finnish-Swedish designer Henrik Wahlforss (whom Papanek met in the course of his SDO encounters), who founded the agency *ErgonomiDesign* in 1969. After joining forces with a group of designers with similar motives in a disused chapel space outside Stockholm called *DesignGruppen*, Wahlforss and his peers generated an approach to design that focused on users: ethnographic and ergonomic research under the conjoined name *ErgonomiDesignGruppen*.[31] Along with a collection of ethnographic artifacts, Papanek kept a study set of designs for disability, including Vennola's *Fingermajig* and the kitchenware designs of Maria Benktzon and Sven-Eric Juhlin (for people with reduced fine-motor skills) manufactured by RFSU Rehab, in Sweden for ErgonomiDesignGruppen. Exemplifying for Papanek the progressive, democratic idea of design for user needs, this group of designers circumscribed a focus on the user that came to be understood more broadly as a Swedish phenomenon from the 1970s onward.[32]

Securing Buckminster Fuller's Ultimate Endorsement

Student projects spanning Papanek's earlier work as a professor at North Carolina State University through to Purdue University and the Scandinavian and Nordic countries formed the backbone of his first major publication, and he duly dedicated the book to them: "This volume is dedicated to my students, for all they have taught me." Notably, the design critic casts the time frame and geographic location of the writing of *Design for the Real World* between1963 and 1971 as "Helsinki—Singaradja (Bali)—Stockholm." Although a somewhat pretentious flourish for a novice author to affect, in some senses this detail offers a realistic overview and timeframe of the amalgamation of sources on which Papanek relied to create *Design for the Real World*. Its contents were largely borne of a combination of earlier design projects rooted in the Cold War experimentation, and those arising from his travels in Scandinavia and Finland, freshly repackaged to suit the students of a postindustrial design culture and ethical consumption lifestyle.

In this respect the choice of Buckminster Fuller, a man whose work was firmly embedded in the politics of the military-industrial complex, to pen *Design for the Real World*'s introduction belies the more anachronistic aspects of the book as a whole. The inclusion of Fuller's contribution, as an endorsement to an unknown author, persuaded the US publisher Pantheon to take on Papanek's breakthrough work. Yet *Design for the Real World* emerged to ride the crest of an early 1970s wave of anticonsumer, anticorporate, and anti-American rhetoric, a rhetoric that focused on establishment figures such as Fuller for its critique.

In 1970, when Fuller was first mooted as an endorsing contributor, the design technocrat had already been targeted by a campaign of mounting disdain in Europe. In 1968 the London-based art group ARse, for example, distributed a spoof pamphlet, the "UN Official Programme," depicting Fuller, geodesic dome spliced onto the crown of his head, and a speech bubble emanating from his mouth with the words "Welcome to the Buckminster Führer show." The reverse side of the pamphlet comprised a photographic collage of Fuller's 1956 design for the US pavilion in Afghanistan, with the sarcastic caption "Domes for U.S. marines, showing the capacity and manageability of 36ft. diameter hemisphere."[33] All the more ironic then, that Papanek had been forced to intellectually prostrate himself before the architect-inventor in order to extract the technocratic designer's opening contribution to his book.

For Papanek negotiating the contract for his first English-language book, Fuller had been the ace up his sleeve. Finally, he could tie the twentieth century's leading maverick designer irrevocably to his relatively unknown projects and ideas, imbuing them with incalculable status—the brand equivalent of a celebrity endorsement. Since first courting a relationship with the inventor in the mid-1950s, (then as a member of the audience at one of Fuller's lectures at MIT) Papanek had finally achieved his goal: to be publicly acknowledged in perpetuity by the great master of twentieth-century design thinking. While his (real or imagined) relationship with Frank Lloyd Wright had served its purpose in the early part of his career, at this point all that stood between Papanek and a step into the world of international design renown was a failure to secure Fuller's written endorsement.

Boasting of the link with Fuller as a friend and ally, Papanek first suggested to his editor at Pantheon that he could persuade the prominent figure to provide a foreword or introduction in 1970. The promise of a piece written by Fuller ensured that the publishers took Papanek's manuscript seriously. There was a problem, however. Having first agreed as an off-the-cuff gesture to pen something, Fuller suddenly reneged, much to Papanek's embarrassment. Having read the manuscript, Fuller stated that he entirely disagreed with its contents. Behind the scenes, and unbeknownst to the author, a correspondence ensued between the publisher in New York City and Fuller's administrative secretary Naomi Wallace in Carbondale, Illinois, trying to establish whether Papanek's claim that Fuller would provide an introductory piece held any veracity. The situation seemed close to resolution when finally Papanek managed to confront Fuller at a lecture, only to have him confirm that he had indeed rescinded his offer of an introduction.

Dejected, Papanek sent his erstwhile role model a heartfelt letter steeped in all the forlorn longing of unrequited love: "Much as I felt both hurt and bewildered when you said that you had read my manuscript and disagreed with what I say; I was relieved when you added that on a far deeper level you feel that we are in agreement," wrote Papanek, clutching at straws, before launching into full-on emotional blackmail. "You see, originally my book was dedicated to you, this dedication I removed since my editor at Pantheon Books felt that it was inappropriate as you were doing the Foreward [sic]. Still, quotes from you abound throughout the book. Often in the past 10 or so years, you and I have shared the speaker's platforms or appeared side-by-side in magazines; always I received the impression that much of what I said seemed at least reasonable to you."[34] Pleading with Fuller to agree to any form of contribution he might deign to make, Papanek wrote with desperation, underlining his words in bold blue marker pen: "write it anyhow and as you see fit."[35] In a final act of acquiescence to Fuller's greatness, Papanek made a rather pathetic attempt at flattery—"I feel that I could gain much insight and maybe even wisdom if you could let me know what elemental flaws my book has in your eyes"—before signing off the letter with contrived intimacy: "all the best of luck wherever you are right now, and much love."[36]

To appease the discomfort of Wallace's embarrassed attempts to forestall Papanek's persistent publisher, Fuller—reluctant but worn down—finally submitted to Papanek's request. Shortly before *Design for the Real World* was due to go to press, his secretary reassured Pantheon that "Dr. Fuller has dictated the introduction to Victor Papanek's book, it has been transcribed, and he has the manuscript with him to rework," before adding tentatively, "I've sent a copy of your letter to him to remind him time is running out."[37]

By the time it went to print, Fuller had provided a surprisingly lengthy introduction, rather than a simple preface or foreword, just over eleven pages long. Beginning with a seemingly genuine gesture to legitimate Papanek's work and their relationship, it was nevertheless carefully crafted to maintain a calculated distance between them: "There are wonderful friendships which endure both despite and because of the fact that the individuals differ greatly in the experiential viewpoints while each admires the integrity which motivates the other. . . . Victor Papanek and I are two such independently articulating friends who are non-competitive and vigorously cooperative."[38] "In this book," continued Fuller, before launching into a summary of his own work biography, largely centering on the impact of the two world wars in transitions of technology and production, "Victor Papanek speaks about everything as design. I agree with that and will elaborate on it in my own way."[39] Describing himself as a "Comprehensive Anticipatory Design Scientist," perhaps the single most salient point of Fuller's introduction (otherwise largely rehashed from his standard lectures) was the observation that Papanek's book "conducts a mass funeral service for a whole segment of now obsolete professionals."[40] Describing MIT as "a vast graveyard of technology," its "rooms full of yesterday's top priority machinery that is now utterly obsolete," Fuller unconsciously harked back to the exact setting in which he had first encountered Papanek—at the mid-1950s MIT Creative Engineering Seminars—where, under the rubric of Cold War US ideologies, both designers began pursuing the potentialities of transdiciplinary design.[41] *Design for the Real World*, as Fuller aptly identified, signaled the emergence of design as a dispersed phenomenon untethered from the machine and intimately connected to all facets of what it was to be human in the late twentieth century. Papanek reiterated the urgency of design's social imperative, following up Fuller's introduction in his preface with an apocryphal statement: "As socially and morally involved designers, we must address ourselves to the needs of a world with its back to the wall while the hands on the clock point perpetually to one minute before twelve."[42]

Despite Papanek's position as an American design professor and critic, and although published by a New York-based publisher, the intended audience of the book was the radically more politicized students and design activists of Europe with whom he had built his reputation and ideas. As the author put it himself, name-checking the world's leading media theorist in the process, "I teach in the United States, I have tried to give a clear picture of what it means to design within a social context. But there is only so much one can say and do, and even in Marshall McLuhan's electronic era, sooner or later one must fall back on the printed word."[43]

California Institute of the Arts: From Utopia to the Real World

By the time Papanek's "printed word" on the politics of design hit the shelves in the fall of 1971, the author had already left behind the Cold War design experimentation of Purdue University for the decidedly more liberal and progressive environs of the recently opened Valencia campus of the privately funded California Institute of the Arts (CalArts). Evidently, Papanek had been actively seeking an alternative to his position at Purdue since at least February 1969, when he received a letter from Fuller (called upon once again to endorse the designer) with a short message, dictated in his absence, confirming his support: "You know how much I admire you, and you can count on me to do my best when the opportunity arises."[44]

Whether or not Fuller played any part in fixing the prestigious professorial position with a recommendation, by fall 1970 Papanek had joined CalArts to head its new design program. Having perfected the art of the unsolicited, hyperbolic, self-promotional letter way back in the 1940s when he first contacted Frank Lloyd Wright at Taliesin, then in the 1960s as a means of securing an entrée into the Nordic and Scandinavian design scene, Papanek applied this technique to the dean of design, Richard Farson, at CalArts in January 1970. Tipped off about a potential faculty opening in the institute by the biomorphic furniture designer Douglas Deeds upon returning from a month-long teaching visit to Sweden and Finland, Papanek pitched his expertise with an extensive CV and lengthy missive to Farson that included copies of eighteen articles written in international design magazines. Before launching into a sustained critique of the paucity of facilities available to pursue the establishment of "an inter-disciplinary team approach of designing for the true needs of man," with typical frankness Papanek explained that, despite having tenure and an unusually high salary, a shift of management at Purdue University had impacted funding in his area of experimental design, leaving him "with an extremely conservative and senile President, who is uninterested in this area and stronly [*sic*] opposed to our view of design."[45]

Having received a standard letter from CalArts administration acknowledging receipt of his application, the ever-persistent Papanek pursued the position with a further three-page letter to Farson, including more references to articles exploring the originality of his design approach, and a further clarification of details of his curriculum vitae. "If all this seems 'pushy' to you," he wrote unabashedly, "it is."[46] Papanek succeeded in impressing the dean and the CalArts team, and astonishingly just a few months after sending his first cold-call letter, he had secured a position in the School of Design and was arranging with the administration the specificities of his family's accommodation requirements. Farson's background as a behavioral scientist and his preexisting interest in social design must certainly have helped sway the appointment of Papanek.

Fortunately for Papanek, in the course of his negotiations with CalArts, *The Environment and the Millions: Design for Service or Profit?* had arrived hot off the press from the publisher in Stockholm; the prospect of a new position then spurred him on to secure a contract for the English version, *Design for the Real World*, with Pantheon Books (see figures 8.3 and 8.4).

There was just one slightly uncomfortable coincidence regarding a major element of both the Swedish and English versions: they both featured an extensive pull-out infographic (the Copenhagen Flowchart) that bore an uncanny resemblance to a wall chart created by Papanek's soon-to-be colleagues at CalArts.

Radical pedagogy framed the mission of CalArts, and just prior to Papanek's arrival, two faculty members, Maurice Stein and Larry Miller, produced *Blueprint for a Counter Education*. A publication-cum-pedagogic kit designed by Marshall Henrichs and comprising curriculum, handbook, wall decoration, and "shooting script," the box featured the dictum: "THE REVOLUTION STARTS HERE." Declaring the traditional university obsolete, the boxed "counter–university" included "faculty" of Herbert Marcuse, McLuhan, Eldridge Cleaver, and Jean-Luc Godard. This initiative and CalArts more generally—first conceived of in 1969 as a creative laboratory with no fixed curriculum—were a far cry from the conservative strictures and Midwest politics of Purdue University, but clearly faculty members Stein (dean of Critical Studies) and Miller (active with the interdisciplinary art collective Fluxus) shared with Papanek a vision of a new politics of design. In fact, by coincidence in the course of his negotiations to join CalArts in 1970, Papanek had been sent a copy of *Blueprint for a Counter Education* to review; shocked to discover he was not alone in generating a provocative infographic he took pains to emphasize that his Copenhagen Flowchart had been generated prior to the release of Stein and Miller's publication.

In fact, despite his later antipathy toward the structureless and creatively anarchic setup of school that would ultimately lead to his resignation from CalArts, in his earliest communications with the dean of design, Papanek expressed an entirely contrary opinion: "Frankly, what turns me on to CalArts most," he wrote to Farson in the months before the School of Design's inauguration, "was [the] explanation of the loose structure of the school as you are planning it. It closely parallels my own attempts at turning our curriculum around here at Purdue, the difference being that a State University tends to slam innovation."[47]

The publication of *Design for the Real World* in 1971 coincided with its author's promotion to dean of the newly formed School of Design at CalArts, a post he occupied from 1971 to 1972, following Farson's resignation, and which lent Papanek the ideal base from which to launch his book and increase its audience exposure. The social designer's arrival at the school also coincided with its taking on the organization of the twenty-first International Design Conference in Aspen (IDCA) under the theme "Paradox." This was the very same conference

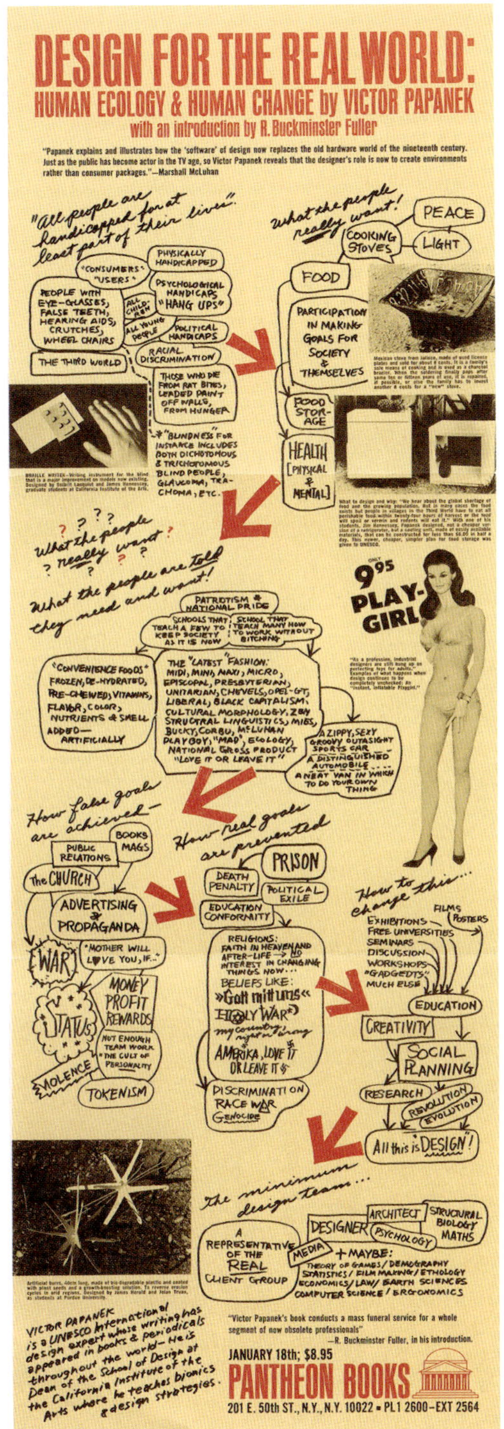

Figure 8.3

Design for the Real World: Human Ecology and Social Change promotional infographic, first English language edition 1971. © Pantheon Books

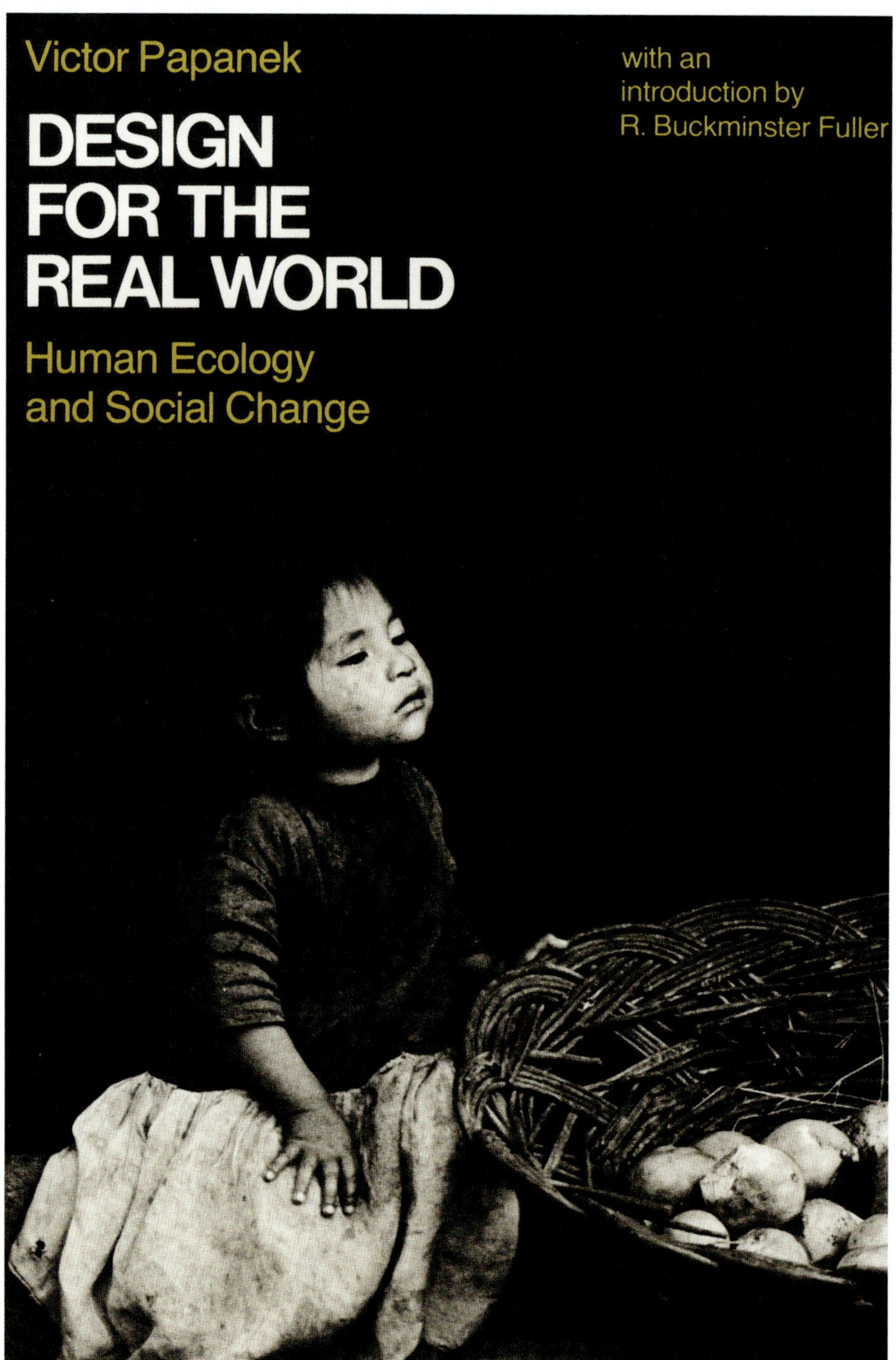

Figure 8.4

Victor Papanek with introduction by R. Buckminster Fuller, *Design for the Real World: Human Ecology and Social Change,* 1971, front cover. © Pantheon Books

body Papanek had castigated in the national design press, in 1968, for its superficiality and endless supply of "dry martinis" and "back-slapping bonhomie."[48] But Papanek had been far from alone in his disapproval of the cultural conservatism of the IDCA. The planning of the forthcoming event was almost entirely framed by the legacy of the previous year's happenings, whereby student activists had actively and physically commandeered the proceedings, taking to task the prominent white male luminaries of the design world and the homogenous IDCA organizing committee itself.[49] Doubt was cast over the entire legitimacy and relevance of the Aspen conference in relation to its ability to effectively address "design problems of the world," and, in an effort to deflect attention from the corporate power base of the organization, CalArts had been invited to devise the 1971 program with the ambition of making it "a turn on for the conferees."[50] Papanek was present at the conference as a dean of the design school, but more significantly as a social designer bringing the "real world" manifesto of his newly launched book to an avid audience of fired-up US design activists.[51] Moving their design maverick double-act from the highly politicized Nordic and Scandinavian design scene of the late 1960s to the Aspen "Paradox" conference of 1971, Fuller and Papanek were reunited on the speakers' platform, with Papanek speaking to the theme of "Revolution in the Third World" and Fuller expounding his theories of design science. Significantly, Farson, who oversaw the initial conception of the event, described the creation of the conference as a "social design problem" in which faculty and students collectively sought to generate participation.[52] The radical restructuring of the IDCA as a participatory, politicized event with an overtly social agenda, and the contemporary movement toward a freed pedagogy in art and design, exemplified by CalArts, came together as an optimal launch pad for *Design for the Real World* as treatise that put the social conscience of designers at the forefront of societal change: "Watching the children of Biafra dying in living color while sipping a frost-beaded Martini can be kicks for lots of people," sneered Papanek, "but only until *their* town starts burning down. To an engaged designer, this way of life, this lack of design, is not acceptable."[53]

Despite his short tenure at CalArts, in an attempt to put his publication into applied action, Papanek initiated the establishment of a Center for the Study of Responsible Design. He was desperately trying to move the School of Design at CalArts away from an elitist artistic laboratory (which he viewed as failing the students) to an educational establishment with real links to its users and industry. "Being Dean of a School of Design that is an hermetic and isolated institution is of no interest to me whatsoever," he informed CalArts President Robert Corrigan, in no uncertain terms.[54] Offering to fundraise a sum to the tune of $100,000 to kickstart the enterprise, Papanek proposed that the Center for the Study of Responsible Design tackle areas of research related to his established expertise in the field (disability, health, educational, developmental, emerging countries, etc.) with one notable exception:

"[one] developmental group will deal with the so-called 'counter-culture' and some of its emerging lifestyles. The concept of 'nomadics' as expressed in the interior fittings or fans, buses, and other vehicles has not as yet been designed rationally, and the needs of neither manufacturers nor young people have been considered."[55]

A couple of years later, Papanek and James Hennessey, his younger colleague at CalArts, would coauthor the first of two books exploring the lifestyle politics of nomadism, *Nomadic Furniture 1*. But the proposal fell flat, despite a pending public relations push for *Design for the Real World* by Papanek's publisher, in which, according to the book's author, "a number of nationwide TV 'talk' shows and interviews" would offer the perfectly timed opportunity to broach industry with the idea.[56]

As a renowned hotbed of counterculture, the creative, utopian CalArts seemed the ideal fit to the antiestablishment stance Papanek espoused at international design festivals, yet it would in fact prove his least successful academic position following confrontations over its "loosened up" educative stance, and progressive approach to gender politics.

Sheila Levrant de Bretteville, a politically engaged design faculty member, had designed an iconic participatory conference program for the IDCA Paradox conference, and earlier in 1970, a visually compelling flyer promoting CalArts' new school of design that incorporated the Papanek-like slogan "If the designer is to make a deliberate contribution to society, he must be able to integrate all he can learn about behavior and resources, ecology and human needs: taste and style just aren't enough."[57] De Bretteville delivered a speech on sexual politics at the Aspen conference, delineating the issue as one that urgently needed be addressed within the design profession.[58]

Active feminists, de Bretteville and her women colleagues were keen to reshape the patriarchal landscape of design pedagogy and education with their proposition of founding the first women's design course. De Bretteville would go on to establish the renowned *Womanhouse* project in 1972 in partnership with the Feminist Art Program run by Judy Chicago and Miriam Schapiro. This celebrated feminist intervention and performance included sardonic inversions of the gender politics around consumerism, including Camille Grey's *Lipstick Bathroom*, which featured lipstick-daubed walls, an installation of hundreds of variations of red lipsticks, and a fur-lined bath. Despite being a self-professed advocate of design for the socially excluded (which by his own definition included women), Papanek strongly opposed de Bretteville's women's design program as a retrograde move toward "ghettoization." Some colleagues suspected this opposition belied a deeper fear of shifting gender power relations.[59]

The tables had turned. Throughout the late 1960s, he stood atop the podium stirring the consciences of crowds of enthusiastic activist students, demanding social revolution; now Papanek was doing his utmost to stamp out even the tiniest hint of that very revolution.

Having ended his preface to *Design for the Real World* with a critique of the defunct rationalist legacy of the Bauhaus, stating that "A philosophy more than half a century old is out of place in a field that must be as forward-looking as this,"[60] in the final months of his tenure Papanek fell back on the words of the arch modernist himself, Walter Gropius, to defend the conservative position he had come to adopt at CalArts.

He began a faculty memorandum of October 25, 1971, titled "Some Thoughts on Starting the Year," with this statement: "Academic training has brought about the development of the great art-proletariat destined to social misery. For this art-proletariat, lulled into a dream of genius, enmeshed in artistic conceit, was being prepared for the profession . . . without being given the equipment of <u>real education</u>."[61] Prior to launching into a six-page memo demanding an immediate change to an overintellectualized, nonvocational, nonapplied culture of design, Papanek reiterated his objection to CalArts elitist avant-gardism by reminding faculty members (and students) that the opening quote was "from a statement Walter Gropius made in 1928, as he tried to define why there was need for a change called The Bauhaus. History doesn't begin with us."[62]

By February 1972, in a further memo to faculty and students titled "Design School Structure," Papanek complained that his office had been "deluged" by students submitting unsolicited proposals for "a much tighter structure of classes, attendance, evaluation procedures and core courses in the Design School."[63] It seemed the students, as well as their dean, were growing weary of the enforced "hippie modernism" that CalArts promoted, craving instead the old-fashioned solidity provided by the type of industrial design programs Papanek himself had hastily left behind in Purdue.[64]

With his first pitch for employment at CalArts, the ambitious social designer had included a bundle of articles covering aspects of his radical design approach, illustrating the unrivalled suitability of his expertise. "Controversial design teacher seeks to stir imagination," read the byline of one such piece exploring the Purdue professor's "shock value" classes.[65] Yet faced with the anarchic artistic free-for-all of CalArts, in a complete turnaround Papanek began holding classes in "Professional Practices" that dealt with the "design fees, retainers, cost plus, royalties, deferred billing, contracts, tax laws, etc."—a desperate attempt to instill in his students an understanding of how to survive as designers in the real world.[66] Next to elective classes offered by more hip-to-it tutors, such as "Cannabis Myths and Folklore' (with a specialty in "19th and early 20th century American hashiology including social and medical uses" and "anti-marijuana laws"), and "Advanced Drug Research (continued)" Papanek's hard-won reputation as dissenting voice within the design profession was waning fast.[67]

Aside from the increasingly contentious situation developing at CalArts, by early 1972 the second edition of *Design for the Real World* (driven by the publishers' high-profile public relations campaign; see figure 8.5) was generating extraordinary levels of attention in the local

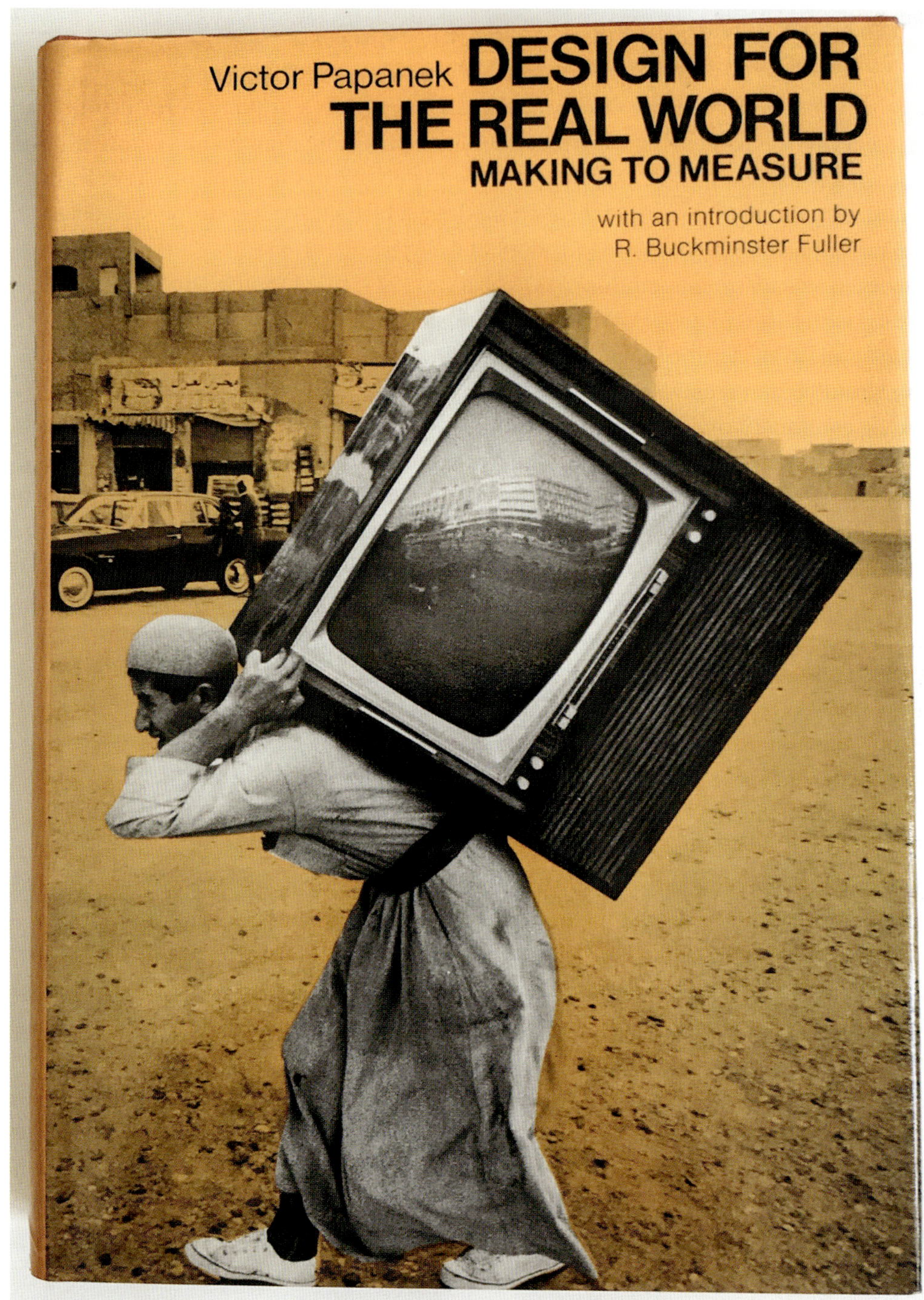

Figure 8.5

Design for the Real World: Making to Measure front cover, second English edition (London: Thames and Hudson, 1972). © Thames and Hudson

and national, mainstream press; some US journalists were decidedly less enthralled by the designer's airy intellectualizing than their European counterparts. "We're deep in the heart of Survivalsville," began a profile article in the *LA Times*. "The consumer-advocate country where the world's clock is always at 10 minutes to 12 and time is running out."[68] Refusing to take Papanek's earnestness too seriously, the piece featured the "elegant, crisp man with the Austrian accent—looking like a fashion designer in his tailored knit shirt-suit" posing in an ergonomic recliner of his own design. One photo shows him sitting upright; the other slightly less dignified, fully reclined with feet pointing to the ceiling.

The *LA Times* journalist cast aside the reverence to which Papanek was accustomed, adopting instead a playful mockery toward the social designer's unerring social conviction and borderline pretentiousness: "You find him seated in his gray, Madison Avenue-come-to-college office surrounded by Marimekko bags ('In Finland it's just a workers' hangout. But here they think it's chic'), and black-and-white photos of Robert Kennedy and children of the Third World ghettos."[69]

Just a few weeks prior, in a rare PR coup, Pantheon Books had managed to secure a review, under the title "Down with Designers?," of *Design for the Real World* in America's premier mainstream journal, *Time* magazine. The reviewer, however, wary of the idealistic naivete gave the book a decidedly lukewarm reception. The author's rhetoric, the reporter explained, "is so extreme at times that he calls corporation executives 'criminals.'"[70] Some critics, the review concluded, might indeed argue that it was Papanek himself, with his utopian vision of design, who had "lost contact with the real world."

The ongoing strife at CalArts, coupled with a less than enthusiastic mainstream reception of the second edition, put a further damper on Papanek's US career as a social designer, making the stream of invitations to teach and guest lecture in Europe increasingly attractive as a long-term proposition. With his ideas still popular with the design schools of the welfare economy nations of the Nordic and Scandinavian region, Papanek moved his attention from the West Coast to Denmark, where he set up a second home in the early 1970s.

In August 1971, Denmark's leading design journal, *Mobilia* (see figure 9.1), featured on its front cover a discarded automobile license plate refashioned into a portable stove on which, Papanek informed readers, a Mexican "family cooks <u>all</u> of their meals."[1] Along with a crank-handle bulk storage refrigerator (see figure 9.2) for distribution in the small Southern African country of Lesotho, codesigned with his assistant from CalArts James Hennessey, the article—"What to Design and Why"—introduced the American social designer to the mainstream Danish design profession. "In deciding what to design and why, the designer redefines his own position in society," espoused Papanek, "instead of 'securely' designing for the boss what that boss tells him to design, the designer emerges as a <u>facilitator</u>. He facilitates between what people need and those who control production."[2]

Having successfully appealed to the broader Danish design community, just a year later Papanek took up residence along with his wife Harlanne and infant daughter in a penthouse apartment on the picturesque Nyhavn, in the center of historic Copenhagen. Taking a temporary leave of absence from his harried position as dean of at the School of Design CalArts, Papanek relished the opportunity to immerse himself in a European design culture that had enthusiastically received *Design for the Real World*. Invited by Erik Herløw, the head of design at the Royal Academy of Fine Arts, to be a professor (in what Herløw identified as the "trend-setting" area of design and environment), he was affiliated with the School of Architecture there and supported by a national fellowship. The intention was that Papanek would found a research study group focusing on "design-based influence in the developing countries" to aid social design policy for Greenland.[3]

Based on his interest in Inuit culture and his participatory design approach, Papanek's research would contribute to a broad reappraisal of the politically contentious issue of housing development in Greenland, Denmark's former colony. In the postwar period, Danish architects had been tasked with installing social housing on the expansive island. Located between the Arctic and North Atlantic oceans, Greenland thus was to be enhanced as a

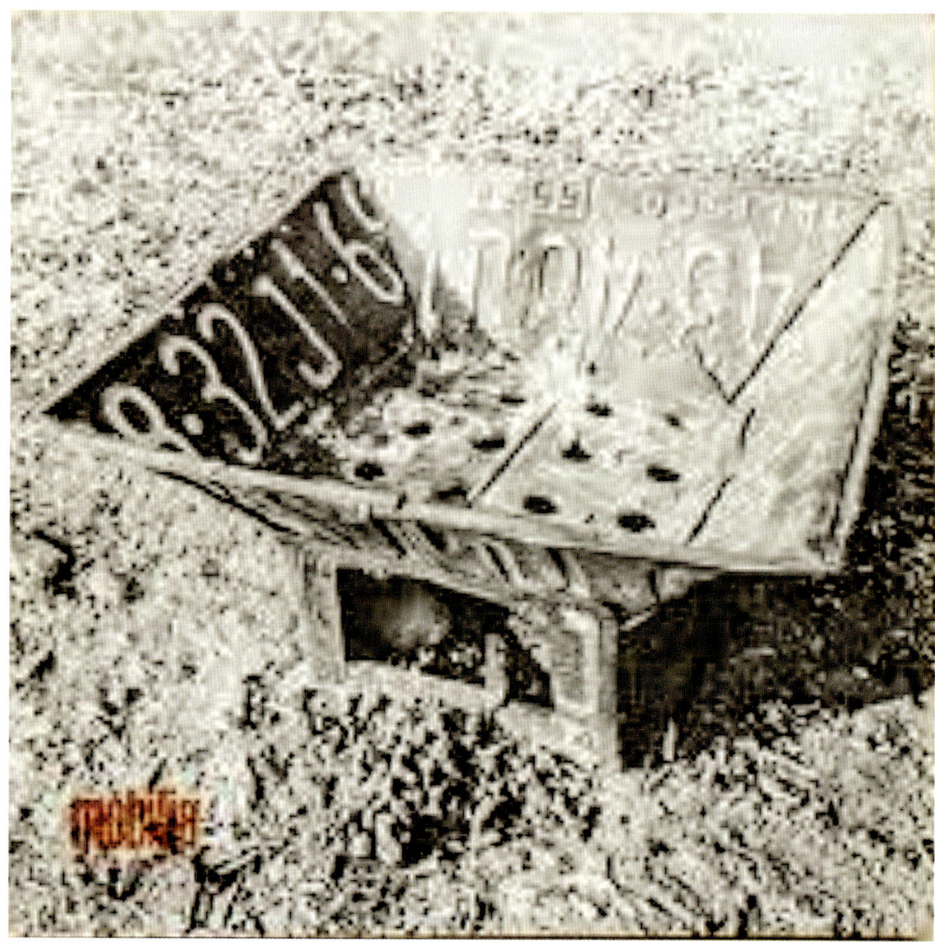

Figure 9.1
Mobilia, front cover, August 1971. © Mobilia, Denmark

modern welfare society in keeping with its postcolonial status as part of the Danish realm.[4]
(The island country achieved full independence in 1979.) By the 1970s, the top-down installation of modernist, rationalist architecture—planned by the Greenland Technical Organization (GTO) and designed functionally to manage the hostile climatic setting—was the subject
of much critical debate in the architectural press, which according to historian Kirsten Birk
Hansen revolved predominantly around "the question of whether people would be able to
adapt and tailor residential construction to the Greenlanders' needs and way of life."[5]

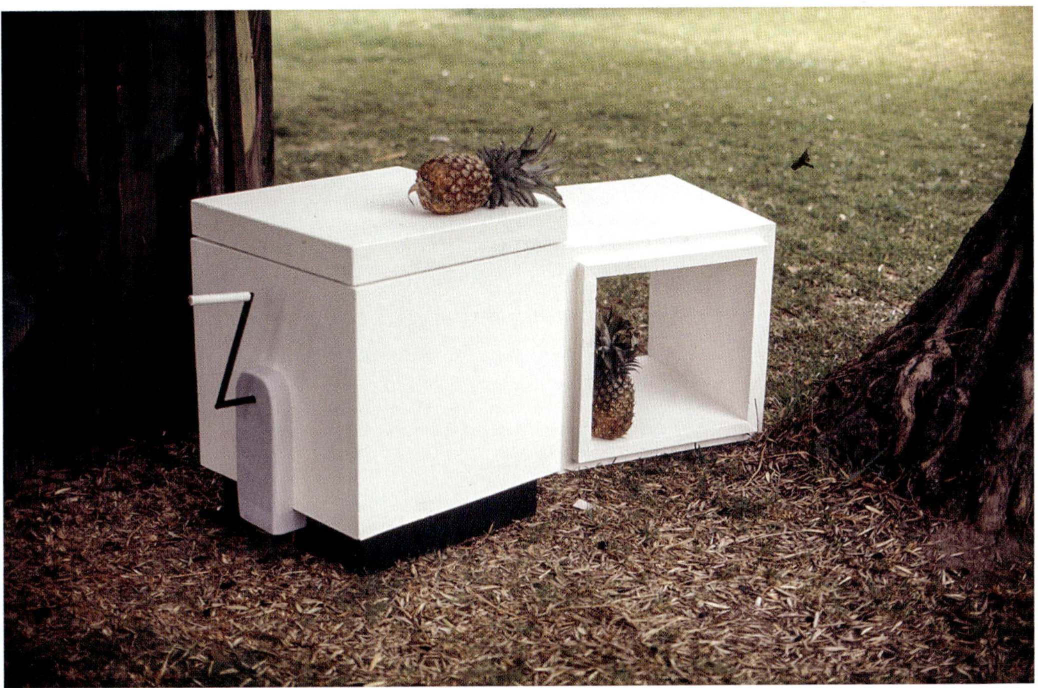

Figure 9.2
Refrigerator for "developing" countries. © Victor J. Papanek Foundation, University of Applied Arts Vienna

Problem Solver and Missionary: Design in a Danish Welfare Economy

Since his earliest work with students in 1968, the Danish press had enthusiastically showcased Papanek's ideas on alternative models of design, and design strategies for the so-called developing world. Indeed, the clarion call of *Design for the Real World* seemed perfectly apposite in addressing the accusations made around the lack of humanity shown in the installation of inflexible "hefty concrete buildings" that the GTO had inflicted on the Greenland community, which some viewed critically as a mode of "welfare colonialism."[6] One architectural design editor, for example, in the leading journal *Arkitekten*, called for the kind of participatory approach Papanek heralded in his own design projects: "What could have been developed instead were forms of industrialized building, where lightweight components . . . could have been used for making new constructions and for the expansion of the original small homes. The education and training of Greenlanders to carry out this kind of industrialized construction would have been a more viable task than educating and training them to be participants in the construction of concrete buildings."[7]

The pertinence of the *Design for the Real World* discourse to Denmark's focus on the ethics of development strategies for Greenland also coincided with the release of the Danish edition, *Miljø for millioner: Design for behov eller profit* (*Environment for Millions: Design for Need or Profit*; see figure 9.3), during the author's guest professorship there. Papanek assured journalists that the Danish version would differ from the previous Swedish and US iterations in its omission of "a fictional story of a design for a tremendously expensive plastic super-woman" after protest and persuasion by "feminist groups."[8] Papanek refers here to the spoof corporate design project of an automated life-size sex doll he first imagined in overtly misogynistic terms in the letters page of the journal *Design Course* at Purdue University. Originally titled "Volita Project," the piece found its way into both the Swedish and US editions of his ethical treatise on the role of design in society as the "Lolita Project."[9]

Feminists were not the only group to throw a broader spotlight on the contradictions of Papanek's preaching. Some local journalists, skeptical of the designer's own ethical standpoint as a privileged Western male academic, were quick to point out the irony of the American professor's views on social equality, questioning how the beneficiary of a well-appointed luxury penthouse sponsored by the National Bank of Denmark could be teaching as a "missionary" of social design at the Academy a few blocks away (see figure 9.4).[10]

Others, however, proved more appreciative of the technical feats with which Papanek regaled his readers. Under the title "Problem Solver and Missionary," one Danish national newspaper showcased the design professor in his fashionable penthouse, seated against the backdrop of his ethnographic mask collection, a staple feature of his traveling "nomadic" interior. Referring to the social designer as a "knight in shining armor" and as a "crusader for a better world," due largely to his distribution of the inexpensive *Tin Can Radio* design to users in Indonesia, the technical journalist praised his policy of extending competencies to the users as a means of empowerment.[11] The figure of "42 million users" was seemingly plucked from the air, but the simple 1960s speculative radio design for UNESCO, designed in the geopolitical information era of the Cold War, had now blossomed into a potent political totem of an emerging appropriate and intermediate technology movement pitted against Western techno-utopian visions.[12]

But the overriding narrative that framed Papanek's stint in Denmark centered on the initial rejection of his *Design for the Real World* manuscript by US publishers prior to its being snapped up by the Swedish and Danish presses. This curried favor with those of an anti-American persuasion and further naturalized the design critic's ideas as befitting the progressive welfarism of the Nordic region generally, and Denmark as his host country more specifically. In this respect, Papanek and his design approach were lent honorary Danish

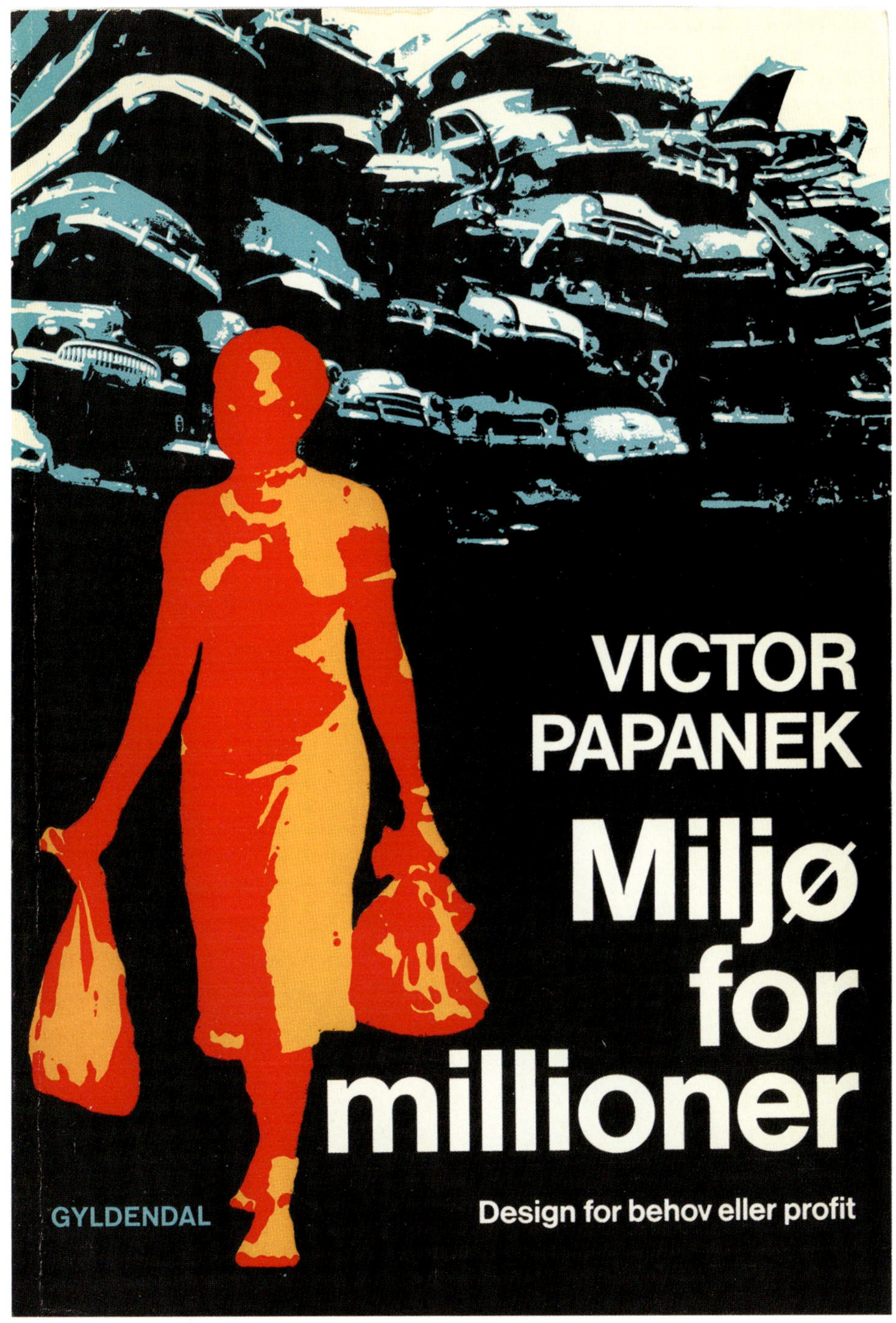

Figure 9.3

Miljø for millioner: Design for behov eller profit (*Environment for Millions: Design for Need or Profit*), first Danish edition, *Design for the Real World* (Copenhagen: Gyldendal, 1972). © Gyldendal. Courtesy Victor J. Papanek Foundation, University of Applied Arts Vienna

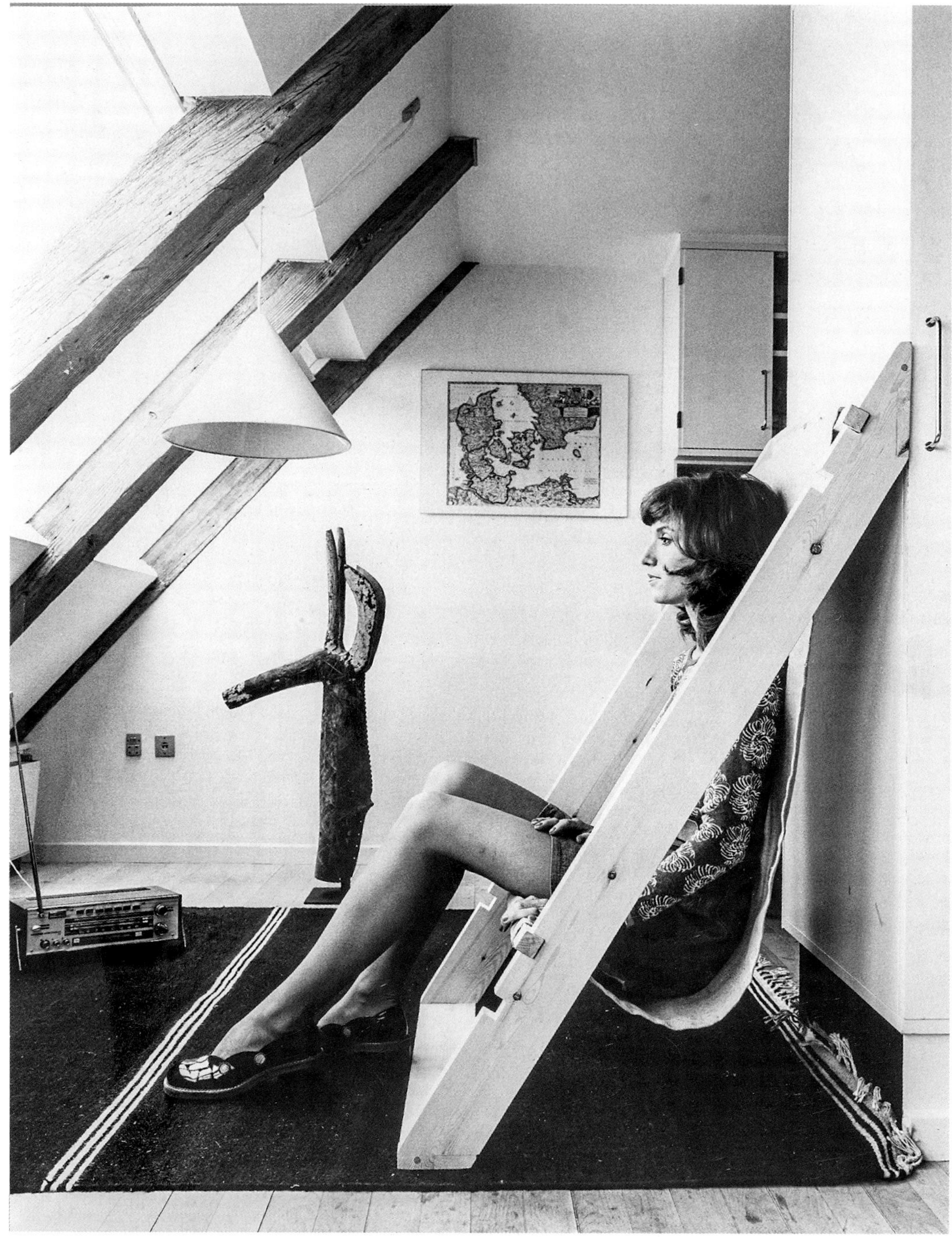

Figure 9.4

Harlanne Papanek, Copenhagen Penthouse Apartment, Denmark, 1971. © Victor J. Papanek Foundation, University of Applied Arts Vienna

status. Drawing parallels between Papanek and the US consumer rights campaigner Ralph Nader, who both provoked the ire of industrialists and harbored grand ambitions to improve the world, a typical newspaper report told how the social designer was a controversial figure in the United States, where his anticorporate ideas were lambasted, but that he relished his success among the Scandinavians. In a newspaper interview, Papanek described Denmark as "fertile ground" for his ideas and expressed an explicit desire to expand his career in the country.[13]

At the time Papanek joined the Royal Danish Academy of Fine Arts, however, it was in a state of extreme upheaval, having abandoned entrance examinations in favor of an open registration system in the wake of student protests. One former student of Papanek from this period, Steen Juhler, described how Danish design and architecture formerly had been controlled by an unspoken hierarchical structure whereby "only a few people and mostly people from architectural families" made it into the Academy, and consequently into the broader national Danish design scene.[14] To Juhler and his dissenting contemporaries, this professional nepotism was inextricably linked to the prolonged observance of an elitist Danish design culture of "good form," a culture that focused on design connoisseurship and the notion of distinction redolent of the mid-century Scandinavian modern style. Juhler and his disaffected group of would-be students, who had initially been rejected from the Academy after failing the entrance exams, envisaged a design culture of collaboration and "real" social engagement beyond aesthetic dilettantism that was perfectly suited to Papanek's agenda.

As a counterpoise to the social-class bias and dynastic design and architecture structure in which the Academy was entrenched, it had been a group of newly forged design activists that had demanded the institution open itself up as a nonhierarchical, interdisciplinary place of learning in which anyone with an interest in design or architecture could participate, regardless of formal qualifications or connections. With the abandonment of the entrance examinations and enrollment procedures, design education was bordering on chaos as nonregistered students moved freely between studios, many of them lacking assigned studios or structured programs. The formerly traditional Royal Danish Academy of Fine Arts was fast descending, by Papanek's measure, into the kind of anarchic pedagogic setup he had been desperate to escape from at CalArts.

Papanek's invitation to join the Academy as a guest professor with a self-proclaimed social design agenda had, in fact, been partly motivated by the faculty's desire to quell the perceived mayhem generated by the informal, open access policy. The social designer's arrival appeased the radicalized student caucus through the introduction of transdisciplinary design programs and hands-on participatory projects. "Design," declared Papanek in one of the opening pages of *Design for the Real World*, "is a luxury enjoyed by a small clique who form the technological moneyed, and cultural 'elite' of each nation"; for young people like Juhler who had felt entirely alienated by the hermetically sealed culture of Danish design, culturally reproduced over generations to ensure the dominance of particular social groups and families, this polemic must have seemed uncannily apt.

With the Academy in a state of flux and political upheaval, design with a social purpose served to transform the institution's agenda, with Papanek relieving pressure on the upper management by tutoring students in design projects with an overtly social agenda: corn mills for Africa, toys for preschool children, exercise equipment for pregnant women, jigsaw puzzles for blind children, low-cost medical apparatuses, nonpolluting stoves and low-tech desk fans for developing countries, play-and-work carts for kindergartens. The formerly disaffected Juhler, for example, worked on a project to design a bicycle wagon for transporting kindergarten-aged children from one of Copenhagen's major squatter tenement communities to a nature hut in the woods, in order to protect them from daily police raids. This project was at once sensitive to the needs of those on the periphery of societal power structures and overtly political because, conversely, kindergarten children could be visibly transported to the squatter tenements, with the intention of forestalling further police raids.

The upheaval at the Royal Danish Academy of Fine Arts was indicative of a general pan-European radicalization of architecture and design schools that had begun in the late 1960s and persisted into the early 1970s, with *Design for the Real World* taken up as a manifesto for the cause.[15] This activism was not confined to the elite of the art schools but rather was directly linked to concerns and priorities of Worker's Unions, ecological groups, pedagogical reform, NGOs, the social landscape movement, ergonomics, alternative and sustainable technologies, community activism, disability rights, alternative transport, health design, occupational therapy, and humanitarian relief.

Toward a Design Anthropology

Danish social anthropologist Mette Bovin, a faculty member at Aarhus University, was most adroit at recognizing the interdisciplinary potential of a new branch of anthropology that Papanek's crusade for a humanitarian design represented, and its potential intersections and application to areas of social activism. As a young student in the late 1960s, Bovin herself had been instrumental in establishing a now well-known relationship with E. E. Evans-Pritchard and Aarhus University, and thus with the British school of anthropology.

In 1973, in a national left-leaning Danish newspaper, Bovin offered a portrait of the peripatetic design critic under the title "Victor Papanek's Ideas and Anthropology" in which she highlighted parallels between his proposed methodologies and those of anthropology more generally: the holistic view of the human being, methods of analysis based on participation, and in-depth cultural awareness.[16] By Bovin's account, the fact that Papanek had spent half of his lifetime outside of his "native culture," and that he used this as a productive method of reflective distancing, critiquing power relations and giving voice to the "client group," made him directly comparable to a contemporary anthropologist. She concluded that his approach contrasted only in that it was not limited to locating "interesting problems" but instead contributed applied solutions through design.[17]

Under the subtitle "Anthropology and the Failures," the social anthropologist reinforced calls for interdisciplinary teamwork, using the example of Papanek's own professional experiences as an agricultural advisor in South East Asia to illustrate her point. Rehashing Papanek's well-worn tale of the importance of fieldwork in respecting Indigenous populations' belief structures (redesigning of plows from steel to plastic, the former material being in conflict with spiritual understandings of the earth and materiality), Bovin reiterated Papanek's point that research conducted by a multidisciplinary team prior to the distribution of the plows would have foreseen this user problem, taking into account the cultural subtleties surrounding even the most basic functional object. While for ethical reasons anthropologists, as a matter of course, would normally prohibit such cultural interventions, denying the possibility of "hands-on" application, Bovin called for a wholesale "de-mystification" of disciplinary boundaries and the rethinking of anthropology's intent.[18] A new configuration of a designer-cum-anthropologist, an expert who combined nuanced cultural insight with practical, participatory skills was deemed to have the potential for effecting positive change by making interventions that operated beyond neo-colonialist gestures (which some had accused Papanek's designs of embodying). Contrary to Papanek's critics, as an anthropologist Bovin identified in the designer's work a genuine intent to pursue an alternative to the tyranny of Western power relations enshrined in the rationalist aesthetic of the modernist design paradigm.

Certainly, Papanek welcomed the input of international designers and the quasi-anthropological problems their projects proffered and as a means of pushing the real-world pedagogy beyond the strictures of the conventional, elitist Western design school. A letter to the president of his home institution CalArts, written from his Danish address in February 1973, spells this out very clearly. In it Papanek emphasized the ambition to internationalize the cohort at CalArts upon his return there: "Maybe because I have written DESIGN FOR THE REAL WORLD (which is now in 9 languages), I find myself increasingly bombarded by letters from students in many parts of the US and the rest of the world who would specifically like to study design with me. . . . I have nine [international] students here [in Denmark] from Germany, Mexico, Indonesia, Canada, USA, India, Kenya and Great Britain."[19]

During his yearlong guest position in Copenhagen, Papanek worked closely with an exchange student from the state-run Institut Teknologi Bandung (ITB) Indonesia, Pak Imam Buchori, tutoring him on the redesign of a kerosene stove and a dual-purpose lampshade used to trap insects. Both projects were born of the anthropological method of participant observation applied here to local Indonesian domestic settings. Buchori's return to Bandung in 1973 coincided with a new, concerted modernization program under the government of President Suharto, with industrial design an integral part of the industrializing policy. Treading a thin line between the wholesale importation of Western ideas (and the model of a liberalized economy that went with it) and a vernacularized, community-based model of design, individuals like Buchori (who would go on to be a professor of industrial design at ITB) were admiring of the ambition of Papanek's "design for the real world" thesis but recognized that it actually relied for its critique on a model of Western design, production, and consumption. "At that time," Buchori later recounted, "Indonesia had no experience of industrialization. . . . We had industry, but it was one that just copied and pasted from the West—when actually there was no Indonesian design involvement whatsoever. So we found it difficult—how does design survive as a profession if there is no industry?"[20] While the *Tin Can Radio* found favor with progressive Danish journalists, keen to remedy the last vestiges of Western colonialism, Buchori found the recycled design faintly patronizing: "Papanek talked about designing a radio with some sort of used materials and lace [decoration]. I found it silly—sorry to say that—but the product is silly and not very good at all!"[21] Yet he and other international students did return to their homelands with alternative modes of design thinking that reached beyond the Bauhaus–Ulm School good-form aesthetic teachings (exemplified by institutions such as the National Institute of Design in Ahmedabad, famously led by Ulm School alumni of Germany). They were formulated and readapted from the hands-on, research-led design philosophy Papanek advocated during this period.

In Denmark, Papanek's work and methods received the strongest reinforcement in terms of their anthropological value, to the point of their being a prototype for contemporary design anthropology that concretized as a discrete discipline at the beginning of the twenty-first century.[22] Papanek, the self-defined anthropologist-cum-designer had finally been offered the unequivocal opportunity to effect social change, as leader of a research group focused on an alternative, participatory design strategy that could radically improve plans for Greenland's housing facilities. Ultimately, however, the project failed to come to fruition. Instead, together with his CalArts colleague James Hennessey, Papanek embarked on an altogether different project: coauthorship of the first volume of a lifestyle furnishing book, *Nomadic Furniture 1,* perfectly suited to the generation of "loosened up" anticonsumers they found themselves teaching on a daily basis.

Nomadic Furniture: Commodity Culture Unhinged

Nomadic Furniture 1 (see figures 9.5 and 9.6), a book project conceived and delivered with lightning speed despite the painstakingly analog method behind its production, drew on

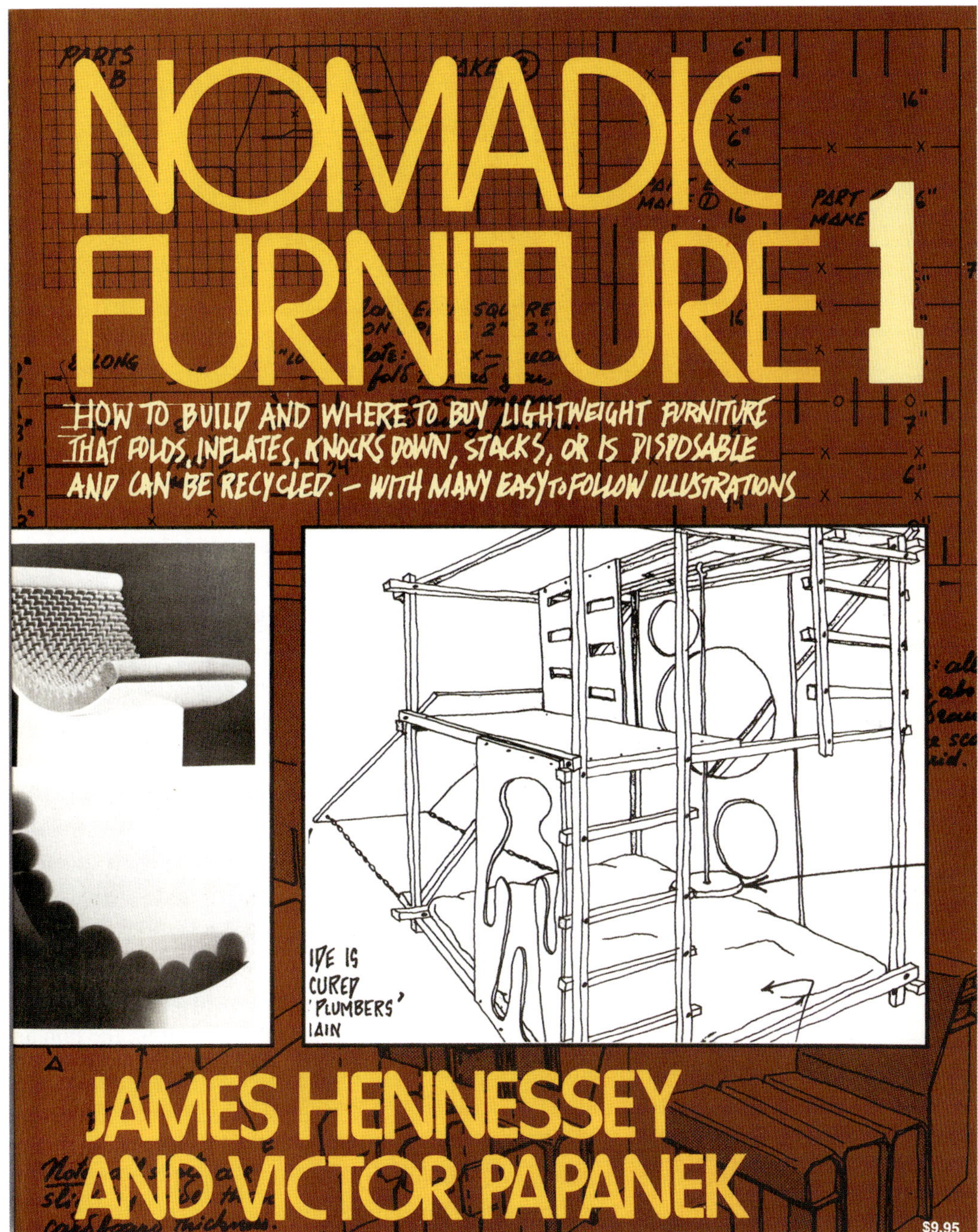

Figure 9.5

Victor Papanek and James Hennessey, *Nomadic Furniture 1*, 1973. (New York: Pantheon Books, 1973). © Victor J. Papanek Foundation and James Hennessey

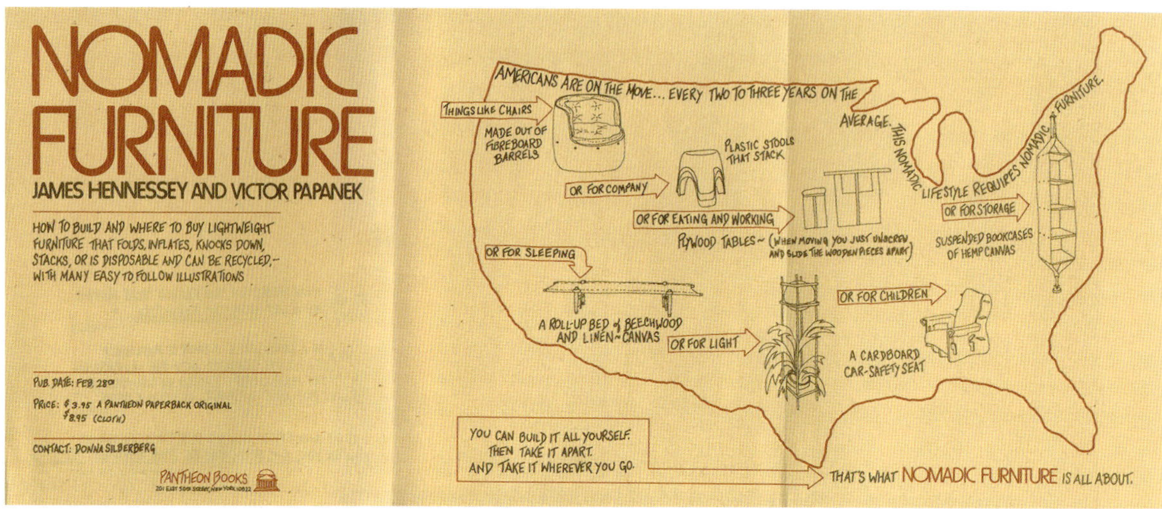

Figure 9.6

Promotional pamphlet for *Nomadic Furniture*. © Victor J. Papanek Foundation and James Hennessey

Papanek's and Hennessey's own personal fascination with the democratic design of portable possessions, as well as the general modishness of the theme. Both having traveled extensively and relocated internationally multiple times, they set about sourcing the ur-nomadic objects that circumscribed a new freed-up lifestyle. Using audiocassettes on which ideas were recorded and mailed to and fro between Denmark and California, Papanek and Hennessey built up a repertoire of items—self-built, purchased, and recycled—that demarked a democratic, self-provisioning design culture with a hint of social conscience. "Most Americans— especially young people—move more and more frequently, over increasingly great distances," explained the back-cover copy, continuing, "Furniture design to suit these contemporary nomads are developing rapidly; many are improvised under the stress of unexpected living conditions."[23]

Loosely based on the *Whole Earth Catalog* concept of consumption for an alternative economy, Papanek conceived the project as an overtly commercial rather than an intellectual venture. Closely following the public relations rumpus stirred up by *Design for the Real World,* the designer even had his publisher, Pantheon Books, sign up *Nomadic Furniture 1* for the Book-of-the Month-Club. As well as signing up with two further books clubs, the release of the book's sequel, *Nomadic Furniture 2* (see figure 9.7), was arranged less than a year later—with the intention of benefiting from the flurry of attention the debut edition had generated.[24]

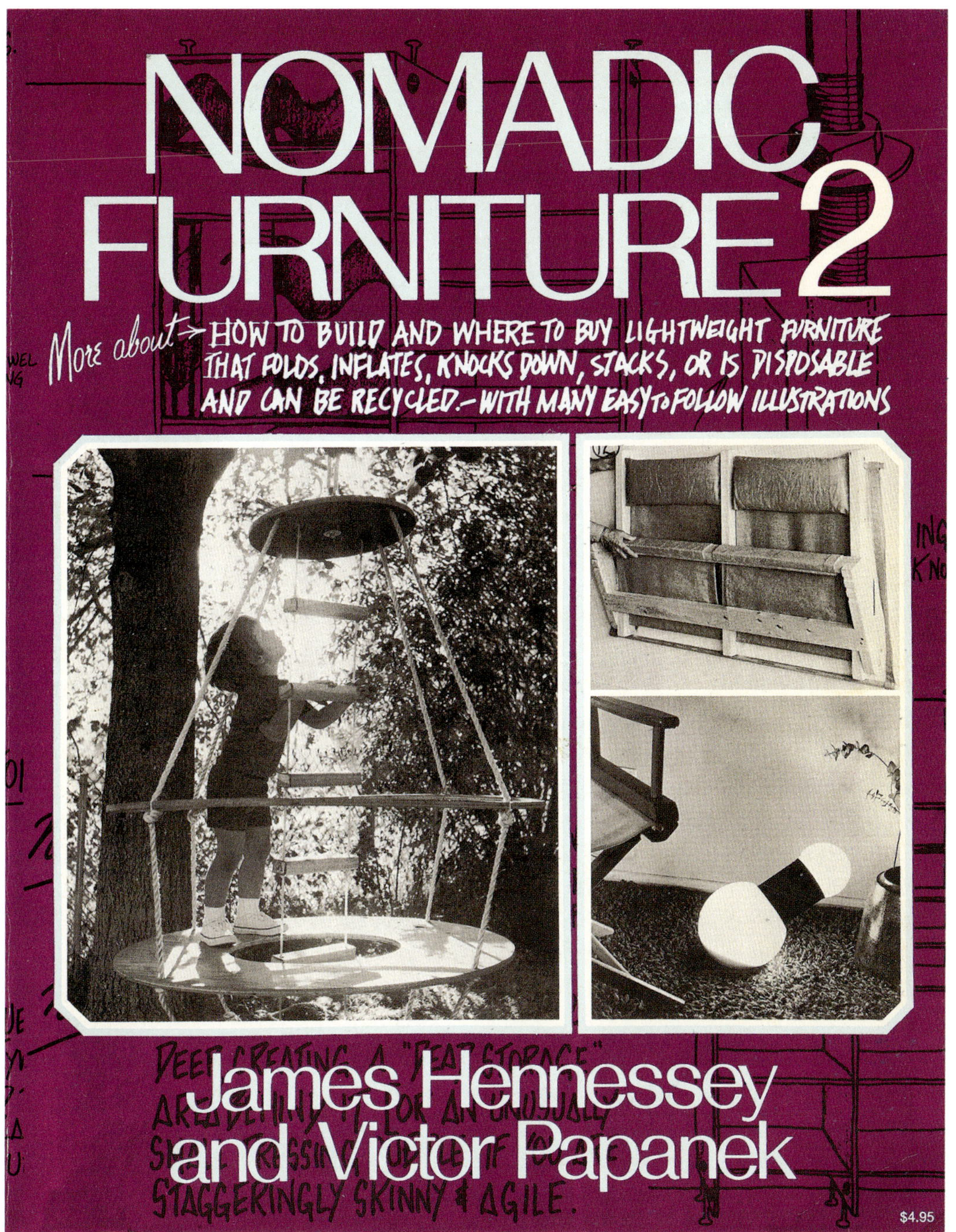

Figure 9.7

Victor Papanek and James Hennessey, *Nomadic Furniture 2* (New York: Pantheon Books, 1974). © Victor J. Papanek Foundation and James Hennessey

With the subtitle "How to Build and Where to Buy Lightweight Furniture That Folds, Inflates, Knocks Down, Stacks, or Is Disposable and Can Be Recycled—with Many Easy To Follow Instructions," the hand-drawn and handwritten book included designs from Hennessey and Papanek's students, design classics (such as the 1968 Sacco beanbag by Piero Gatti, Cesare Paolini, and Franco Teodoro and the 1938 butterfly chair by Antoni Bonet, Juan Kurchan, and Jorge Ferrari-Hardoy), and even one of Papanek's earliest customizable designs from the 1940s, the Transite Table. The *CP-1 Cube*, first designed by a transdisciplinary design team overseen by Papanek in Finland in 1968 as a play environment for children with cerebral palsy, featured as a prototype piece of nomadic furniture with a variety of applications: entertainment cube, children's cube, relaxation cube, and work cube.

The cube designs drew heavily on designer and architect Ken Isaacs's late 1960s *Beach Matrix* installation (1967) and his earlier work, the *Living Cube* (1954), structures made, as Isaacs explained, as part of "new prototypical systems in architecture, living equipment, fabricating means and communications" that rejected preexisting forms and their "visible allegiance to value-saturated historical models."[25] Self-build items such as Isaacs's *Superchair* (1967)—adaptable, sustainable, and portable—anticipated his later work *How to Build Your Own Living Structures*, whose release coincided with Hennessey and Papanek's *Nomadic Furniture 2* in 1974. Rooted in an overarching matrix theory developed during his time as a graduate student at the Cranbrook Academy of Art (which itself had parallels with Buckminster Fuller's Dymaxion Principle), Isaacs challenged the postwar American idea of interiors, architecture, and furnishing as static symbols attesting to social status.

An influence even closer to home for Papanek's *Nomadic Furniture* project was Design Group, comprising Papanek's CalArts colleagues Peter de Bretteville, Toby Cowan, and Jack Reineck, who generated a flexible "dormitory furniture system" for students, for use in the CalArts dormitory with the instructions that "that each resident must determine his or her own needs and arrange the elements accordingly."[26]

Nomadism, and its auxiliaries self-build and DIY (Do It Yourself), formed part of a broader gamut of decentering practices familiar to the late 1960s and early 1970s counterculture movement ranging from the "quintessential US countercultural project" Drop City in Colorado (1965) to the Italian *Global Tools* collective (1973–1975).[27] Tracing the pervasive interest in the nomadic phenomenon in the visual culture of this period, historian Silvia Bottinelli observes that the term *nomadism* came to be used to define "an archetypal and dialectic pattern, in which the nomadism is seen as a fluid, dynamic, antagonistic force in contrast to the establishment, and that's one that undermines pre-existing power structures."[28] In addition to sketching out a political sensibility that upended bourgeois values of stability and fixedness, nomadism signaled a broader discontent with commodity culture, and a cultural revolution of things that first ignited in the late 1960s.

The Death of the Object

In 1968, the year of the Paris student riots and Papanek's first intervention in pan-Scandinavian design activism outside Helsinki, a revolutionary book titled *The System of Objects* offered a theoretical interpretation of society's newly insatiable appetite for consumption. Penned by the Marxist French philosopher and sociologist Jean Baudrillard, its pages described the process by which late-capitalist consumer culture stripped away meaning from concrete symbols through a hyper-abstraction of signs and signifiers, and the dislocation of things from any semblance of utilitarian value. The sociologist offered a particularly compelling argument regarding the generational changes of the structures of interior design, and the pronounced shift to a kind of modern, multifunctional nomadism that Hennessey and Papanek's *Nomadic Furniture* celebrated at the start of the 1970s. With previous generations, Baudrillard asserted, "things" were interpreted as a blueprint of social structure in which the "arrangement of furniture," for example, stood as "a faithful image of the familial and social structures"—bourgeois interiors emphasizing "unifunctionality, immovability, imposing presence and hierarchical labeling." The emergence, however, of a newly mediated system of objects, in which "real things" were rendered redundant, created a qualitatively different, dislocated relationship between people and things in the new nomadic, information-led, technologized society.[29] Because the primary function of furnishing, Baudrillard argued, had been to personify human relationships, "the modern object liberated in its function" (new modes of transient and foldable, collapsible, and mobile furnishings) had upended the traditional patriarchal bourgeois interior allowing for "a greater openness in . . . social relationships."[30]

From the Paris riots to the ransacking of the 14th Milan Triennial in May 1968, dissent toward consumer culture and the perceived erosion of authentic labor politics lay at the core of protest in Europe. Baudillard's articulation of a historical and ontological shift in the meaning of things was coupled with the publication of the cult book *The Revolution of Everyday Life* (*Traité de savoir-vivre à l'usage des jeunes generations*, 1967), a mainstay of the 1960s student activist bookshelf. Its author, philosopher and Situationist Raoul Vaneigem, decried the death of the working class, and in its place the rise of the consumer, whose only power resided in shopping. "Purchasing power is a license to purchase power," Vaneigem declared, continuing in a tone not dissimilar to that of Papanek himself; "With Volkswagen your problems are over! The man with good taste is savvy too: he chooses Mercedes-Benz!"[31] "The old proletariat sold its labor in order to subsist" whereas "the new proletarian'" was now reduced to trading theirs "in order to consume."[32] *The Revolution of Everyday Life* would reach Anglophone youth several years after it had kindled the May 1968 Paris student uprisings. But along with fellow Situationist Guy Debord's *The Society of the Spectacle* (*La Société du*

Spectacle, 1967) it popularized a neo-Marxist critique of commodity fetishism that inspired the wholesale rejection of consumer culture as a political act. Like Baudrillard, Vaneigem and Debord identified a momentous shift from an era of authentic social life to a contemporary state in which the commodity had colonized every aspect of daily life, leaving in its wake a generation of depoliticized dupes wallowing in a state of false consciousness. Philosopher Wolfgang Haug—in his widely read polemic *Kritik der Warenästhetik* (*Critique of Commodity Aesthetics*, 1971), released the same year as Papanek's *Design for the Real World*—also took up the anticonsumer mantle, specifically castigating designers as the handmaidens of consumer capitalism, operating in wanton collusion with advertisers to generate the insatiable desire that drove modern consumerism.

It would be easy to assume this burgeoning neo-Marxist discourse around consumerism, despite its clear ties to the polemic underpinning *Design for the Real World* and *Nomadic Furniture,* is far removed from Papanek's terms of reference as an American designer trained in the Cold War period. Yet the understanding of his work and teaching in Europe was entirely framed by this genre of radical anticonsumer politics, a fact that became increasingly problematic to Papanek in the latter stages of his interactions with students in Denmark.

Far from being confined to the coffeehouse banter of Left-leaning intellectuals, the influence of anticonsumerist philosophers and activists had spread to mainstream discourse and even to the upper echelons of the US design establishment. In the same year that Baudrillard further debunked the model of consumption as a needs-based phenomenon in his critical-theoretical *The Consumer Society* (1970), the committee of the IDCA invited him to attend its conference as one of a collection of theorists dubbed the "French Group."[33] The conference kicked off with cocktails served against the spectacular backdrop of sunset amid a mountain range with Charles and Ray Eames, George Nelson, and their ilk mingling, confident in their position as the old guard of design, content to purvey corporate design to an eager consumer public. Theirs was the same cozy scene of "back-slapping bonhomie" Papanek had critiqued in his 1968 article praising, by contrast, the activist-based design conference of the pan-Scandinavian students in Helsinki.[34] The arrival of the French Group, and the chosen conference theme of "Environment by Design," had been a failed attempt by the IDCA's executive committee to appease growing student unrest, as witnessed by the organizers of Italy's doomed Milan Triennale two years previously, when students had trashed the fair in protest of designers' role in fueling consumer capitalism.

But when the US activist group Ant Farm arrived at Aspen 1970 and defiantly erected a vinyl nomadic inflatable "atop the sacrosanct landscape" designed by Bauhaus luminary Herbert Bayer, it was evident that the Aspen conference would result (as Reyner Banham would later put it) in "a guaranteed communications failure."[35] Historian Greg Castillo describes how, after being denied access to main conference events, "the Ant Farm contingent lurked at the margins of the event as self-proclaimed 'rabble-rousers.'"[36]

Ultimately IDCA 1970 descended into an acrimonious confrontation involving several groups: design outsiders, student activists, and counterculture and environmental groups that together ridiculed the ingenuousness of the corporate organizers' attempt to address the political and social implications of design in society. As previously mentioned, this event became a watershed in US design culture, with the resignation of the IDCA committee and the consequent shift to CalArts agreeing to take on the follow-up conference with Papanek (as new dean of the new School of Design) overseeing a radical new format in which conference attendees, not the organizers, steered the on-the-ground proceedings.

Despite this, Papanek was far removed from interventionary radical activism, and even further removed from the intellectual neo-Marxist theorists of the French Group and politically motivated ideas of "life without objects" espoused by Italian radical designers Superstudio.[37] But without doubt, he was fully aware of the sea change afoot among the generation of students with whom he interacted on a daily basis. Before leaving CalArts to take up a post as guest professor at the Royal Danish Academy of Fine Arts, and following the counterculture-dominated, anticonsumerist 1970 and 1971 IDCA conferences (both of which he took part in), he proposed the establishment of the Center for the Study of Responsible Design, with its specific research focus on emerging lifestyles of "nomadics." This was the first inkling of what would become his second major research and publication project.

Both *Nomadic Furniture 1* and *2* proved successful, coinciding as they did with a spate of exhibitions and installations around similar themes of anti-commoditization and sustainable consumer culture. In 1973, for example, the International Design Center (IDZ) Berlin featured an exhibition titled *Design It Yourself: Möbel für Grundbedarf des Wohnens* (*Furniture for Basic Living*), including designs by Papanek. This DIY spontaneous design aesthetic challenged the supremacy of mass-manufacturing standardization with the intended effect of democratizing design. As part of a broader discourse of alternative culture, and anti-commoditization sentiment, the exhibition promoted the idea of self-empowerment by the adoption of low-impact appropriate technology: a set of basic design instructions and a simple set of tools. Self-assembly furniture and the new "Low-Tech-Kultur" were meant as an overtly political statement regarding the overturning of hierarchies of taste and design authorship.[38] In response to a less than favorable review of *Nomadic Furniture 1* in the Danish design journal *DanskForm* (which singled out the unviable design solution of the cardboard child's safety car seat for particular criticism), Papanek cited the book's influence on the IDZ Berlin *Design It Yourself* exhibition in defense of its relevance to "an *entirely new* section of people: housewives, workers, the elderly, unemployed" as well as "young revolutionary students."[39]

In 1974, the Gallery of Contemporary Art, Zagreb, offered Papanek one of his first forays into the socialist federation of Yugoslavia, with an invitation to exhibit the work he had shown at the IDZ 1973 installation. This marked the beginning of a broader interest in Papanek's work that emanated from Eastern Europe and the Soviet Union and came to further fruition in the 1980s, largely replacing his interactions with Scandinavia and the Nordic region.

By the time of his engagement with Zagreb as a nomadic designer in early 1974, Papanek had moved on from Copenhagen to a position in the design school of Manchester Polytechnic. His departure from the Royal Danish Academy of Fine Arts was far from frictionless—with Papanek weaponizing the national newspapers to publicly denounce the school of design as little more than a facilitator of romantic bourgeois students, who exercised a weaker work ethic than the infants of a Danish kindergarten.

Bourgeois Romantics and Naive Idealists: Severing the Scandinavian Connection

In the final months of Papanek's guest position in Copenhagen, under the provocative title "Promises of More Bullets against 'Lazy Students,'" the journalist of a weekly national newspaper recounted the biting criticism Papanek had fired at the Academy for the reign of chaos that pervaded its classrooms, and the laziness of its posse of bourgeois Marxist students. Rumor had it that the "world-famous American designer" had sought a renewal of his contract as guest professor, only to be offered hourly paid work to teach an international cohort of students.[40] Clearly perceiving the offer as snub, Papanek had launched into a tirade, offering an interview to a national newspaper under the extraordinary headline "Unfair Comparison of the Academy of Fine Arts with a Kindergarten: Still, Kindergartens Have a Minimum of Discipline and Some Work Gets Done There."[41] Claiming that the school was "chaired by a group of so-called Marxists, who actually perform applied fascism," Papanek accused the institution of failing its students and wasting taxpayers' money due to the profound lack of knowledge or competence within its faculty. "Imagine," exhorted the disgruntled designer, "you yield yourself to a brain surgeon, who spent five years studying in a way, in which one studies here. It would be better to shoot a bullet through your forehead—in any case, this would be less painful."[42] He and his wife had at first envisaged settling in Denmark, he went on to say, but after finding the work ethic of British design students faultless by comparison with that of their Danish contemporaries, he had decided that his professional interests must come first, accepting a position at a design school in England, where basic professional standards, rather than political ideals, were maintained. Lamenting the institution's lack of motivation to turn around a system that had for decades relied on the notion of good form, ultimately Papanek criticized the reluctance of the Royal Danish Academy to educate students in the basics: "They do not know anything about the philosophy behind design. Isn't the absolute minimum requirement for a designer, that he is able to formulate his ideas on paper and with models? A tremendous part of them are unable to do these very things."[43]

Finally, confirming his complete disassociation from the radical Scandinavian design activism that had helped forge the ideas behind *Design for the Real World*, Papanek argued that good designers should be socially aware but remain objectively depoliticized. Erik Herløw, leading faculty member of design at the Royal Danish Academy, was forced to defend the cohort in a national newspaper interview, explaining away the year-long upheaval as an experimental phase. He made assurances that the turmoil and insecurity were merely part of a more general youth upheaval that would soon "settle down."[44]

By the end of 1973, Papanek had taken up a temporary position as principal lecturer at the Manchester Polytechnic design school, where he remained for just under two years as head of postgraduate studies; the family relocated from their trendy penthouse apartment in Copenhagen to a conventional mock-Tudor house in an exclusive suburb of Cheshire in Northern England. That same year, the hand-drawn 1968 Copenhagen Flowchart delineating the politics of design, which had kicked off Papanek's career in Denmark, was made into a full-size wall chart. Dubbed the "Big Character" poster and sold in fashionable outlets across Scandinavia, it placed the social designer alongside revolutionary heroes like Che Guevara, as a counterculture pinup.

Professor of the Real World: Pragmatism before Politics

The British design community lapped up the social design message Papanek brought with him from Denmark, and its comparatively rigid design educational structure perfectly matched the ambitions the "Professor of the Real World" (as one UK magazine described him) had to tutor a motivated, rigorous cohort keen to apply the ideas to functioning prototypes.[45] With the UK edition of *Design for the Real World* released in 1971 and sales of *Nomadic Furniture* booming, Papanek had an eager new audience at his disposal, all set to admire what one journalist described as "the stark and unalloyed quality of Papanek's ideas."[46] Stung by the resentment that had started to fester among the students and faculty of the Royal Danish Academy over his conventional, apolitical stance (some even suspected him of working for the US intelligence service), Papanek began strategically promoting his reputation as a *simpliste* and pragmatist. "The political sophist will find a complex enemy in Victor Papanek," reported the UK's foremost agenda-setting glossy consumer magazine *NOVA*. "He has little taste for or patience with lengthy dilations on economic or political imperatives. He is essentially a designer."[47] Papanek's steadfast reluctance (and inability) to engage with the complex issues underpinning the role of design in society, which had so frustrated the Scandinavians and Continental Europeans, was repackaged as a deliberate ploy on his part. He was the authentic pragmatist, not the elitist intellectual, who used design to serve those without a voice whatever the system in place: "All systems—private capitalism, state socialism and mixed economies," Papanek stated in his magazine interview, riffing on Buckminster Fuller, "are built on the assumption that we must buy more, consume more, throw away more, and consequently destroy Liferaft Earth."[48]

Photographic portraits showed him (once again) flanked by his collection of nomadic furniture and ethnographic objects, and his ideas became notably retuned to the growing environmental and ecological discourse: "[Papanek] considers that much of the ecological chaos and economic waste of the modern world is the direct responsibility of industrial designers who are at best 'pimps for sales departments,' at worst criminals," commented the *NOVA* profile.[49] Describing an encounter with Papanek as "something like attending a global village fete," due to his reeling off design anthropological feats from fieldwork with the Navajos to the Inuit, the British interviewer made passing reference to Papanek's status as a former refugee escaping the Austrian *Anschluss*.[50] Now aged forty-eight and based in England, Papanek allowed for the first time the story of his émigré refugee past to be incorporated into the larger part of his career biography, alongside the less reliable but ever-persistent account of his working relationship to Frank Lloyd Wright.

By the time of his departure from the UK at the end of his limited contract, the designer had reinvigorated his reputation as an agent-provocateur of design, recast as a radical pragmatist whose ideas had impact in the real world: "As author of *Design for the Real World*, he has been disliked and even loathed by his design contemporaries: he has been sneered at for his preoccupation with the non-profitable needs of third-world people; he has been accused of the single-handed subversion of design schools which previously had a reputation for industrial submissiveness."[51]

He left behind him in Manchester one of his most successful codesigns, a remarkable tetrakaidecahedral portable playground structure with integrated audiovisual components designed to empower children (see figure 9.8), which would feature in his final book, *The Green Imperative*, in 1995. Despite his ambition to relocate to Europe permanently, Papanek left England, never to return to Europe in any substantive academic capacity again in his career.

Design for Need: The Real World Agenda Goes Mainstream

In April 1976, Papanek was reinstalled in North America as a visiting guest professor in architecture and industrial design at Carleton University, Ottawa, but he did return to the UK to

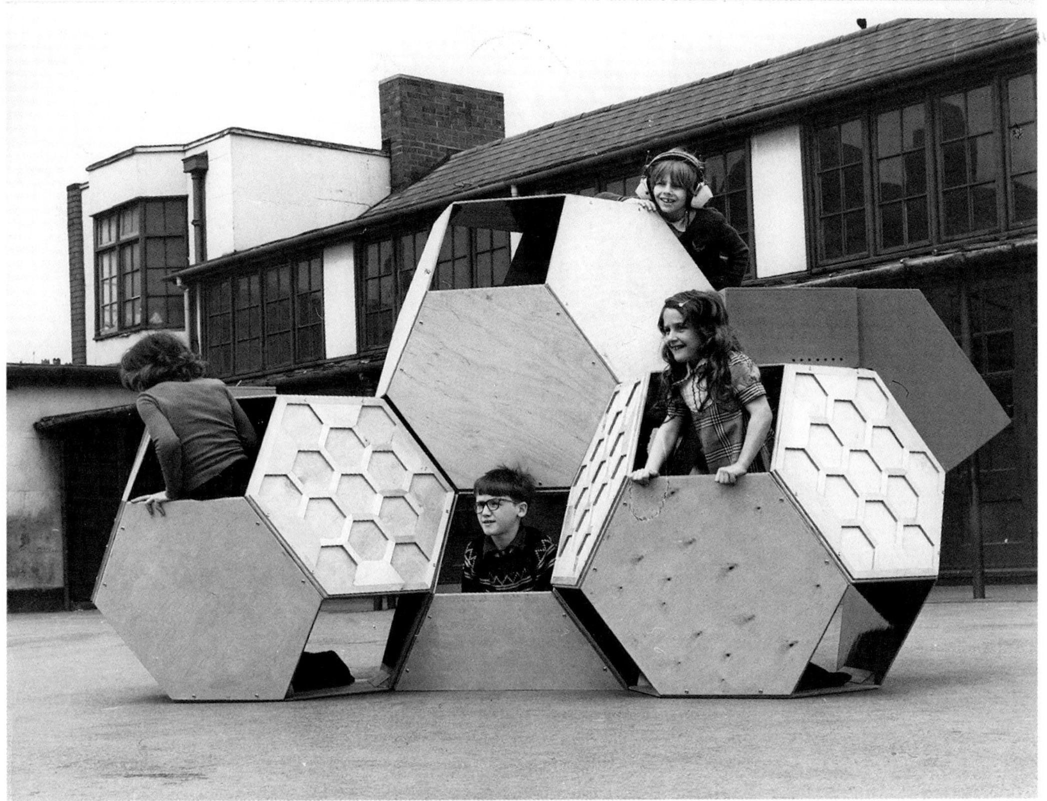

Figure 9.8
Tetrakaidecahedral portable playground. © Victor J. Papanek Foundation, University of Applied Arts Vienna

contribute to a groundbreaking event that his ideas had helped spawn. The hosting of the exhibition and symposium *Design for Need: The Social Contribution of Design* by the prestigious Royal College of Art, London (in cooperation with the renowned Imperial College of Science and Technology), marked the unequivocal impact of Papanek's ideas on the British design scene. Sponsored by a bevy of the highest-profile design concerns, ranging from the Design Council through to the International Council of Societies of Industrial Design (ICSID), Papanek joined an impressive lineup of expert speakers drawn from the kinds of interdisciplinary backgrounds his 1968 "Minimal Design Team" infographic (part of the Copenhagen Flowchart) had sketched out.

The year of his contribution to the *Design for Need* symposium, MoMA in New York exhibited an experimental design for a vehicle for the disabled overseen by Papanek as a consultant

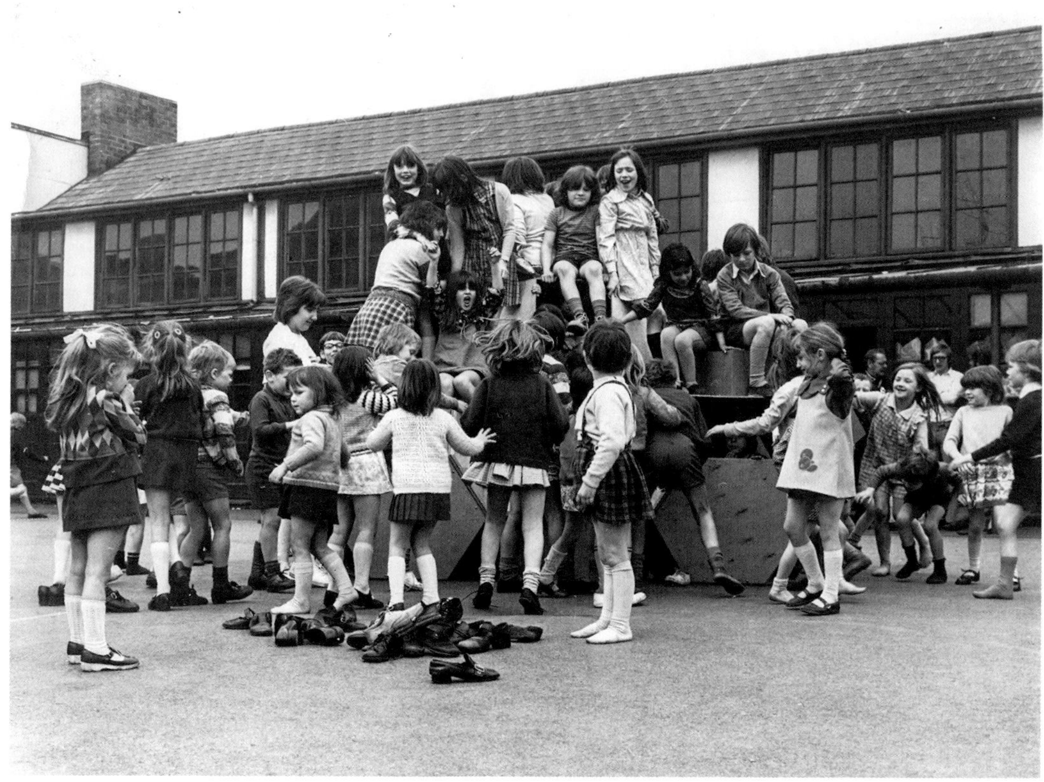

Figure 9.8
Continued

with the Swedish auto company Volvo. A review feature in *Domus* magazine, "The Taxi Project: Realistic Solutions for Today," praised the adaptability of the design for wheelchair use.[52] Papanek had successfully worked in an interdisciplinary team to get the vehicle to prototype stage.

As an established social designer, Papanek stood in good stead to address the theme "Because People Count: Twelve Methodologies for Action." Yet he was a minor player in the symposium, where he met up with figures from some of the most formative stages of his career, among them Yrjö Sotamaa (Finland), Goroslav Keller (Yugoslavia), and his erstwhile nemesis and critic, Gui Bonsiepe (Argentina). The proceedings, including Papanek's talk, were published in an edited volume, offering a concrete contribution to interdisciplinary design covering the fields that *Design for the Real World* had identified as vital areas for concern over a decade and a half earlier.[53] The symposium tackled issues of design for disability, eco-housing, organic waste experiments, recycling, design for the elderly, playgrounds and structures, low-technology devices, design fieldwork research, and design for developing, all of which Papanek had been teaching, writing, and lecturing about globally decades before the design establishment had taken heed.

Despite Papanek's reputation as design subversive, *Design for Need* marked a new phase in his career as an international consultant and expert in policy making for the Third World. By 1974, he had become a member of Working Group IV of the influential design body ICSID, the social designers meeting up quarterly to further the council's "investigative interest in 'developing countries.'"[54] Chaired by Paul Hogan, manager of the Irish Export Board's development aid program, and including Jörg Glasenapp, advisor to United Nations Industrial Development Organisation (UNIDO), Amrik Kalsi, industrial design and pedagogue in Kenya, and Knut Yran, head of Philips's industrial design office in Eindhoven, Netherlands, Working Group IV represented the design establishment's visible endeavor to balance earlier top-down genres of design development doctrines, following demands in development theory concerning the politics of intervention and the incorporation of a broader, more representative contingent. Yet, despite his pivotal and globally respected position as an expert in this arena, Papanek had in fact been reduced to haggling and lobbying in order to gain a position in the working group.

The original ICSID Working Group IV, devoted to "problems of design in developing countries," had been led by Papanek's key adversary, Gui Bonsiepe, who had been invited to act as its founding coordinator from its inception in 1970, and who four years later would provide a condemning critique of *Design for the Real World* in the Italian design journal *Casabella*.[55] In fact, it was only with Hogan's role as the new coordinator in 1974 that any suggestion of the social designer's involvement with Working Group IV arose. Records show a marked reluctance to include Papanek at all: "'a number of Council members are opposed to Papanek, as they no longer agree with his philosophy in dealing with developing countries," read the minutes of one particular meeting.[56] Correspondence between ICSID members shows evidence of a small counter-force of support largely based on a perceived obligation to recognize the significance and public profile of Papanek's impact on the area of design and development: "Victor Papanek is world-wide renowned, and the fact that we asked him to be part of Working Group IV is a recognition of his value," wrote the ICSID General Secretary herself, Josine des Cressonnières, to Indian design professor Sudhakar Nadkarni.[57] Despite the reluctance of some ICSID members, Papanek would eventually join Working Group IV, recounting his experience in a design journal some years later as the foundation for this ambition to form an international design school for peripheral countries, and producing graduates with real world experience.[58]

In 1976, the disbanding of Working Group IV, however, signaled the end of Papanek's formal involvement with ICSID—this despite his ongoing and persistent attempts to remain as a participant in its activities and policy making. In January 1978, for example, Papanek contacted fellow Austrian designer Carl Auböck (in Auböck's capacity as leader of the organization's developing countries framework) expressing his disappointment at being "cut off from ICSID" in spite of his desire to contribute to its efforts.[59] In 1981, he revisited the issue of his exclusion head-on in a chagrin-filled missive to ICSID's chief executive, Hélène de Callataÿ, complaining that he had not been invited to deliver a paper at the latest ICSID congress for the first time in ten years.[60]

Despite his integral role in Working Group IV, by 1979, when ICSID's involvement in exploration of the potentialities of design in the peripheral economies culminated in a large-scale conference in India, and the signing of the famed ICSID-UNIDO Ahmedabad Declaration policy document on design and development, Papanek's ICSID role had, humiliatingly, been reduced to that of a tokenistic gesture: handing over the trophy to the winner of a design prize at the congress's closing ceremony. With a keen following in India, though, he delivered a public lecture in association with the National Institute of Design (NID) Alumni Association in a vast auditorium, under the title "Victor Papanek, USA," which explored "new directions in design for various levels of development."[61] Beyond the realms of the design establishment, Papanek retained his reputation as a groundbreaking social designer, and as the figure who had done his utmost to challenge narrow parameters of a design culture geared toward corporate profit.

The Ahmedabad Declaration: Design for the Real World and Global Policy Making

When Arthur J. Pulos, president of ICSID, delivered an opening speech entitled "The Profession of Industrial Design" at the organization's congress in Mexico City on October 14, 1979, he made explicit reference to design's overlap with a development agenda: "In the century to come, the design professions, with industrial design in the vanguard, will rededicate their efforts toward the final emancipation of all humans from drudgery and social and economic subjugation . . . human beings in a new Renaissance will, once again, become the masters of their environment as the race achieves, finally, that ultimate form of equilibrium known as peace."[62]

Anthropologist Arturo Escobar has drawn attention to the prescience of 1970s design critiques, such as Papanek's, written "as industrialism and US cultural, military and economic hegemony were coming to their peak" and to the significance of understanding design as part of a broader historiography of development policies.[63]

Escobar's work in establishing the historiography of development studies is particularly useful in casting light on the ways in which the hegemonic vision of the "Third World" forged in the postwar period—through policies that held up the industrialized nations of North America and Europe as the appropriate models for imitation—framed the competing design discourses in which Papanek's work was taken up.[64] By the close of the 1970s, the initial top-down economic development theories of the 1950s had devolved into a "basic human needs approach" that "emphasized not only economic growth per se as in earlier decades but also the distribution of the benefits of growth," appealing in particular to appropriate technology and design movements concretized in Papanek's *Design for the Real World* polemic.[65] As criticism mounted regarding the false premise of established postwar development theory, that riches would eventually trickle down to the poorest, a basic-needs approach evolved that embraced a mode of development policy aimed at the so-called poorest people of society.

With the student uprisings and critiques of the suit-and-tie corporate designers that dominated professional design bodies such as the IDCA, organizations such as ICSID risked seeming similarly anachronistic if they failed to act on the broad-ranging critique of design's role in a Western modernizing project. Whatever critics might have identified as its failings, *Design for the Real World* had been part of a countermovement that had drawn attention to the top-down expansionist policies of the so-called Western nations. Such was the political backdrop to the joint ICSID/UNIDO Design for Development Congress of 1979, strategically staged at the National Institute of Design (NID), Ahmedabad, and the Institute of Technology, Bombay, in recognition of India's pivotal role in the Cold War modernization paradigm.

Following Nehru's death in 1964, Indira Gandhi's Congress Party had taken power. By 1979, the replacement coalition government of opposition parties established after the so-called Indian Emergency, in which Gandhi's party was accused of corruption, had collapsed. In the context of India's shifting domestic politics and an emergent basic-needs rhetoric in international development, the Ahmedabad congress can be understood as emerging from this disjuncture with the unrealized Nehruvian vision of design as a catalyst of change. It constituted a major diplomatic undertaking aimed at bolstering Western relations, after a period of considerable rupture in Indian politics that had unsettled preexisting Cold War cultural diplomatic strategies, up until Indira Gandhi's reelection with a newly honed pro-foreign policy approach by 1980.[66]

The congress, the first of its kind, took place over the course of ten days in January 1979, bringing together 130 delegates (ninety-eight of whom were representatives of India) from twenty-five nations. The conference firmly positioned India at the center of a design and development policy-making agenda. The self-proclaimed mission was to address "the role of industrial design in the development and diversification of a developing country's industrial production by designing new products and re-designing old ones," also identifying "ways and means" of satisfying needs through "co-operative and bilateral arrangements."[67] The historic meeting was timed to coincide with the auspicious Hindu festival *Makar Sankranti* and with Gujarat's cultural highlight, the dramatic and picturesque Kite Flying Festival that would be watched by an array of social policy experts, design leaders, and NGO representatives. India acted as a case study regarding the opportunities for industrial design in the context of a mixed economy—with particular attention paid to design education, the upgrading of skills for small- and medium-scale industries, and the potential for craft and village industries.

Romesh Thapar, left-leaning author of *India and Transition* (1956), delivered the opening keynote address titled "Identity in Modernisation." An outspoken critic of Nehru, Thapar was a member of the Club of Rome global think tank that in 1972 published the influential *Limits to Growth* report on sustainability and expansionism—a publication frequently cited together with *Design for the Real World* as part of the canon of environmental critique in the 1970s. As a measure of their desire to appeal to a newly mainstream movement of social design, an aide-memoire circulated within UNIDO four months prior to the Ahmedabad congress stated clearly that the "role of industrial design in related areas of social need, including design for the handicapped" would be clearly demonstrated.[68]

A political showcase, the result of a decade-long series of working group meetings that had engaged with extensive issues of design in peripheral economies to which Papanek had contributed, the Design for Development theme had been preceded in 1977 by the signing of a memorandum of understanding between UNIDO and ICSID, as well as in-depth discussions later that same year at the tenth ICSID congress in Dublin, Ireland. The Dublin ICSID meeting was of particular significance, as it was on this occasion that Ashoke Chatterjee, executive director of NID, accused the ICSID membership of exclusively representing the industrially advanced economies "at a time when only few developing lands [had] felt the need for design as a motive force in their economic improvement."[69]

The astute selection of NID as the location for the ICSID 1979 Congress placed India at the forefront of the policy debate, while drawing on the Cold War legacy of the West's cultural-political intervention in Nehru's modernizing agenda. In 1977, NID had been awarded the ICSID–Philips Award for industrial design in developing countries. The following year, a UNIDO pilot report conducted by John Reid, former president of ICSID, exploring *The State of Industrial Design in Developing Countries*, had confirmed India, and NID under the directorship of Chatterjee specifically, as the optimal venue for the commencement of the congress. Chatterjee himself had introduced Reid, acting as a UNIDO consultant, to the design culture of NID, and Gujarat more generally. Reid had presided over the ICSID executive board from 1969 to 1971, during a period in which affirmative action in recruiting a broader constituency outside developed countries was a priority for the society. He began his report with a definition of design as a socially bound practice: "the development of industrial design . . . despite its enormous technical and scientific content, is an *art*, not a science. It is concerned with people—their hopes, needs and aspirations."[70] The dialectic between preserving design as an embodiment of national identity and authenticity, and the explicit drive to innovate new designs fit for export to a Western market, was a defining theme of the design development agenda. Early into the section of the report dealing with India, Reid encapsulated this search for the authentic, echoing the Eameses' fetishization of the *lota* in their famed *India Report*.[71] "It is sad," observed Reid, "that the first chair I saw was of Scandinavian design in itself 'derived' from an American original."[72] A rather clumsy slogan coined in the making of this UNIDO pilot report summarized a freshly honed design and development agenda that made Papanek's work seem radically politicized: "Industrial Designing Is about Caring for People."

The few historical accounts of the congress generally identify the Ahmedabad Declaration as a golden moment, a crucial turning point in the recognition of the social potential of industrial design in the Third World, developing, or peripheral economies. Such accounts, in some instances penned as academic articles by original attendees themselves, frame the declaration in the context of the legacy of the Eameses' famed 1958 *India Report*, which led to the establishment of NID in 1961. Ahmedabad's unique claim to the national design heritage of India, as the site of Mahatma Gandhi's first *ashram* and the political legacy of the *swadeshi* policy, is also noted as the backdrop to the declaration. The Eameses' involvement in India, in itself the result of a decade of Cold War politicking, arguably fueled the rhetoric of a modern democracy of design, underpinned by the neocolonial Modernist pedagogy of Ulm school tutors and their ilk, the end of which was marked by the 1979 Ahmedabad congress.

Design Dilemmas and Environmental Solutions

The Ahmedabad congress consolidated a postcolonial concept of design for social usefulness, epitomized by the emergence of what art historian Saloni Mathur has described as a "design in an Indian idiom," with proposals for adapting Indigenous forms from the tiffin lunchbox to the automated rickshaw entering conventional professional design media. These hybridized designs recognized the social potentialities of design while seeking to remedy the contradiction of development discourse—and followed the genre of vernacularized designs created by Papanek and his students from the dung-powered *Tin Can Radio* to the calabash cassette player. This renunciation of Western approaches to ideas in design, reiterated in Thapar's Ahmedabad congress keynote address under the title "Identity and Modernization," had called for a renewed offensive against vulgarity and the upholding of traditional aesthetic values, thus marking a clear turning point in understanding the relations between design and development policy and the objects born of this paradigm.

The development design typology generated by the 1970s within the parameters of the intermediate and appropriate technology ethos first emerged in Papanek's "real world" model as the material embodiment of the dialectics of the development discourse.[73]

The archetypes of a "design for development" material culture that Papanek and his students had generated, and which by the time of the Ahmedabad congress in 1979 had become part of the mainstream, stood as shorthand for a broader set of ideological intents that sought to challenge the authoritarian logic of design as the "central political technology of modernity" and its environmental consequences.[74]

How Things Don't Work: Technology and Its Discontents

Following the ICSID-UNIDO Design for Development Congress, Papanek felt he had been ousted from a design profession that had subsumed the key rhetoric of his life's work into the policies of their major institutions, with scant acknowledgment of his influence. His final book project of the 1970s (coauthored with his trusted friend, design engineer Hennessey) entirely avoided reference to any facet of the design for development theme. Evocatively titled *How Things Don't Work* (1977) (see figure 9.9), which might at this point have summed up Papanek's general disappointment with the design world he had idealistically sought to transform, it provided a witty, pragmatic bookend to a decade that had opened with the critical optimism of his *Design for the Real World.* Now chair of the Department of Design, Kansas City Art Institute, Papanek used the book to launch a new weekly radio show, *By Accident or Design*, drawing on a format he had first pioneered in his TV broadcasts in Ontario in the late 1950s.[75]

Whereas his opening treatise of the decade had provided a blanket condemnation of the industrial design profession, *How Things Don't Work* still critiqued bad design but found pragmatic technological solutions that minimized environmental impact, obsolescence, and the acceleration of a neophiliac consumer culture.

Nomadic Furniture 1 and *2* had mimicked the *Whole Earth Catalog* with their handwritten text and illustrations, offering ideas for low-cost, portable furnishings and storage for an alternative lifestyle. The latest coauthored volume rejected the counterculture aesthetic for a style reminiscent of a technical handbook, offering the general public and design community a guide to suitable technology solutions and, most innovative of all, models of new ownership that broke the mold of conventional consumer patterns, banishing the need for all but the most essential technological goods. Communal sharing, recycling, and the

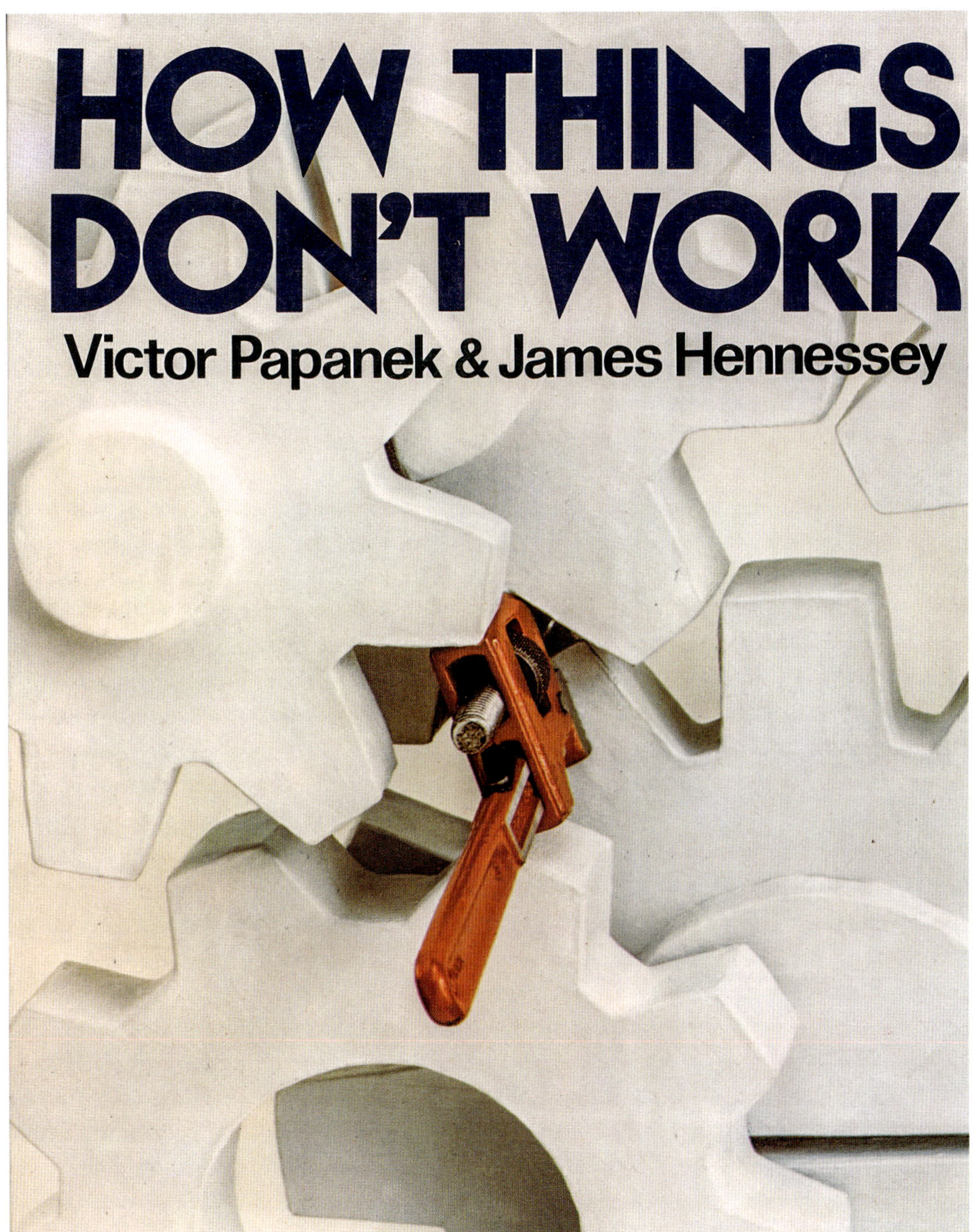

Figure 9.9
Victor Papanek and James Hennessey, *How Things Don't Work* (New York: Pantheon Books, 1977). © Victor J. Papanek Foundation and James Hennessey

purchasing of do-it-yourself kits were featured alongside critiques of gimmickry familiar to Papanek's earliest tirades against electric nail-polish dryers and Japanese full-person washing machines.

Chapter titles drew on the format of one-line witticisms first developed in his *Design Dimensions* TV broadcast series for WUNC-TV (a typical segment on his 1950s series was called "The Chrome-Plated Marshmallow"). The chapter "No Roast Tonight—the Lights on My Carving Knife Need Realignment" regaled readers with the horrors of the environmental impact generated from the packaging and creation of one typical, overly designed basic product.[76] Another, "Is Your Hi-Fi Out of Fashion? Is Your Mink Coat Obsolete?," began with a quotation from degrowth economist Ezra Mishan's book *Making the World Safe for Pornography* (1973): "Just as rapid technological innovations entail unforeseeable ecological and health hazards, so also can rapid cultural innovation produce unforeseen social hazards," before embarking on an intriguing mini-essay on the sociology of consumption.[77] "High-Fidelity systems and high-fashion fur coats are similar in many ways," warned Papanek in his recognizably sardonic fashion. "While it may be somewhat difficult to listen to an ocelot coat, and the warmth generated by a solid-state amplifier is negligible, they are alike in how they are acquired, maintained, disposed of and, eventually, replaced by newer versions."[78] At first glance Mishan, author of *The Costs of Economic Growth* (1966) and an early critic of expansionist economics whose de facto thesis calling decommodifying a source of well-being preempted the rise of the environmental movement, is a predictable choice of reference for Papanek and Hennessey's critique of the over-technologized culture of design. Yet the citation is extracted from one of Mishan's more controversial works, in which he argued that economic and technological growth led to unchecked hedonism and permissiveness (generating cultural innovations such as psychedelic drug use and pornography) as opposed to furthering the welfare of humanity. In light of Papanek's harried departure from and public critique of what he considered to be Denmark's lazy, Marxist romantic hedonists, four years prior to the release of *How Things Don't Work*, however, the reference is a subtle attack on the type of ideological design activism the social designer had, once and for all, left behind.

Instead, practical solutions such as the pooling of lawn mowers between neighbors (promoting good community relations and alleviating the need for duplication of rarely used machinery), along with the advocacy of adaptable, multifunction designed vehicles (the ambulance-cum-fire engine) formed the basis of a humorous book that, according to the promotion spiel on the dust jacket, offered an alternative to a dysfunctional product world: "Poking fun at the nonsense and near-nonsense products that engulf us, Victor Papanek and James Hennessey pose the fundamental question whether some products are needed at all and suggest alternative ways of making and owning things."[79]

Most significantly, the newly honed pragmatism of his latest publication tied Papanek to the rising alternative technology and environmental movement and its "prophet" and "secular guru" British economist and German-émigré E. F. Schumacher, who had died the year of *How Things Don't Work*'s release.[80] Schumacher, author of *Small Is Beautiful: A Study of Economics as If People Mattered* (1973), himself a student of political scientist Austrian-émigré Leonard Kohr (author of *Development Without Aid: The Translucent Society*, 1979), stood as the key figure of the burgeoning degrowth, decentralization, and intermediate technology movement.

Papanek's own work had started to be included in the canon of intermediate technology at least as early as 1974, during his time at Manchester Polytechnic. His essay "Areas of Attack for Responsible Design" was included in the anthology *Man-Made Futures*, which featured leading voices including British socialist politician Tony Benn, US professors architectural theorist Christopher Alexander and sociologist Daniel Bell, and Ivan Illich, author of the highly influential books *De-schooling Society* and *Tools for Conviviality*.[81] The title, *Man-Made Futures*, was inadvertently revealing, given its exclusively male list of contributors and the entire avoidance of feminist discourse and gender analysis around development economics, degrowth, and technology.[82]

How Things Don't Work, which according to Hennessey was the least successful book ever published by Pantheon, marked the end of Papanek's decade-long reign at the forefront of design activism. One particularly influential reviewer, the former editor-in-chief of *I.D.*, J. Roger Guilfoyle, described Papanek in the damning article "Still Naive But More Reasonable and Responsible" as "glib and exasperating, but his heart is in the right place in his controversial new book."[83] Observing that the authors had simply "grafted the do-it-yourself approach of *Nomadic Furniture* to the intellectual construct of *Design for the Real World*," the high-profile review appeared in *I.D.*, the industry's mouthpiece that had regularly featured Papanek's own articles throughout the earlier stages of his career. It concluded condemningly, and somewhat sardonically, that the book would have "benefited greatly if the positive language of *Nomadic Furniture* could have muted the tone of *Design for the Real World*."[84] More than just a stinking review, the piece read as an editorial on Papanek's life's work from the perspective of an embittered US design establishment. Reluctantly acknowledging the success of Papanek's first sole-authored book in Europe, it then dwelt on his notoriety closer to home: "It is difficult in retrospect to determine what enraged the profession more: Papanek's temerity in criticizing the design quality of products or the fact that he was one of its own," the review noted, before poking fun at some of the well-meaning suggestions for good design put forward in the pages of *How Things Don't Work*.[85] "How about a bathtub foam-lined for safety?" it teased, before commenting on the author's "peculiar blindness to basic human nature" because "like safety airbags in cars—it's a great idea but nobody wants them."[86]

After a decade-long stint as premier design activist, the lukewarm reception of *How Things Don't Work* marked the beginning of an era in which Papanek increasingly turned to the anthropological, holistic, and spiritual aspects of design as an antidote to problems of spreading globalization, and newly emerging questions of cultural identity and consumer culture. And yet, quirky and idealistic as it may be, his penultimate work holds an enormous resonance in its call for the abandoning of ownership in favor of collectivized share schemes, preempting the present day sharing economy by decades.

A decade and a half after the first publication of *Design for the Real World,* the currency of Papanek's ideas were waning, with one leading design journalist of the 1980s writing off the social designer as nothing more than "a cult figure while ecology was fashionable during the early seventies."[1] Papanek's arch critic, Stephen Bayley, was a posturing, designer-suit-wearing figure who cast himself as the key advocate of 1980s' designer-led consumerism and middle-class taste-making (promoting costly "design icons" from Brompton fold-up bicycles to Alessi's notoriously nonfunctional Philippe Starck lemon squeezer). Affiliated with the gentrification project of London's Docklands area (the controversial development was widely accused of culture washing through the large-scale displacement of working-class communities), Bayley took up the post of director of the privately owned Design Museum when it first opened there in 1989. Set up in the fashionably postindustrial setting of a former banana warehouse, the new museum represented a movement toward a style-over-substance model of design that Papanek and his followers had spent decades challenging. The Design Museum, the first of its kind, was bankrolled by the Conran design retail empire; the institution's major purpose was to promote good-taste commercial design to the general public in the setting of a flagship "yuppie" neighborhood comprising high-end restaurants and river-view apartments.[2] As flag-bearer for the all-out consumer culture and free-market economy that arrived with Thatcherism in the 1980s (the UK prime minister Margaret Thatcher famously declaring "there was no such thing as society"), Bayley came to personify the deliberate jettisoning of the previous decade's politicized design idealism. "Green design," he scorned, "is a foolish idea, something created for and by journalists."[3] Regularly appearing in glossy magazines touting the latest status symbols and must-have designer items, the yuppie design guru and cult-object worshipper might easily have stepped from the pages of *Design for the Real World* as a Papanekian parody of some future design dystopia.

One of the principal tenets of Papanek's groundbreaking book of the 1970s had been that "design must be independent of concern of the gross national product" to facilitate its socially responsible and accountable role in society.[4] Nothing could be further removed from the subsequent decade's model of design culture in which the individual quest for the accumulation of stuff, at any cost, was viewed as the key driver of the economy and individual well-being. As Thatcher herself put it in her opening speech for the inauguration of the Design Museum in 1989: "The things we buy and the jobs we do are really the essence of the life in which we live, and so more and more a part of the sense of community, and that we wish in fact to enjoy the things we buy and know more and more about them."[5] After praising the burgeoning privatization of museums and galleries sector more generally, Thatcher went on to describe the project as an "Exhibition Centre and not a museum," making patently clear that commerce, not historical knowledge, education, or critical understanding, occupied center stage in the new society-less paradigm. Buying things, not societal relationships, was the glue that would bind people together.

Taking on the Designer Decade: The Death Knell of Design Activism

Prior to this onslaught, as a regular writer for the British design press, Papanek had attempted (in opposition to figures such as Bayley) to take up the mantle of critic of "designer" culture with a newly honed conviction, eager to reclaim the enormous popularity his ideas had enjoyed in the 1970s. The October 1980 issue of *Design*—the highly regarded and widely read journal of the British Design Council—even featured articles by Bayley and Papanek alongside each other, perfectly encapsulating the contradictions of an emerging "designer decade." While Bayley sycophantically praised the luxury Swedish automaker Saab for its ingeniousness in steering the market toward recognition of its new high-end brand identity, by contrast, affecting his typical sardonic aplomb, Papanek ripped into the superficiality of fashion designer labels and their origins in the environmentally disastrous cult of the status-seeking automobile: "I went into a Gucci shop the other day," he recounted. "The luggage, handbags, jewellery and shoes were quite impressive. So were the belts; fine leather, careful handstitching, superb craftsmanship. The prices are absurd of course," he continued, launching into an apocryphal tale, "but the quality is unusually high as well. One of those professionally cheerful young women employed by the shop showed me some pigskin belts. 'I really like them,' I said, 'but do you have one with [a] plain buckle.' The place went quiet, as if I had sneezed in church. But you see I really don't want a belt buckle the consists of two intertwined letters G. . . . My initials don't match. So, in a manner of speaking, I was thrown out of Gucci's."[6]

Sounding somewhat like a disparaging parent, at a loss to understand the proclivities of a younger generation, Papanek proceeds to blame the "name game" phenomenon as a throwback to the machinations of the postwar US automobile industry that he and fellow American consumer critic Vance Packard lambasted in the early 1960s: "Then there are 'designer' jeans; Fiorucci, Klein, Vanderbilt, even, from Australia, Bogarts. Why anyone would pay up to £30 extra for perfectly ordinary jeans just for the sake of some slight variations in the stitching on hip pocket and name tag, must be counted one of the great mysteries of the age," he ranted. "Like much else, I really think it began with automobiles. Most cars have small chrome-plated designers' labels just like jeans do. They say things like 'Air-Conditioned by Carrier,' 'Fuel-Injected,' 'Cruise-O-Magic,' 'Five-speed,' and 'Body by Fisher.' These are the signs that let the folks in the car behind you know that you have spent more on your gas-guzzler than they have on theirs."[7]

In the 1980s, applying his trademark pop-sociology style, Packard, like Papanek, also continued to tackle the corrosive effects of an exploded consumer culture. His contribution to the discourse came in the form of two polemical works: *Our Endangered Children: Growing Up in a Changing World* (1983), which warned how the US preoccupation with sex, money, status, and power put the real needs of future generations in jeopardy; and the follow-up volume *The Ultra Rich: How Much Is Too Much?* (1989), a semiethnographic exposé on the excesses of real-life multimillionaires.

Both Papanek and Packard, consumer critics belonging to a previous generation, decried the rise of the wanton and wasteful consumerism they had so passionately sought to warn their audiences of over the course of the previous two decades. Reiterating with renewed urgency the moral and ethical imperative of curbing the rise of corporate and advertising power, as manifest in 1980s ubiquity of the logo, Papanek lamented the loss of authenticity. "Can it be that we no longer expect, or even value, decent materials, natural colours, simple yet elegant design, and good craftsmanship?" he pleaded. "Do we look for decoration, in the forlorn hope that the emptiness of what we own will somehow become magically transformed if we print the logo of some corporation, the cynical advertising slogan of a company, or the facile smile of some syndicated [McDonalds'] clown across our belongings?"[8]

Having apparently fallen out of fashion as a "cult figure" by the early 1980s, Papanek cut an increasingly lonely figure representing an increasingly redundant analog, human-needs approach to design emphasizing inclusivity through ergonomics and the moral necessity of manufacturers to provide easy-to-read diagrams and instruction manuals.[9] In the age of late-capitalist post-Fordism and the postmodern mentality, the modes of design Papanek had dedicated his career to policing had largely dematerialized, reemerging in the form of entities such as brand and identity management and design consultancies. As a globalized phenomenon, "the neoliberal object" of design no longer adhered to the fixed linearity of design, production, advertising, and consumption on which most of Papanek's "design for the real world" paradigms relied. The shift from manufacturing to a service-based economy manifestly shifted the priorities of designers.[10]

Coca-Colonialism, McDonaldization, and the Cultural Object: The Anti-Consumerist Legacy

Yet the anticonsumerist rhetoric the design critic had forged over a decade aptly paved the way for a new wave of critical cultural studies focused on the degenerative effects of American culture in the form of "McDonaldization," and what Papanek described as "Coca-Colonialism."[11] Buoyed by a surge of postmodern interest in the objects and meanings of consumer society, from shopping malls to brand identities, quite unexpectedly Papanek's ideas found a new lease on life in the Scandinavian country that had first embraced him. His treatise on a better future through social design, titled *Miljön och Miljonerna: Design som Tjänst eller Förtjänst? (Environment and Millions: Design for Service or Profit?)* had been published in 1970. More than a decade on, a team of Swedish filmmakers traveled to Kansas, tasked with making a feature-length documentary, titled *Jorden vi ärvde (The Earth We Inherit)*, which followed a day in the working life of the renowned design critic and academic.[12] The extraordinarily compelling film opens with a scene of Papanek broadcasting his critical design show, *By Accident or Design,* on Kansas City Radio, earnestly instructing the listening audience on the hazards of uninformed consumer behavior, design's role in the impending environmental crisis, and the continued rise in industrial pollution. Along with the documentary, a transcript of the show reveals how Papanek had lost none of the hubris or acerbic bite he had displayed in his earliest lectures delivered to 1960s design activists: "Times are tough. Money is short and people are out of work," he exhorted. "We're in a recession. And still we buy and consume useless trash. We listen to glittering commercials on TV and absorbedly read 4-color advertisements in magazines, and then go out and spend. We buy things that don't work, for money we haven't got to impress others who don't care."[13]

Yet, unlike his early design politicking with the impassioned students and designers of Scandinavia, Papanek's rhetoric now took on the tone of American self-help books, providing a practical six-step guide to avoiding gimmickry. Step 2 posed the question "Does each product feature have a purpose?" A warning promptly followed: "Electric peanut-butter [making] machines (runaway best sellers during the 1975 Christmas shopping spree), automatic apple corers and similar devices are silly, although their essential bankruptcy is disguised by topnotch industrial design."[14]

Despite his more accessible public face, none of the seriousness of Papanek's design pedagogy had dissipated. The film offers a rare insight into the social designer's interactive studio teaching methods, with his Kansas design students shown integrating ethnographic objects into their contemporary design thinking, and taking part in "bisociation creative iteration" exercises inspired by Austrian émigré writer Arthur Koestler's 1964 seminal work *The Act of Creation*, which Papanek much admired. Interspliced with dramatized footage of fast-food waste (polystyrene cartons piled one on top of the other until the heap topples over), the documentary then turns from the studio to a fly-on-the-wall view of the academic's doomed trip to the local shopping mall: his mission, to find a minimally packaged, non-synthetic shirt. The film culminates in the intimacy of Papanek's own Kansas living room, where he is shown half-naked (standing against a backdrop of African and Inuit masks), clumsily unpacking the shirt from its reams of polythene wrapping. He then tries on the garment while addressing the camera face on, explaining that over-packaging is not a phenomenon unique to North America. "I've wrestled with shirts in Sweden, I've fought shirts in Indonesia and Africa," Papanek tells the viewers before offering a solution: "It's true all over the world. What we have to look for as end-users and consumers are things that are well designed, beautiful and pleasant to handle, and perform their function very, very well. And it's up to us; to the people who design and to the people who buy things, to assert our will in the marketplace and make sure that the things that we are offered are really very nice."[15]

Whether or not the social designer intended to offer so anodyne a conclusion to a feature documentary exploring his critical environmental stance, ultimately his final statement condoned a simplistic free-market paradigm in which humankind (as consumers) controlled its own fate through the implementation of rational, instrumental choice. What the documentary reveals, however, is Papanek's continued, or renewed resonance in a Scandinavian nation whose celebrated postwar welfare structure was threatened by the specter of American consumerism, and global monetization policies set on market deregulation, denationalization, the boosting of consumer spending, and the paring back of previously collectively owned social resources.

If the overpackaged, synthetic shirt vignette failed to impress his followers, during the same period of the early 1980s Papanek began to engage with the increasingly popular field of material culture studies within anthropology, art history, and philosophy.

"The Cultural Object," an essay written for an innovative 1980s journal exploring architecture in the developing world, offered far more sophisticated insight into the subtle shift in Papanek's thinking around design and material culture during this period.[16] Drawing on the work of philosopher Hannah Arendt and art historian George Kubler, the extended article abandoned the reductionist dualism of the West and undeveloped nations on which so much of his former design criticism and practice had relied. Instead, Papanek embraced an expanded understanding of designed things as socially embedded entities, and the objectification of the social relations that bound people together: "Objects are made and then projected into the real world. Their profound fixity stabilizes life. The ephemeral is made visible through the built object—tables, hoes, beds, cups or fish hooks."[17] Status seeking, fashion, and style are shown to be cross-cultural, transhistorical phenomena rather than merely the instruments of a dysfunctional Western consumer culture gone awry, with just a variation in the nature of their temporality. "Fashion and "style" exist in *all* societies," he conceded. "They are subject to the same pendulum swings and cyclical changes as in technologically advanced countries. The pendulum merely swings more slowly."[18] Peppered with vignettes illustrating the appropriation of solid, holistically meaningful Indigenous artifacts by a market economy driven by profit, the essay extended the "real world" polemic into a broader critique of modernity. For example, traditional Guatemalan clay water jars had been replaced by more expensive, yet inferior, mass-produced Taiwanese synthetic versions: "In the [plastic] containers, water is tepid and unappetizing—some authorities think that these containers also make water carcinogenic. The Taiwanese vulgarities sell for about (US) \$8.50 each. It is a sad comment on the loss of cultural pride, engendered by world wide Coca-Colonialism, that most Guatemalan farmers hope to sell enough of their lovely clay-vessels to enable them to buy one plastic pot."[19]

This search for a remedy to the loss of authenticity in the face of modernity's onset contained in the "The Cultural Object" essay was uncannily redolent of the basic premise of Bernhard Rudofsky's semi-anthropological 1944 MoMA exhibition *Are Clothes Modern?* that had so heavily influenced Papanek in his formative years as a young émigré designer in New York establishing his first practice, the Design Clinic. Perhaps, in some ways, at this point in his life Papanek had begun to revisit his earliest attempts at making sense of design's social and cultural relevance beyond the aesthetic and rationalist paradigms of modernism. Certainly, this continued attempt to make sense of the human dimensions of design inspired his first, and last, major writing project of the 1980s.

Design for Human Scale

Design for Human Scale (see figure 10.1), Papanek's second sole-authored treatise on design of 1984, has remained a little-known and scarcely cited contribution to the field. An extension of *Design for the Real World*'s attempt to challenge Western definitions and understandings of design, it drew on his work with students in the UK, Denmark, and the Kansas City Art Institute, where he had been based in the years prior to its publication. Now that he was a professor in the School of Architecture, University of Kansas, the dust jacket promotional blurb shifted emphasis from his work as a designer to that of a transdisciplinary pedagogue who came to "industrial design from the disciplines of architecture and anthropology."[20] His profile constructed him as a player in a newly globalized design field, with positions as a "permanent Senior Design Consultant" at the World Health Organization, "Volvo in Sweden (developing a taxi for the handicapped and work enrichment programs for Volvo's works)," and, perhaps most surprising of all, "the government of Papua New Guinea."[21] In 1982, his life's commitment to challenging the corporate mechanisms and profit-driven motives of

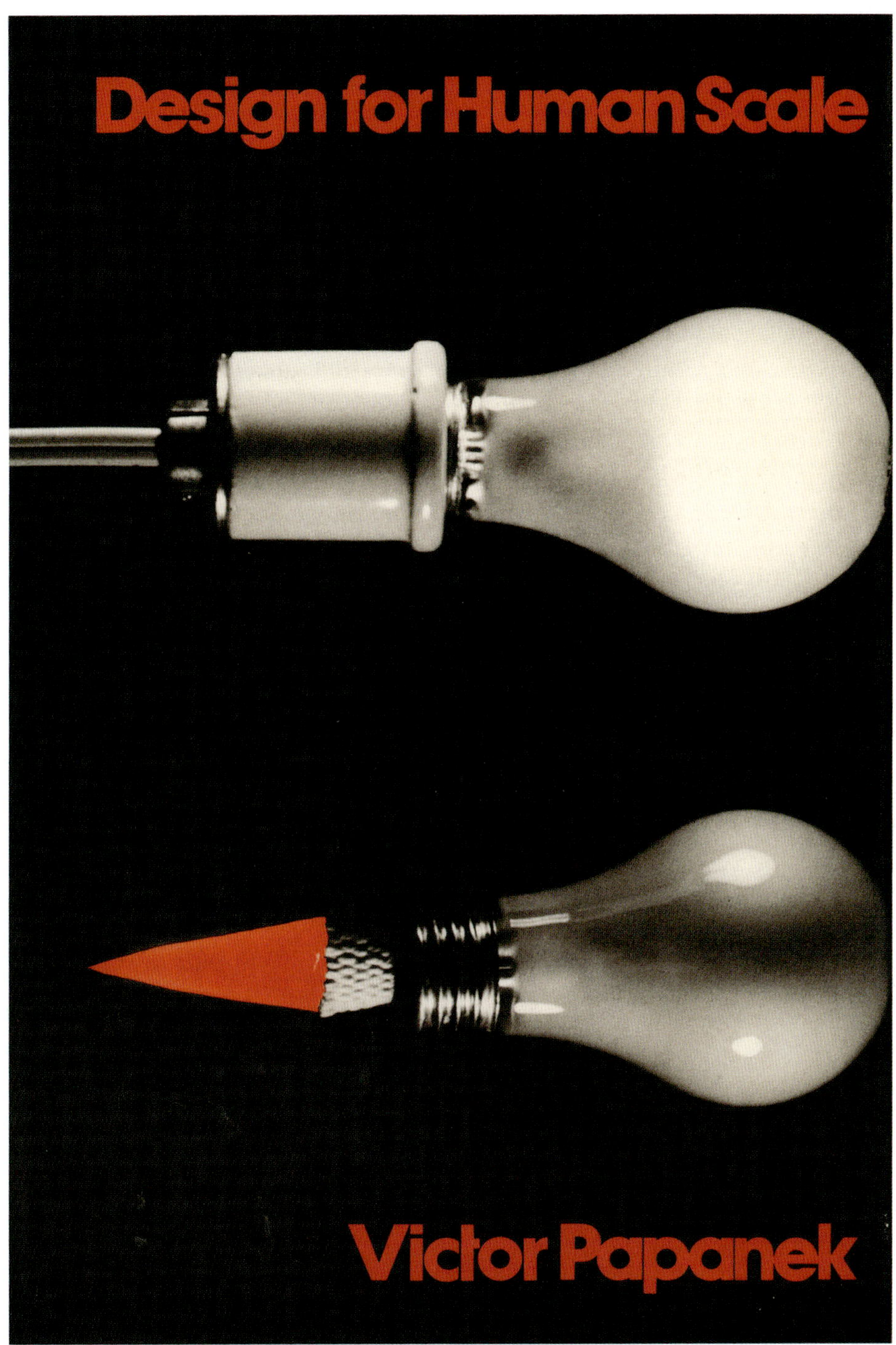

Figure 10.1

Victor Papanek, *Design for Human Scale* (New York: Van Nostrand Reinhold, 1984), front cover. Courtesy Victor J. Papanek Foundation, University of Applied Arts Vienna

the industrial design industry brought him the prestigious nomination for the *The Right Livelihood Award*—commonly known as the "Alternative Nobel Prize." First initiated in 1980 by a Swedish philanthropist, the award is bestowed in recognition of social and ecological innovation, challenging the narrow remit of the conventional Nobel Prize and its preoccupation with the industrialized nations, and as such was an enormous honor for Papanek to have been nominated for.

While *Design for the Real World*, the *Nomadic Furniture* editions, and *How Things Don't Work* had preempted a shift within design toward environmentalism and sustainability with their rejection of conventional consumer culture and object obsolescence, and the advocacy of self-made design, systems of shared ownership, and recycling, none addressed green design head-on. But by the 1980s, as ecological design and ethical consumerism moved firmly into the mainstream, *Design for the Real World* reemerged as a classic of its time rather than a radical intervention into the field. *Design for the Human Scale*, in contrast, showed Papanek's increasing concern with "ecology, ethology, and the environment" and broader interactions between design practice and ecology that overtly drew on the work of E. F. Schumacher and the appropriate and alternative ecology movements.[22]

Seeking a remedy to the alienation of modernity, Papanek gradually moved his analysis away from anticonsumerism toward a broader discourse of human scale and the sensorial realm (drawing from hermeneutics and phenomenology) as means of humanizing the designed world. From his position within a school of architecture, this newly honed approach turned to broader questions of dwelling and planning. Working toward a holistic understanding of design practice, Papanek described the "biotechnology of communities" (a network of complex intersecting factors from the environmental to the psychosociological) as a means of counteracting the contemporary shift to making "design process more systematic, scientific, and predictable, as well as computer-compatible."[23] Incorporating the field of biogenetics, he argued against the idea of designers and architects as rational planners working with a tabula rasa: "Just as birds choose the ideal location for their nests with their strongly developed siting instincts and without help of design consultants, so do certain human groups."[24]

In response to the shifting sands of the social design terrain, Papanek's principal polemical standpoint in the 1980s became the acknowledgment of an ever-greater need for the humanization of design. His ideas still held sway with the Danish design fraternity to the point that in the early 1980s, he was invited by the editors of *Mobilia* to provide an extensive overview of contemporary design from his expert perspective. Under the subtitle "Humanizing Design," Papanek reflected on the increasing difficulties of codesign in an era in which "the dividing line between tangible and intangible products and services sometimes appears blurred," making "the social, societal and individual problems of end users . . . difficult to work with."[25] Headings, such as "Self-Reliance," "Design Participation," and "Design Decentralization" promised to deliver commentary of contemporary issues of dematerialized design in a global context; ultimately, however, the polemic merely summoned up well-worn examples or resorted increasingly to quasi-anthropological analysis of objects considered to be holistic design, ranging from Papua New Guinean fish hooks to Mesoamerican decorative clay vessels from his private ethnographic collection.

Indeed, anachronistic as it seemed, Papanek's iconic early 1970s "designs for development," originally forged in the heat of Cold War US product design experimentation, were still accruing favorable attention. In 1981 the designer was awarded the ICSID Kyoto Honors Award for his *Batta-Kōya* communications device for Tanzania and Nigeria—the type of device condemned, as far back as a 1974, by fellow designer for the peripheral economies Gui Bonsiepe as morally repugnant objects of US military sponsorship and neocolonial intent. In 1983, Papanek went on to receive an award for "Outstanding Design for Developing Nations" from UNESCO. While both awards offered retrospective acknowledgment for contributions Papanek made in the 1970s, they also signaled his unmitigated acceptance into the fold of the design establishment, and in so doing, his waning relevance to a new generation of designers. This is not to suggest that demand for Papanek's genre of design criticism entirely disappeared in the later decades of his career. On the contrary, correspondence reveals a steady flow of offers to lecture internationally (including in New Zealand, Japan, Czechoslovakia, and Portugal) and regular invitations to act as a design award jury member.[26] His work also reached new audiences, and even new economic and political regimes.

The Soviet Connection

The publication of *Design for the Human Scale* coincided with the unanticipated success of Papanek's work in the last years of the Soviet Union. Papanek had befriended Yuri Soloviev, a leading Russian figure within ICSID who served as vice president with the council from 1969 to 1973. Founder of the Soviet Design Council, the All-Union Scientific Research Institute of Industrial Design (VNIITE), Soloviev first met Papanek during the social designer's involvement with the ICSID Working Group IV of design and development in the 1970s. The 1975 ICSID congress took place in Moscow, bringing the two designers further into contact. Despite this active connection with VNIITE and the Soviet design community, Papanek's first publication in the socialist state came about under the auspices of a piece of orchestrated pro-US propaganda.

In 1983, an edition of the Russian-language journal *Amerika* (*Америка*), which had been published by the US Department of State as a glossy and fully illustrated propaganda vehicle since the start of the Cold War in the 1940s, ran a feature on *Design for the Real World* and the ideas formulated in Papanek's most recently released book.[27] The article's most notable aspect was its overt appeal to environmentalist concerns. "The main design challenge is to establish a link, as tight as possible," wrote Papanek, "between design and the need of humankind. . . . Designers pay too little attention to such social problems as environmental pollution, depletion of natural resources, and the fate of objects after the expiry of their functional life."[28]

Just three years prior to the breakup of the Soviet Union in 1988, Papanek was invited by Soviet designers to contribute to *Tekhnicheskaya Estetika*, the journal of the Soviet Design Council, VNIITE. Under the title "Ecological Design: Quest, Results," Papanek reiterated the environmental imperative of design in a stronger tone than ever before: "Many contemporary designers like polystyrene that can be used, for example, in the manufacture of low-cost disposable coffee cups. However, polystyrene directly leads to the expansion of the ozone-layer hole over Antarctica, causing a sharp increase in ultraviolet radiation throughout the globe, which consequently leads to the increase in the number of skin and blood cancers and various congenital anomalies." Papanek emphasized that "this should not be taken as a warning call to *inaction*. Rather, I'm trying to convince designers that they will always be facing different choices in their work, and their decisions will be leading to the most extensive environmental consequences."[29]

This article, aimed at a new generation of designers searching for a socially relevant ethos in the changing climate of the Soviet Union, also marked a turn for Papanek directly toward ecological and environmental concerns, which would form the focus of his final book, published shortly after his divorce from his fifth wife, Harlanne, and three years before his death in Kansas, following prolonged illness, in 1998.

The Green Imperative: The Last Stand

It was *The Green Imperative: Natural Design for the Real World*—published in 1995, twenty-five years after the release of *Design for the Real World*—that stood as a testament to the author's lifelong attempts to champion socially responsible design, through the shifting of his attention to the contemporary issues of environmentalism. The book appeared in the UK the same year with an alternative subtitle to the US version, *Ecology and Ethics in Design and Architecture*, clearly pitching to a European audience more familiar with the intersections of ecology and design practice. Written over a period of four years during fellowships and residencies in locations as diverse as Bali, the Schumacher College in the United Kingdom, and rural Spain, it was supported by a National Endowment for the Arts "Distinguished Designer Award" and an "Outstanding Design Award" from the IKEA Foundation in the Netherlands—a far cry from the ostracism he had experienced from the mainstream industrial design fraternity following the publication of his debut book, *Design for the Real World*. Based ostensibly on his lifelong collection of teaching materials, lectures, and articles, and formulated over the period of 1984 to 1989, the original title was drawn from the acceptance speech he delivered in 1986 after receiving an honorary doctorate at the University of Zagreb in recognition of his work "humanizing technology through industrial design."[30] Papanek's original intention for his next sole-authored book had been to include the more ephemeral aspects of his design journalistic work, along with his Zagreb speech, to address the unabated, market-driven environmental destruction of the world's resources.

Instead, *The Green Imperative,* Papanek's last foray into the politics of design, proved to be a more developed book project. It began life as an unpublished, incomplete manuscript titled "The Edifice Complex: Thoughts on Architecture and Design in an Age of Greed." A play on "Oedipus Complex," it made direct reference to the contemporary currency of the term *edifice complex*, coined to describe politician and former First Lady of the Phillipines Imelda Marcos's predilection for appropriating public funds to finance architectural showpiece projects as a form of political propaganda. The title "Edifice Complex" was transmuted into *The Green Imperative* after the publisher, Thames and Hudson, recognized the resonance of Papanek's ideas with those of newly reinvigorated ecological and green politics. Ambitious in its scope, the original plan for the book was to have environmentalist James Lovelock (famed theorist of the since-disputed Gaia hypothesis of terrestrial self-regulation) provide the foreword but, sadly for Papanek, he declined. Whereas the first edition of *Design for the Real World* had benefited immeasurably from design maverick Buckminster Fuller's specially commissioned introduction, Papanek's final polemic was published without the benefit of celebrity expert endorsement.

Whether or not Papanek envisaged *The Green Imperative* as his final book is unclear. But the work itself was clearly conceived as the culmination of his lifelong interest in the transformation of design from a practice of corporate profiteering to one of humanitarian significance. Covering topics that had been the mainstay of Papanek's work since the 1970s—including the anthropological, spiritual, and cultural aspects of design as well as the significance of biomorphics and the natural world—*The Green Imperative* also revealed the changing emphasis within the design industry itself, which, in the decades since Papanek's first book, finally acknowledged the prescience of environmental and social issues.

Although the book retained the anticonsumerist vitriol of Papanek's earlier work, its major premise centered on scaling back design to a spiritualized, joyful process that would nurture intuitive connections to the environment, communities, and their values rather than stoking unbridled consumption.

Some critics, like US designer Ken Isaacs (an activist and early pioneer of self-build environments whose work had heavily influenced Papanek and Hennessey's *Nomadic Furniture*), could barely conceal their contempt for the hippie-hangover idealism. "Does one arrive at the spiritual life by meditating in the check-out lane with a Swatch (while wishing for a Rolex)?" Isaacs scoffed in his scathing review of the book, before concluding that the intellectual naivete and untimeliness of Papanek's approach was tantamount to "whistling in the cemetery."[31] Despite its eminently marketable title, *The Green Imperative*'s optimistic treatise on design's capacity to effect social transformation ultimately proved as anachronistic as its critics had predicted. The indelibly blurred distinctions between Western and Indigenous cultures that defined the über-globalized worldview of the mid-1990s made Papanek's borrowing of native authenticity look suspiciously like old-school cultural appropriation.

Toward the end of his life and career, and just prior to the publication of *The Green Imperative*, Papanek started to embrace a spiritualized version of the Schumacher "small is beautiful" philosophy that emerged with the establishment of the Schumacher College for alternative economics in Devon, England, in 1990. In 1992, he contributed the article "Sensing Dwelling" to its journal, *Resurgence*.[32] Far from pragmatic, the designer had at this point turned his hand to considering the "non-dimensional" aspects of design, arguing "we can taste, touch, smell, and see and listen to a building and derive joy from it when it is built with an inborn sense of harmony."[33]

Ultimately, however, this approach proved anachronistic to a new generation of readers. Its rhetoric belonged to a more idealist era of theoretical debate that predated the rise of the designer superstar and omnipresent brand culture redolent of the 1980s and early 1990s. As such, it received an ambivalent reception when it appeared in 1995. Papanek's well-worn meditations on the holistic design of authentic cultures, from Inuit ice dwellings to Mongolian yurts, coupled with warnings of the detrimental effects of mass marketing, struck a discordant note in an era in which aggressive free-market culture was a stark reality.

Set within the context of a mid-1990s design criticism increasingly looking to nonobjects, interactions, and software design—rather than traditional manufactured product design and architecture—the didactic tone and content of *The Green Imperative* fell on deaf ears. The battles it described had already been lost; designers were equipping themselves with alternative theoretical apparatus to manage an ever more complex design environment, reliant on new interactive technologies and less visible sources of political power. Nevertheless, *The Green Imperative* consolidated Victor Papanek's international reputation as a key proponent of an alternative, non-market-driven, and critical approach to design. In the year of its publication, 1995, Papanek became the recipient of the International Lewis Mumford Award for Development.

And yet certain aspects of the social designer's last testament remain pertinent: the exploration of bionics, the role designers play in accelerating global climate change, and the pressing question of what form design's socioenvironmental accountability might take. Moreover, design remains, as Papanek asserted in the heyday of radical politics, a force as likely to inflict harm as to remedy it. Tasked with shaping our material and immaterial existence, should designers be held answerable to broader humanitarian goals, or merely to the whims of their techno-ideological paymasters? Just how far does their responsibility extend to the after effects of their work, be it a driverless vehicle or a disposable diaper? Although the global efficacy of much of contemporary social design is debatable, Papanek's clarion call to action and the feverish polemicizing of his ideas are still prescient. Recent developments in the decolonizing design and transition design movements as well as the normalization of shared-ownership and rental schemes around a range of commodities from lawnmowers to handbags, are all premised on debates engendered in Papanek's collectivized notion of design. After more than half a century of readership, the English-language version of *Design for the Real World* has never fallen out of print, and new language translations (including Ukranian, Catalan, and Mandarin) are constantly initiated.

Prior to his death in 1998, Victor Papanek, ever dedicated to the figure who first inspired a young Viennese refugee's interest in the social aspects of architecture and design, requested that his ashes be scattered at Frank Lloyd Wright's Taliesin studio in Spring Green, Wisconsin.[34]

Concluding Observations

Victor Papanek's life, ideas, and his widely read and cited publication, *Design for the Real World: Human Ecology and Social Change*, remain totemic within a socially responsible design movement that has come to impact large swaths of contemporary design practice. Yet, the historical and geopolitical origins of many of these ideas remain, on the whole, critically unexplored within design. As twenty-first-century social designers in the Euro-American context adopt quasi-anthropological techniques, identifying subjects worthy of their intervention (from newly arrived refugees to the vulnerable elderly in nursing homes), they form part of the design industry's neoliberal role as facilitators of social services in the context of declining state provision. The term *social* is all too often left unproblematized, and the real politics of design's interventions into the lives of potential users and its troubled relation to the Global North's project of modernity and development, is left critically unexamined on the macrolevel.

Through the conduit of biography, the tracing of the career, design work, writings, teachings, cultural and intellectual networks, and personal life of its subject, this monograph has sought to reintroduce historical complexity into the origins of the social design movement. As a refugee outsider traumatically cut loose from his life in bourgeois Vienna, Papanek's early observations of a full-blown consumer culture led to his engagement with some of the most radical ideas and experiments around social and racial exclusion that could be found in 1940s North America. The early exposure to the popular culture critiques of community psychiatrist Fredric Wertham and his condemnation of violent racial stereotypes in comic books, as well as the pioneering social psychiatry of the Lafargue Clinic, supported by and serving African American communities in Harlem, shaped the would-be social designer's early critiques of the discontents of free-market-driven consumer culture. Immersed in the creative crucible of 1940s New York City, Papanek mapped these ideas onto the practice of design as a healer of social ills, with the establishment of his first studio, Design Clinic, in 1946; he already envisaged a human-centered mode of design that countered the industrial rationalism of his modernist forbearers.

Yet, by the late postwar period, far from being just an idealistic offshoot of progressive counterculture activism, the new emphasis on the social dimensions of design and the primacy of the local user was inextricably tied to shifting postwar geopolitics of decolonialization, US military interventions, and the construction of alternative antisocialist paradigms of welfare and development as a bulwark against communism. That Papanek's work was intricately interwoven into theses indices of Cold War politics—meaning he was simultaneously overseeing transdisciplinary graduate research groups bankrolled by the US military, while travelling across Europe codesigning sensory environments for disabled children and delivering antigovernment and antimilitary speeches as an advocate of 1960s and 1970s design activism—does not negate the significance of the discourse he came to represent. On the contrary, his works (many coproduced with students or local groups and users), and contemporary responses to them, allow insight into the inherently contradictory discourses and renderings around the notion of the social as manifest in industrial design in postwar Euro-American culture. Unlike his ally, architect and techno-maverick Buckminster Fuller, Papanek did not flaunt his connections to corporate, government, and military sponsorship; in fact, these collaborations remained largely unacknowledged during his lifetime. For it was of course these connections that sustained the broader practice of transdisciplinary and humanitarian design that he came to advocate, with designs such as the celebrated dung-powered *Tin Can Radio* for nonliterate communities of the Global South operating as both socially inclusive mechanisms of media dissemination in the new "global village" and as efficient dissemination tools for US military propaganda.

Perhaps perversely, *Design for the Real World*, and Papanek himself, acted as shorthand for an approach to design that respected and incorporated Indigenous cultures and otherwise excluded groups—among them he identified Black Americans, people of color, women, children, the disabled, the working classes, people of the Global South, and the white rural poor. Yet, a crucial question remains pertinent: To what extent did the invisible labor and cultural appropriation of such groups generate the contents of *Design for the Real World* and create Papanek's reputation as the premier social designer of the twentieth century? In 1985, on the occasion of the re-release of the second edition of the seminal work, one reviewer aptly summed up the shift in sentiment toward design for development with the heading "Victor Papanek: The Pioneer of Patronising Design."[35]

Despite this shift in context, as a manifesto for social change through the rethinking of design's role in constituting and perpetuating social exclusion, *Design for the Real World* remains the most globally far-reaching book on design of the twentieth century. Somewhat anachronistically, the recent Mandarin Chinese translation of the book introduces the polemic, and the once radical ideas that underpin it, to the world's most rapidly modernizing and polluting nation—with a new generation of readers desperately searching for an alternative, sustainable model of benign social progress shaped through design—and so its legacy continues.

Notes

Introduction

1. Victor Papanek, preface to *Design for the Real World: Human Ecology and Social Change* (New York: Pantheon, 1971), xxi.

2. Satish Kumar, interview with Victor Papanek, Schumacher College, 1991, film, Victor J. Papanek Foundation, University of Applied Arts Vienna.

3. Papanek, *Design for the Real World*, 44–45.

4. Papanek, 44–45.

5. Papanek, 45–46; emphasis in original.

6. Victor Papanek, "What Is Contemporary?," *La Mer*, August 1950.

7. See Peter Lang and William Menking, eds., *Superstudio: Life without Objects* (Milan: Skira, 2003).

8. Alessandro Poli, cited in Lang and Menking, *Superstudio*, 226.

9. See Ross Elfine, "Superstudio and the 'Refusal to Work,'" *Design and Culture* 8, no. 1 (2016): 55–77. Elfine suggests that the radical Italians transformed the reality of mass unemployment among 1970s Italian architects and planners into an avant-gardist politicized stance of the "refusal to work" in a postindustrial, late-capitalist economy.

10. Papanek, *Design for the Real World*, xxvi.

11. See chapter 9, this volume.

12. See chapter 9.

13. Barbara Ward and René Dubois, *Only One World: The Care and Maintenance of a Small Planet* (London: Andre Deutsch, 1972).

14. Victor J. Papanek, "Design in a World Perspective," *Industrial Design* (April 1969): 84.

15. Simon Sadler, "A Container and Its Contents: Re-reading Tomás Maldonado's *Design, Nature and Revolution: Toward a Critical Ecology* (1970, trans. 1972)" (2013): 52, available at UC Berkeley eScholarship, *Room One Thousand*, https://escholarship.org/uc/item/25f1r08h.

16. Tomás Maldonado, *Design, Nature and Revolution: Toward a Critical Ecology* (New York: Harper & Row, 1970, trans. 1972), 30. Papanek's personal copy of *Design, Nature and Revolution: Toward a Critical Ecology* is held within the designer's personal library at the Victor J. Papanek Foundation, University of Applied Arts Vienna.

17. Papanek, *Design for the Real World*, xxvi.

Chapter 1

1. "Flight to the Valley of the Sun," *Harper's Bazaar* 76 (January 1942): 40–47.

2. Papanek states that he "found Mr Wright's building more attractive to me (at the time), than my girlfriend!" (author's translation). In Victor Papanek, "Mit Bleistift und Pflugschar: Erinnerungen und Anekdotisches zu FLW" ("The Pencil and the Plow: Personal Memories of Frank Lloyd Wright"), *Architektur und Bauforum* 147 (1991): 46, 46–48.

3. Lewis A. Coser, *Refugee Scholars in America: Their Impact and Their Experiences* (New Haven: Yale University Press, 1984).

4. For an example of the plotting of career intersections and émigré networks in design, see Donald Albrecht, *Designing Home: Jews and Mid-Century Modernism* (San Francisco: Contemporary Jewish Museum, 2014).

5. The compulsory questionnaire (*Fragebogen*) Helene Papanek was forced to complete as part of the National Socialist processing of Jews exiting Vienna reveals that all personal possessions, including her wedding ring, were confiscated through the assets (*Vermögen*) assessment process. The family gourmet food and wine business was confiscated without payment. The permission to leave was granted with the support of a reference given by a paper stationery business owner, situated close to the Papanek family grocery store. The cost of the administration of the forced emigration was born by the victims through a tax levy. The administrator of the Israelitische Kultusgemeinde Wien (IKG) notes in the document: "The processing of documentation for the departure of Frau Papanek, and especially her young 15 year son is <u>extremely urgent</u>" (author's translation). Shut down by the Nazis with the Anschluss of the Third Reich in 1938, the IKG was reopened to process the emigration documents for the Central Office for Jewish Emigration (*Zentralstelle für jüdische Auswanderung*) as part of a Nazis program to accelerate Jewish deportation. "Fragebogen, Fürsorge-Zentrale der Israelitische Kultusgemeinde Wien: Auswanderungsabteilung," Helene Papanek, 1010, Opernring 15, Wien, no. 51284, 10 March 1939. Austrian National State Archives, Vienna.

6. Victor Papanek and his mother Helene arrived at Ellis Island on the SS *Penland* ship from Vlissingen, Holland. Their contacts in New York City were the Strauss family; Edward Strauss was Helene's uncle on her maternal side. "Manifest of Passengers for the United States Immigrant Inspector at the Port of Arrival," New York, 3 April 1939, www.libertyellisfoundation.org.

7. Victor Papanek, "Family History," date unknown. Victor J. Papanek Foundation, University of Applied Arts Vienna. Members of the Papanek family are not recorded as "von Papanek" in any official Austrian documentation, including birth certificates, prior to their departure to the United States, despite Victor Papanek's suggestion otherwise.

8. The gourmet food and wine store was the main branch of a chain belonging to a great-uncle, Alois von Spitz-Straus. See Al Gowan, *Victor Papanek: Path of a Prophet* (Cambridge, MA: Merrimack Media, 2015).

9. For broader contextualization of the Ringstrasse and its relationship to the Jewish community, see Elana Shapira, "Moses and Hercules: Jewish Patrons Fashioning the Ringstrasse and the Prater," in the exhibition catalog *The Jewish Ringstrasse*, ed. Gabriele Fritz-Kohlbaer (Jewish Museum Vienna, March 2015), 162–188.

10. On Lazarsfeld and the politics of consumption in early twentieth-century Vienna, see Oliver Kühschelm, "(Mis) Understanding Consumption: Expertise and Consumer Policies in Vienna, 1918–1938" in *Émigré Cultures in Architecture and Design*, ed. Alison J. Clarke and Elana Shapira (London: Bloomsbury, 2017), 45–61.

11. Richard Papanek died on March 18, 1935, of unspecified heart disease and was buried in the Jewish section of the *Wiener Zentralfriedhof* with the title of "Kommerzialrat" engraved on his gravestone.

12. School report, February 13, 1937, Realgymnasium XIX Wien, Strassegasse 39, des Vereines "Österreichchisches Landerziehungshiem," catalog no. 19 (1936–1937).

13. Helene Papanek's widowed mother Maria Spitz and her sister Bertha Lederer and Bertha's husband Alfred Lederer sought refuge in Prague. But in March 1939, a year after the Anschluss of Austria, the Czech Republic was also annexed to Nazi Germany. In November 1941, Helene's mother, sister, and brother-in-law were relocated to Theresienstadt Ghetto (Terezín), where Helene's mother was murdered in September 1942; Bertha was deported to Treblinka and died there that same month; the fate of Alfred Lederer is unknown. Victor Papanek's uncle, Joseph Papanek, was held in Terezín concentration camp and murdered on September 2, 1944. I thank my colleague Elana Shapira for related research on these details.

14. Richard Papanek and Helene Spitz were married on June 3, 1919, in the synagogue in the sixth district, Schmalzhofgasse 3, Vienna. Victor Papanek's birth was registered with the Israelitic Community, Vienna (*israelitische Kultusgemeinde in Wien*), January 17, 1924.

15. The archive of the Victor J. Papanek Foundation holds a collection of Victor Papanek's empty wooden *Sachertorte* boxes.

16. Lisa Silverman, "Leopoldstadt, Judenplatz, and Beyond: Rethinking Vienna's Jewish Spaces," *East Central Europe* 42 (2015): 249.

17. Lisa Silverman, "Repossessing the Past? Property, Memory, and Austrian Jewish Narrative Histories," *Austrian Studies* 11 (2003): 138–153.

18. Silverman, "Repossessing the Past?," 140.

19. Silverman, 141.

20. M. Jeffrey Hardwick, *Mall Maker: Victor Gruen, Architect of an American Dream* (Philadelphia: University of Pennsylvania Press, 2004), 91.

21. Harriet Freidenreich, *Jewish Politics in Vienna* (Bloomington: Indiana University Press, 1991), 208–209. Freidenreich identifies the nostalgia invoked by Viennese Jews as an integral part of their émigré biographies. Also see Hardwick, *Mall Maker*, 91.

22. See M. Jeffrey Hardwick, "A Viennese Refugee and the Re-Forming of American Consumer Society," in *Economic History Yearbook (2005), 2,* ed. Jörg Baten et al. (Cologne: Akademie Verlag, 2005), 93, 89–114. See also Hardwick, *Mall Maker*.

23. Daniel Horowitz, "The Émigré as Celebrant of American Consumer Culture," in *Getting and Spending: European and American Consumer Societies in the Twentieth Century,* ed. Susan Strasser, Charles McGovern, and Matthias Judt (Cambridge: Cambridge University Press, 1998), 149–166.

24. See Hardwick, "A Viennese Refugee"; Hardwick, *Mall Maker*.

25. Jan Logemann, "European Imports? European Immigrants and the Transformation of American Consumer Culture from the 1920s to the 1960s," *Bulletin of the German Historical Institute* 52 (Spring 2013): 113.

26. The Hannah Lavanburg House, a Jewish charitable hostel, was equipped with a synagogue, a swimming pool, and a gym as well as meeting rooms and extensive other facilities. The Sixteenth Census of the United States, New York, 1940, records Papanek as a resident, with no source of employment noted.

27. Lisa Silverman, "Leopoldstadt, Judenplatz, and Beyond: Rethinking Vienna's Jewish Spaces," *East Central Europe* 42 (2015): 250.

28. The Lower East Side was described as the "capital" of Jewish America at the turn of the twentieth century. See Peter Buhle, *From the Lower East Side to Hollywood: Jews in Popular Culture* (New York: Verso, 2004), 19.

29. News Syndicate Co. Inc; The New York Times Co.; Daily Mirror Inc.; Heart Consolidated Publications Inc., *New York City Market Analysis* (1943), courtesy of CUNY Center for Urban Research, http://www.1940snewyork.com/.

30. The fair was held by President Franklin D. Roosevelt at the occasion of the 150th anniversary of the inauguration of the first president of the United States.

31. See Robert Kargon, Karen Fiss, Morris Low, and Arthur P. Morella, "Whose Modernity: Utopia and Commerce at the 1939 New York World's Fair," in *World's Fairs on the Eve of War: Science, Technology and Modernity, 1937–1942* (Pittsburgh: University of Pittsburgh Press, 2015), 57–83.

32. Kargon et al., "Whose Modernity."

33. Papanek is quoted as saying, "I would like to jump in here with a story. I once worked for Raymond Loewy in New York." From "How to Work with a Package Designer," *Canadian Packaging* (February 1957): 43.

34. The autobiography was part of a broader and sustained promotional campaign and revolved around a rags-to-riches-type émigré story despite Loewy's upper-class French origins.

35. Despite this critique of "skyscraper furniture" designers, in 1935 Paul T. Frankl's own autobiography makes mention of Raymond Loewy as a client. Frankl specifically notes his "meager" contribution of furnishings to Loewy's Palm Springs retreat: his few armchairs and a sofa designed "to add to the comfort of [Loewy's] illustrious guests watching the never-ending sunsets in the sky and the beguiling beauties in the pool." The inclusion of the small commission for Loewy appears to have been advantageous to Frankl as in the ensuing paragraph he also "name-drops" Alfred Hitchcock and Cary Grant as high-profile Hollywood clients. Christopher Long and Aurora McClain, eds., *Paul T. Frankl Autobiography* (Los Angeles: DoppelHouse Press, 2013), 128–129. See also Christopher Long, *Paul T. Frankl and Modern American Design* (New Haven, CT: Yale University Press, 2007).

36. Boris Artzybasheff, "Designer Raymond Loewy: He Streamlines the Salescurve," *Time*, October 31, 1949, cover illustration.

37. Loewy's comments were reprinted as "Aesthetics in Industry," *Architectural Review* 99 (January 1946): iv. See also Jeffrey L. Meikle, *Twentieth Century Limited: Industrial Design in America, 1925–1939* (Philadelphia: Temple University Press, 2001), 134.

38. Loewy created a single window decoration for Macy's working all night on his first day on the job that—according to his recollection—his employer disastrously misunderstood, and Loewy quit before he could be fired. See Raymond Loewy, *Never Leave Well Enough Alone* (New York: Simon and Schuster, 1951), 56–58. Loewy went on to work for Wanamaker's and Saks department stores.

39. Jake Gorst, "Designing Loewy: Elizabeth Reese Remembered," *Modernism Magazine* (Winter 2007): 55.

40. One exception is the work of Jan Logemann, who positions Loewy as a European migrant illustrating the "transatlantic connections that shaped the 'golden age' of American commercial design." Logemann, "European Imports?," 125.

41. Loewy summarized his background thus: "Father, whom I consider a reasonable prototype for a gentleman, was born in Vienna, son of a family of educators, physicians, and other members of the professions." Loewy, *Never Leave Well Enough Alone*, 18.

42. Loewy, 32.

43. Loewy, 173–174; compare Meikle, *Twentieth Century Limited*, 89–90.

44. Victor J. Papanek, "Phylogenocide: A History of the Industrial Design Profession," in *Design for the Real World: Human Ecology and Social Change* (New York: Pantheon, 1971), 30–31.

45. Papanek, "Phylogenocide," xxi.

46. Papanek, xxi.

47. Loewy, *Never Leave Well Enough Alone*, 95.

48. Loewy, 128.

49. See Andrew S. Gross, "Imagining the Interstate: Henry Miller, Post-Tourism, and the Disappearance of American Place," in *Public Space and the Ideology of Place in American Culture*, ed. Miles Orvell and Jeffrey L. Meikle (Amsterdam: Rodopi, 2009).

50. Henry Miller, *Varda: The Master Builder* (Berkeley, CA: Circle Edition, 1947), Victor J. Papanek Foundation. Related to the author in a personal interview with Papanek's former colleague and creative collaborator James Hennessey, Netherlands, December 2012. Papanek was resident in a Jewish Charitable Hostel in 1940, but he could have been referring to the efficiency apartment his mother rented near Broadway (145 East 26th Street, New York), which he later took over as his address.

51. Victor J. Papanek was inducted into the US Army on November 8, 1943, with the home address of 514 Schenck Avenue, Brooklyn. He was issued an "Honorable Discharge Certificate," dated December 14, 1944; stated reason for discharge was "Below minimum physical standards for induction." His "Report of Separation" record of the same date shows his highest held rank as that of private, and states that he did not receive any decorations, citations, or military qualifications (such as marksmanship). His character was noted as "Excellent." On this report, his home address is listed as 127 East 15th Street, New York, and his "Permanent Address for Mailing Purposes," as 514 Schenck Avenue, Brooklyn—the same address as his bride Anna Lipschitz, as stated on their marriage certificate of June 30, 1944. Victor J. Papanek Foundation, University of Applied Arts Vienna.

52. Herbert A. Strauss, "The Immigration and Acculturation of the German Jew in the United States of America," *The Leo Baeck Institute Year Book* 16, no. 1 (January 1971): 63–94, 90; emphasis in original.

53. Marriage certificate, dated June 30, 1944; Anna Lipschitz's address is given as 514 Schenck Avenue, Brooklyn, NY; the groom, Victor Papanek, as "US Army." Victor J. Papanek Foundation, University of Applied Arts Vienna.

54. The union may have been a marriage of convenience. However, the address on Papanek's naturalization papers (669 Schenk Avenue), which granted him US citizenship on June 25, 1945, is the same street declared for his bride Anna Lipschitz on the marriage certificate of June 30, 1944, namely, Schenck Avenue, Brooklyn. According to Victor's naturalization papers, his mother, Helene Papanek, lived two blocks away from Papanek's Brooklyn address in November 1944.

55. Hardwick, *Mall Maker*, 16.

56. See Mario Tedeschini Lalli, "Descent from Paradise: Saul Steinberg's Italian Years (1933–1941)," *Quest: Issues in Contemporary Jewish History* 2 (October 2011).

57. Victor Papanek remarried in 1951, to Winifred Nelson.

58. 145 East 26th Street.

59. Featured in Victor Papanek's design scrapbook and notated by Papanek as having appeared in *Interior* magazine in December 1944. However, there is no evidence the design ever appeared in any edition of *Interior* magazine.

60. The Nelsons' apartment was located on Park Avenue and 83rd Street, Manhattan.

61. The Nelsons' interior with the streamlined bassinet featured in a New York *Sun* article under the title "Nurseries Use Plastics for Decoration," (March 31, 1941). See http://www.nyhistory.org/exhibit/bassinet, accessed March 3, 2017.

62. As well as publishing the popular book *How to Build Your Own Furniture* (1949), Bry developed a niche style designing postwar urban space-saving designs displayed in his Manhattan showroom.

63. Stanley Abercrombie, *George Nelson: The Design of Modern Design* (Cambridge: MIT Press, 1994), 238, quoted in John Harwood, "The Wound Man: George Nelson and the 'End of Architecture,'" *Grey Room* 31 (Spring 2008): 90–115; 92.

64. Harwood, "The Wound Man."

65. Ben Schnall, "Design's Bad Boy," *Architectural Forum: The Magazine of Building*, no. 2 (February 1947): 88–91; 88.

66. Schnall, "Design's Bad Boy," 138.

67. See Andrea Bocco Guarneri, *Bernard Rudofksy: A Human Designer* (New York: Springer, 2003).

68. Prior to his arrival in the United States, Bernard Rudfosky had traveled and lived in Santorini, Greece; Capri and Procida, Italy; São Paulo, Brazil. In his memoirs Rudofsky recounted that the scholastic merits of his classical European training, far exceeding those of a less rigorous postwar American education, played to his advantage; see Guarneri, *Bernard Rudofksy*, 95. For an overview of Rudofsky's early work and formative architectural training in Vienna, see Monika Platzer, ed., introduction to *Lessons from Bernard Rudofsky: Life as a Voyage* (Basel: Birkhauser, 2007), 12–35.

69. See Harwood, "The Wound Man," 23n19. For the MoMA context, see Eliot F. Noyes, *Organic Design in Home Furnishings* (New York: Museum of Modern Art, 1941).

70. For a broad discussion of Rudofsky's experience of New York see Platzer, *Lessons from Bernard Rudofsky*.

71. See MoMA press release, "Museum of Modern Art to Open Exhibition *Are Clothes Modern?*," 1944, https://assets.moma.org/documents/moma_press-release_325447.pdf?_ga=2.68668825.986082888 .1594105582-107489561.1587391895, accessed July 15, 2020.

72. Bernard Rudofsky's later exhibitions and publications—including *Architecture without Architects: A Short Guide to Non-Pedigreed Architecture* (1964) and *Streets for People: A Primer for Americans* (1969)— paved the way for Papanek's *Design for the Real World* with their defiance of the professionalization of innate human needs and activities. They embraced a cosmopolitan, humanist design agenda whose legacy is evident in the alternative design movement of the 1970s, and some aspects of contemporary design studies.

73. For a discussion of Rudofsky, see Felicity D. Scott, "Bernard Rudofsky: Not at Home," in *Émigré Cultures in Design and Architecture*, ed. Alison J. Clarke and Elana Shapira (London: Bloomsbury Academic), 221–236.

74. Elana Shapira, "The Question of Gender: Kiesler, Rudofsky, Papanek, and Surrealism," in *Émigré Cultures in Design and Architecture*, ed. Alison J. Clarke and Elana Shapira (London: Bloomsbury Academic), 141–158; 141.

75. Victor Papanek collected and collated a series of business cards from this period, providing an invaluable "footprint" of his early informal design connections and sources including the Manhattan showroom of Finstevn.

76. Exhibition catalog for *Alvar Aalto: Architecture and Furniture*, Museum of Modern Art, New York, March 15–April 18, 1938, 12.

77. Edgar Kaufmann Jr., *Prize Design for Modern Furniture from the International Competition for Low-Cost Furniture Design*, Museum of Modern Art, New York, NY, 1950.

78. MoMA press release, May 1, 1950, https://www.moma.org/d/c/press_releases/W1siZiIsIjMyNTcyOSJdXQ .pdf?sha=5392d22fc4b7df42, 1, accessed March 17, 2017.

79. See MoMA, "What Was Good Design? MoMA's Message, 1944–56" exhibition, May 6, 2009–January 10, 2011, https://www.moma.org/calendar/exhibitions/957, accessed June 26, 2020.

80. Kaufmann Sr. met Lazlo Gabor in Vienna (who worked with Bernard Rudofsky in 1936), and the latter would become the Kaufmann art director. See Richard Cleary, *Merchant Prince and Master Builder: Edgar J. Kaufmann and Frank Lloyd Wright* (Pittsburgh: Carnegie Museum of Art, 1991), 27.

81. The Cooper Union School of Art "Student Registers" of November 18, 2008 (p. 46), and November 21, 1947 (pp. 90–155), in the Cooper Union Library and Archive, New York, show Victor Papanek (resident at 145 East 26th Street, New York) listed as a student enrolled in the first and second year of an evening class in architecture, respectively. The 1984 Cooper Union Alumni Directory lists Papanek as alumnus of the class of 1950, which would seem to confirm he took part in classes at the institution from 1946 to 1950 (the standard duration of a four-year evening course at Cooper Union during this period). There is no formal certification or record of Papanek having graduated from the Cooper Union School of Art as either architect or designer. However, on March 28, 1987, on the occasion of the Cooper Union Alumni Association Founder's Party, held at the Waldorf-Astoria Hotel, New York, Papanek was awarded the "Augustus St. Gaudens Award for outstanding professional achievement in industrial design and education" and represented as a graduate of 1950.

82. The Cooper Union School of Art prospectus "Free-Day and Evening Courses" for 1946–1947 (p. 5), Cooper Union Archive and Library, New York.

83. "Dr. Paul Zucker Is Dead at 82; Many Years at Cooper Union," *New York Times*, February 16, 1971, 36, http://www.nytimes.com/1971/02/16/archives/dr-paul-zucher-is-dead-at-82-many-years-at-cooper -union-architect.html.

84. "Michael Radoslovich: A Designer of the Fair," *Forest Hills-Kew Gardens Post*, May 5, 1939, Hofstra University Library Special Collections: Digital Collections, "1939 WF 1–8 clipping 1," http://omeka .hofstra.edu/items/show/519.

85. "Michael Radoslovich," *Forest Hills-Kew Gardens Post*.

86. See Bruce I. Kodish, *Korzybski: A Biography* (Pasadena, CA: Existential, 2011).

Chapter 2

1. The store was originally located at 209 East 14th Street and later moved to a ground-level store at 212 East 14th Street, renamed "Movie Star News."

2. Victor J. Papanek, J. L. Constant Distinguished Professor, School of Architecture and Urban Design, Kansas University, letter to *Weekly* journalist Karen Essex in Los Angeles, dated November 5, 1991, Victor J. Papanek Foundation, University of Applied Arts Vienna. This letter was written in response to Essex's article "In Search of Bettie Page," *LA Weekly*, October 10–17, 1991, 22–28.

3. Margaret Talbot, "From the Stacks: Chicks and Chuckles," review of *Bettie Page: The Life of a Pin-Up*, by Karen Essex (with James L. Swanson), *New Republic*, September 8, 1997, https://newrepublic.com/article/116626/margaret-talbot-bettie-page.

4. Talbot, "From the Stacks."

5. Papanek, letter to Essex.

6. The Victor J. Papanek Foundation library contains a section on "erotica"; Papanek used this collection as reference material for his lectures. His ambivalent approach to gender politics—on the one hand praising the liberated women of 1960s Finnish women, and on the other criticizing the creation of Sheila Levrant de Bretteville's women's design program in his capacity as dean at CalArts in the early 1970s is all the more intriguing in light of this formative encounter with America's famed pin-up model.

7. Jeanne C. Hertle was born on October 1, 1926, in Queens, New York City, *New York, New York, Birth Index, 1910–1965* database.

8. *Cooper Union Yearbook, Advancement of Science and Art*, 1948, Cooper Union Library and Archive, New York City.

9. For an extensive inquiry into the postwar censorship and debate around mass culture and juvenile delinquency, see Bart Beaty, *Fredric Wertham and the Critique of Mass Culture* (Jackson: University Press of Mississippi, 2005). The Senate Subcommittee on Juvenile Delinquency investigating the role of mass media as a motivational factor of youth crime, established April 1953, chaired by Senator Robert Hendrickson (R-New Jersey), took place in April 1954.

10. See Whitney Strub, "The Rediscovery of Pornography: The Emergence of a Cold War Moral Panic," in *Perversion for Profit: The Politics of Pornography and the Rise of the New Right* (New York: Columbia University Press, 2011), 11–42.

11. See Laura Fermi, *Illustrious Immigrants: The Intellectual Migration from Europe, 1930–1941* (Chicago: University Chicago Press, 1971); Beaty, *Fredric Wertham*, 8.

12. Beaty, 10.

13. Beaty, 156. Beaty defines the "New York Intellectuals" as "a loose-knit collection of journalist and public intellectuals affiliated with small political and cultural magazines such as the *Partisan Review*, the *New Republic*, and *Commentary*" (52).

14. Regarding Paul Lazarsfeld's role in the Kefauver Hearings, see Beaty, *Fredric Wertham.*

15. Beaty, 48.

16. Beaty, 11.

17. Dennis A. Doyle, *Psychiatry and Racial Liberalism in Harlem, 1936–1938* (Rochester, NY: University of Rochester Press, 2016), 121.

18. Ralph Ellison, "Harlem Is Nowhere," in *The Collected Essays of Ralph Ellison*, ed. John F. Callahan (New York: Modern Library, 1995), 320. Cited in Badar Sahar Ahad, *Freud Upside Down: African American Literature and Psychoanalytical Culture* (Chicago: University of Illinois Press, 2010), 96. The clinic was named in honor of the Cuban-born black French physician Paul Lafargue, who married Karl Marx's daughter Laura. For an extended insight into Ralph Ellison's depiction of the Lafargue Clinic, see Carlo Rotella, *October Cities: The Redevelopment of the Urban Literature* (Berkeley and Los Angeles: University of California Press, 1998), 223.

19. Doyle, *Psychiatry and Racial Liberalism,* 12.

20. For an excellent and extensive historical overview of the Lafargue Clinic initiative, see Gabriel N. Mendes, *Under the Strain of Color: The Promise of an Antiracist Psychiatry* (Ithaca, NY: Cornell University Press, 2015). Mendes describes Wertham as generating for himself a "reputation as a self-important nuisance" (160). See also Dennis Doyle, "'A Fine New Child': The Lafargue Mental Hygiene Clinic and Harlem's African American Communities, 1946–1958," *Journal of the History of Medicine and Allied Sciences* 64, no. 2 (2009): 173–212.

21. The author would like to express sincere thanks to the Professor Bart Beaty for his sharing of resources, and insights regarding the possible relationship between Fredric Wertham and Victor Papanek. His rigorous and detailed historical analysis of Wertham and his writings has proved formative in positioning Papanek in the intellectual and critical milieu in 1940s New York.

22. Al Gowan, *Victor Papanek: Path of a Design Prophet* (Cambridge, MA; Merrimack Media, 2015), 22.

23. CalArts Special Collections & Institute Archives, Valencia, California.

24. Gowan, *Victor Papanek,* 26.

25. See Ernst Papanck with Edward Linn, *Out of the Fire: A Poignant Account of How an Eminent Educator Helped Save Jewish Children from the Hitler Onslaught* (New York: Morrow and Company, 1975).

26. See obituary, Dr. Ernst Papanek, of Wiltwyck School, *New York Times*, August 8, 1973, http://www.nytimes.com/1973/08/08/archives/dr-ernst-papanek-of-wiltwyck-school.html.

27. Dennis A. Doyle, "Black Celebrities, Selfhood, and Psychiatry in the Civil Rights Era: The Wiltwyck School for Boys and the Floyd Patterson House," *Social History of Medicine* 28, no. 2 (May 2015): 330–350.

28. Doyle, *Psychiatry and Racial Liberalism,* 109.

29. Doyle, 110.

30. For a detailed and excellent discussion of this conceptual rift between the Lafargue Clinic and Wiltwyck School protagonists, see Dennis A. Doyle, "Criticizing The Quiet One: Color-Blind Psychiatry Divided in New York City," in *Psychiatry and Racial Liberalism in Harlem*, 108–112.

31. Victor Papanek, "Mit Bleistift und Pflugschar: Erinnerungen und Anekdotisches zu FLW," *Architektur und Bauforum* 147 (1991): 46–48; 46.

32. Victor Papanek, "The Pencil and the Plow: Personal Memories of Frank Lloyd Wright," original typed manuscript (1991), 1, Victor J. Papanek Foundation, University of Applied Arts Vienna.

33. Papanek, "Mit Bleistift und Pflugschar," 46.

34. Papanek, 46.

35. Papenek, 47.

36. Papanek, 47.

37. Papanek, 47.

38. Edgar Tafel, *Apprentice to Genius: Years with Frank Lloyd Wright* (New York: McGraw-Hill, 1979), 143.

39. Olgivanna was a follower of mystic G. I. Gurdjieff as well as theosophical teachings.

40. The brochure was entitled "Application for Fellowship, the Taliesin Fellowship Spring Green Wisconsin, 1932."

41. See *A Girl Is a Fellow Here: 100 Women Architects in the Studio of Frank Lloyd Wright*, directed by Beverley Willis, (2009; Beverly Willis Architecture Foundation), https://www.bwaf.org/film-a-girl-is-a-fellow-here/. See also Lois Davidson Gottlieb, *A Way of Life: An Apprenticeship with Frank Lloyd Wright* (Melbourne: Images Publishing Group, 2006).

42. Papanek, "Mit Bleistift und Pflugschar," 46.

43. Papanek, 46.

44. Papanek, 48.

45. Papanek, 48.

46. Frank Lloyd Wright, letter to Victor J. Papanek, dated Easter 1949, Victor J. Papanek Foundation, University of Applied Arts Vienna.

47. Papanek, letter to Wright.

48. This unverified color slide of the Taliesin breakfast is part of the holdings of the Victor J. Papanek Foundation, University of Applied Arts Vienna.

49. Victor Papanek, "Mit Bleistift und Pflugschar: Erinnerungen und Anekdotisches zu FLW," *Architektur und Bauforum* 147 (1991): 46–48.

50. Letter to Victor Papanek from William Wesley Petters, Chairman of the Board, Frank Lloyd Wright Foundation, Taliesin West, Arizona, November 28, 1985. Victor J. Papanek Foundation, University of Applied Arts Vienna.

51. See Donald Leslie Johnson, *The Fountainheads: Wright, Rand, the FBI and Hollywood* (Jefferson, NC: McFarland, 2005), 62–65.

52. See Bosley Crowther, "The Screen in Review: Gary Cooper Plays an Idealistic Architect in Film Version of 'The Fountainhead,'" *New York Times*, July 9, 1949, http://www.nytimes.com/movie/review ?res=9902E7DA113EE03BBC4153DFB1668382659EDE.

53. "Victor Papanek: Gadfly to a Changing Profession," *Design and Environment* 1, no. 1 (1970): 35.

54. Papanek, "Pencil and the Plow," 6.

55. Affidavit for License to Marry, Victor J. Papanek and Ada M. Epstein, February 25, 1949, Municipal Archives, City of New York Records. The author wishes to acknowledge Michael K. Foster for his genealogical skills in uncovering Papanek's second marriage.

56. In 1950, Ada M. Epstein married a Bernie Horowitz, in Manhattan. She died in New Haven, Connecticut, on March 1, 2008. Source: US Social Security Death Index, 1935–2014. Jeanne C. Hertle boarded the *Queen of Bermuda*, May 20, 1950 for a two-week sojourn to Bermuda. Source: The National Archives, Washington, DC; Series Title: Passenger and Crew Lists of Vessels and Airplanes Departing from New York, New York, 07/01/1948–12/31/1956; NAI Number: 3335533; Record Group Title: Records of the Immigration and Naturalization Service, 1787–2004; Record Group Number: 85; Series Number: A4169; NARA Roll Number: 76.

57. In June 1925, the Lipschitz family lived at number 131 Sheridan Avenue, and the Epstein family at 135 Sheridan Avenue, Brooklyn, New York City. Source: Enumeration of the Inhabitants, Brooklyn, Kings County, New York City, June 1 1925. Index provided by Ancestry.com.

58. *U.S., School Yearbooks, 1880–2012*; Yearbook Title: *Elms 61*; Year: 1948. US, School Yearbooks, 1880–2013 (database).

59. It is uncertain whether a formal annulment of Papanek's first marriage to Anna Lipschitz took place. However, Papanek's completion of the form "Affidavit for License to Marry," for his marriage to Ada M. Epstein in 1949, states that the annulment was finalized on April 4, 1947 without a court appearance, attorney, or waiver—which suggests it took place by default, or not at all. Affidavit for License to Marry, Victor J. Papanek and Ada M. Epstein, February 25, 1949, Manhattan, Municipal Archives, City of New York Records.

60. His 1973 and 1974 coauthored publications *Nomadic Furniture 1* and *2* and his famed diagrams showing the politics of design all drew on this distinctive free-hand writing style possibly informed by his work with comic books.

Chapter 3

1. Victor's mother, with whom he maintained a close relationship until her death in 1967, accompanied him to San Francisco, financing herself and supplementing his income through work as a live-in nanny (personal communication with Nicolette Papanek, September 12, 2017). Winifred Nelson Higginbotham (née Nelson) was born in 1923 in Norfolk, Massachusetts, to a moderately wealthy "self-made" family. Although it is unclear how she and Papanek met, according to the 1930 US Census her own father Thatcher Nelson described his occupation as "designer," although they lived on a small homestead. By 1940, he ran his own advertising business (US Census 1940, Wayland, Middlesex, Massachusetts, roll T627_1621, p. 9A, enumeration district 9–608; US 1930, Wellesley, Norfolk, Massachusetts, roll 937, p. 14A, enumeration district 0129, FHL microfilm. 2340672; http://www.ancestry.com, accessed 2012.

2. The "Preview Invitation" for Studio 44, c. 1951, shows the address as Fillmore 3044, San Francisco. Victor J. Papanek Foundation, University of Applied Arts Vienna.

3. Letter to Victor Papanek from V. C. Morris, July 23, 1951. Victor J. Papanek Foundation, University of Applied Arts Vienna.

4. Headed paper Studio 44, San Francisco. Victor J. Papanek Foundation, University of Applied Arts Vienna.

5. Information based on San Francisco City Directory 1951–1953, San Francisco Public Library. Courtesy of Michael Pierazzi of San Francisco, local historical mapping consultant. In fact, the listings reveal that Studio 44 shared the same telephone number and address as the Grunewald Furniture Store, though it is unclear whether this denoted anything other than an expedient business arrangement for sharing overhead costs.

6. A business card for the Pacific Shop, Sutter Street, San Francisco, remained in Papanek's personal papers from this period. Victor J. Papanek Foundation, University of Applied Arts Vienna. See Wendy Kaplan and Staci Steinberger, "'It Has to Be Sold': The Dissemination of California Design, 1945–1965," in *Living in a Modern Way: California Design 1930–1965*, ed. Wendy Kaplan (Cambridge, MA: LACMA and MIT Press, 2011), 299.

7. "Amberg-Hirth of San Francisco," *Design* 45, no. 9 (1944): 18.

8. Victor Papanek, CV CalArts 1970, Gump's Painting Exhibition "Group Show" Gump's San Francisco, 1952.

9. "Gump's Goes Modern," *Time*, May 30, 1949, 78.

10. Christina Klein, *Cold War Orientalism: Asia in the Middlebrow Imagination, 1945–1961* (Berkeley and Los Angeles: University of California Press, 2003), 23.

11. For an extended discussion of the origins of the term, see Kevin Starr, "Baghdad by the Bay: Herb Caen's San Francisco," chap. 4 in *Gold Dreams: California in an Age of Abundance, 1950–1963* (Oxford: Oxford University Press, 2009), 88–131.

12. The original nineteenth-century Gump store was founded in 1865 by Abraham Livingstone Gump. Starr, *Gold Dreams,* 97.

13. Pat Kirkham, "At Home with California Modern, 1945–65," in *Living in a Modern Way,* ed. Kaplan, 164.

14. Kirkham, 156.

15. The Samisen chair was featured in Mary Sullivan Simons, "Fresh Sculptural Approaches," *Interiors* 115, no. 7 (1956): 116–117; 117. Carlo Mollion's biomorphic chair design was featured in Olga Gueft, "Mollino's Flexible Fantasies," *Interiors* 112, no. 9 (1953): 89. Designed as an aesthetic statement, and purportedly proportioned to relieve weight from resting on the spinal column, Papanek later used the Samisen chair in *Design for the Real World* to illustrate how an ergonomic design could so easily fail. In its description, he stated: "Designed by author, while still a student. The chair sold successfully, but was eventually withdrawn from the market by the designer on the grounds that it was ugly and expensive." Victor Papanek, *Design for the Real World: Human Ecology and Social Change* (Pantheon Books: New York, 1971), 267.

16. Klein, *Cold War Orientalism,* 2003.

17. Kaplan, introduction to *Living in a Modern Way,* 28.

18. Henry Dreyfuss, "California: World's New Design Center," *Western Advertising,* June 1947, 60, cited in Kaplan, 28.

19. See Jarrell C. Jackman and Carla M. Borden, eds., *The Muses Flee Hitler: Cultural Transfer and Adaptation* (Washington, DC: Smithsonian Institution Press, 1983).

20. See, for example, Dorothy Lamb Crawford, *A Windfall of Musicians: Hitler's Émigrés and Exiles in Southern California* (New Haven, CT: Yale University Press, 2011).

21. Reinhold Grimm, *Bertolt Brecht: Poems and Prose* (New York: Continuum, 2003), 100.

22. Detlev Claussen, *Theodor W. Adorno: One Last Genius* (Cambridge, MA: Harvard University Press, 2008), 116.

23. Victor Papanek included Berthold Brecht's 1967 *Gesammlte Werke* in the bibliography of the first Pantheon English-language edition of *Design for the Real World* (1971) under the subheading "Personal Statements by Designers and Others" (330).

24. See Christopher Long, "Becoming American: Paul T. Frankl's Passage to a New Design Aesthetic," in *Émigré Cultures in Architecture and Design,* ed. Alison. J. Clarke and Elana Shapira (London: Bloomsbury, 2017), 79–91. 80.

25. Long, "Becoming American," 80.

26. Long, 80.

27. Starr, *Gold Dreams,* 52.

28. *Nomadic Furniture 1* and *2* (New York: Pantheon, 1973 and 1974) were coauthored with James Hennessey.

29. *La Mer* 2 (August 1950): 25–26. Published in Newport Beach, California.

30. *La Mer* 2, 25.

31. *La Mer* 2, 26.

32. "What Makes the California Look?," *Los Angeles Times* Home magazine, October 21, 1951, cited in Kaplan, introduction, 27.

33. Pat Kirkham emphasizes the point of the practical reality of the "California Look" in "At Home with California Modern, 1945–65," in Kaplan, *Living in a Modern Way*, 148.

34. Promotion brochure for the Art League of California, 1952, Victor J. Papanek Foundation, University of Applied Arts Vienna.

35. *La Mer* 2, 25.

36. The Herman Miller version of the table originated from a 1939, design by Isamu Noguchi, in rosewood and glass for Anson Conger Goodyear, first president of the Museum of Modern Art, New York.

37. Charles (Bob) Nelles, "Let Me Recommend: 'Goodbye Mr. Chippendale,'" review, *Ottawa Citizen*, February 17, 1945, 8.

38. Long, "Becoming American," 88.

39. Richard Gump, *Good Taste Costs No More* (Garden City, NY: Doubleday, 1951).

40. Victor J. Papanek, curriculum vitae, 1970, CalArts Archive. Two decades afterward Papanek described his teaching at Art League of California, San Francisco, and this may have been subject to revision in accordance with his application for a teaching position at CalArts, 1970.

41. Victor J. Papanek, "A Short Appraisal of the Latest Art Books," *San Francisco Chronicle*, May 18, 1952, 21. *The City Lights Journal* did indeed feature articles on the subjects of popular culture and comic books during the early 1950s, none of which, however, were penned by Papanek. The author wishes to thank to Michael Pierazzi for further contextual research on the *City Lights Journal* during the 1950s, and for discovering Papanek's art criticism featured in the *San Francisco Chronicle*.

42. See Victor J. Papanek, "Art and Sculpture Find a Logical Setting in Modern Architecture," *San Francisco Chronicle*, August 3, 1952, 18; Victor J. Papanek, "The Bauhaus Helped German Artists Grasp the Twentieth Century," *San Francisco Chronicle*, August 24, 1952, 21.

43. The 1954 Pasadena city directory lists Victor Papanek's address as 790 Neldorhe A, Pasadena, and his profession "designer."

44. Based on the art collection of A. E. Gallatin, the Gallery of Living Art at NYU (1926–1943) was renamed the Museum of Living Art in 1936 and predated the Museum of Modern Art, New York, the Whitney Museum of American Art, and the Solomon R. Guggenheim Museum of Non-Objective

Painting. See "A. E. Gallatin's Gallery and Museum of Living Art (1927–1943)," *American Art* 7, no. 2 (Spring 1993): 47–63.

45. See Judith Definer, "The Art of California Counter-Cultures in the 1950s," trans. Olga Garlic, *Perspectives: Actualité en histoire de l'art* 2 (2015): 1–77.

46. The design for the new gallery should be considered in relation to the development boom of the Orange County area in the 1950s. According to Starr, *Gold Dreams*: "Between 1950 and 1961, Orange County population (220,000 in 1950; 704,000 by 1960) witnessed the construction of 2,500 housing tracts containing 144,000 building lots. Developers transformed Newport Beach from yachting resort into residential city" (12).

47. There is a possibility that one source, based on audiotape recordings made of Victor Papanek in 1996 and shortly before his death in 1998, may have mistranscribed the name "Briffault" as "Beaufort" because Papanek designed an interior for a hunting lodge for an "A. E. Briffault" shortly after his design for the Gallery of Living Arts in Corona Del Mar, California. While there may be some truth to this claim, it is more likely that Papanek embellished or fabricated the reasons for his swift departure to the West Coast in 1950, in order to avoid mention of the more complicated aspects of his personal and financial situation, including two undisclosed New York City marriages (Anna Lipschitz, 1944; Ada M. Epstein, 1949). The same source (based on uncorroborated oral history) suggests the following: "During one of his freelance jobs, Papanek met Robert Beaufort, an elderly, wealthy advertising agency owner with a beautiful young wife. Who sought to ask him to design and build a contemporary Art Museum in his home town of Carona Del Mar [*sic*]." Al Gowan, *Victor Papanek: Path of a Design Prophet* (Cambridge, MA: Merrimack Media, 2015), 25.

48. As relayed by personal communication with Nicolette Papanek, Victor's first daughter with Winifred Nelson, September 21, 2017. The author wishes to thank Nicolette Papanek for her invaluable conversations regarding her father's life and works.

49. Anthropologist Edmund Snow Carpenter, an abiding influence on Papanek, makes this comparison between early 1950s Toronto and Moscow, emphasizing the city's bleakness and unlikeliness as the setting for intellectual innovation around global media theory. Harald E. L. Prins, "On Marshall McLuhan and Explorations," oral history recording, 2011, https://www.youtube.com/watch?v=t6HyFyMjlXA.

Chapter 4

1. Matthew Wisnioski, "Why MIT Institutionalized the Avant-Garde: Negotiating Aesthetic Virtue in the Postwar Defense Institute," *Configurations* 21, no. 1 (Winter 2013): 85–116; 90.

2. For a "case study" in the agency of institutions in the Cold War North American political economy and culture, see Rebecca Lowen, *Creating the Cold War University: The Transformation of Stanford* (Berkeley: University of California Press, 1997).

3. The author thanks archivist Scott Hillis, Visual Resources and Special Collections, Dorothy H. Hoover Library, OCAD University, Toronto, for this contextual information from internal institutional reports. Personal communication, May 23, 2014.

4. Scott Hillis, personal communication, May 23, 2014.

5. John E. Arnold, *Creative Engineering: Promoting Innovation by Thinking Differently*, ed. William J. Clancey (Stanford Digital Repository, [1959] 2016), https://stacks.stanford.edu/file/druid:jb100vs5745/Creative%20Engineering%20-%20John%20E.%20Arnold.pdf.

6. Peter Justin Kizilos-Clift, "Humanizing the Cold War Campus: The Battle for Hearts and Minds at MIT, 1945–1965" (PhD diss., University of Minnesota, 2009), 6.

7. Kizilos-Clift, "Humanizing the Cold War Campus," 6.

8. Kizilos-Clift, 6.

9. Kizilos-Clift, 215.

10. William M. Kays, Peter Z. Bulkeley, and Carlton A. Pederson, "Memorial Resolution: John E. Arnold (1913–1963)," Stanford University Department of Mechanical Engineering (1963), https://web.archive.org/web/20150619203510/http://historicalsociety.stanford.edu/pdfmem/ArnoldJ.pdf, accessed August 1, 2017. Arnold's publication came out shortly after he was recruited by Stanford University to be the director of the design division within the department of mechanical engineering in 1957.

11. Arthur Pulos, *The American Design Adventure* (Cambridge, MA: MIT Press, 1988), 185.

12. Pulos, *The American Design Adventure*, 185.

13. Roster of Registrants, Massachusetts Institute of Technology, Office of the Summer Session, Special Summer Program 2.739, Creative Engineering and Product Design, Monday, June 20, through Friday, July 1, 1955.

14. John E. Arnold, "History of Creative Engineering at M.I.T." (Creative Engineering Summer School opening address, June 21, 1954, in "Seminar Notes 1954," M.I.T., Cambridge 39, Massachusetts).

15. Laura Micheletti Puaca, *Technocratic Feminism and the Politics of National Security, 1940–1960* (Chapel Hill: University of North Carolina Press, 2014), 2.

16. Puaca, *Technocratic Feminism and the Politics of National Security*, 2.

17. Henry Dreyfuss, *Designing for People* (New York: Simon & Schuster, 1955), 29, quoted in David Serlin, "Engineering Masculinity: Veterans and Prosthetics after World War Two," in *Artificial Parts, Practical Lives: Modern Histories of Prosthetics*, ed. Katherine Ott, David Serlin, and Stephen Mihm (New York: New York University Press, 2002), 65. For an expansion of Serlin's broader thesis on the body and Cold War engineering, see also David Serlin, *Replaceable You: Engineering the Body in Postwar America* (Chicago: University of Chicago Press, 2004).

18. Serlin, "Engineering Masculinity," 65–66.

19. Letter from John Dixon, assistant to R. Buckminster Fuller, to Victor Papanek, August 3, 1955. The letter appears to suggest that Papanek's first encounter with Fuller had not been at MIT but earlier in May of the same year at a lecture Fuller delivered at Syracuse University, New York. Victor J. Papanek Foundation, University of Applied Arts Vienna.

20. See Allen Walker Read, "Formative Influences on Korzybski's General Semantics," in *Alfred Korzybski: Collected Writings, 1920–1950*, ed. M. Kendig (International Non-Aristotelian Library; Englewood, NJ: Institute of General Semantics, 1990), xxi–xxv.

21. Letter to Mrs Tenkuu from Victor Papanek, including résumé, written November 18, 1966; personal collection of the late Liisa Tenkuu, Finland. Permission courtesy of the owner.

22. R. Buckminster Fuller, introduction to Victor Papanek, *Design for the Real World: Human Ecology and Social Change* (New York: Pantheon Books, 1971), xvi.

23. Newspaper clipping, c. 1955, Victor J. Papanek Foundation, University of Applied Arts Vienna.

24. Papanek, *Design for the Real World*, 200.

25. Felicity D. Scott, *Outlaw Territories: Environments of Insecurity/Architectures of Counterinsurgency* (New York: Zone Books, 2016), 248.

26. Business card and stationery, c. 1954–1959. Victor J. Papanek Foundation, University of Applied Arts Vienna.

27. *Domus* 305 (April 1955): 58.

28. International Design Conference, Aspen, Colorado, June 23–July 1, 1956: (i) "Management and Design," June 24, 1956; (ii) "Education and Design," June 30, 1956, Getty Archives, box 3, fol. 4.

29. "How to Work with a Designer," *Canadian Packaging* 10, no. 1 (February 1957): 41.

30. "How to Work with a Designer," 43–44.

31. "How to Work with a Designer," 42.

32. "How to Work with a Designer," 43.

33. Avrom Fleishman, "Design as a Political Force: Part 2," *Industrial Design* 4, no. 2 (1957): 37.

34. Fleishman, "Design as a Political Force," 37.

35. Fleishman, 37.

36. Fleishman, 47.

37. Fleishman, 46.

38. Fleishman, 46.

39. Fleishman, 50.

40. V. J. Papanek, "A Bridge in Time," in *Explorations: Studies in Communication* (1957), republished in *Explorations: Verbi- Voci- Visual* (New York: Something Else, 1967), 1–10.

41. See Frances Stonor Saunders, *The CIA and the Cultural Cold War* (London, Granta, 1999); Kathleen McCarthy, "From Cold War to Cultural Development: The International Cultural Activities of the Ford Foundation, 1950–1980," *Daedalus* 116, no 1 (Winter 1987): 93–117.

42. Marshall McLuhan, *Understanding Media* (New York: McGraw-Hill, 1964), 159.

43. Edmund Carpenter, *Oh, What a Blow that Phantom Gave Me!* (New York: Holt, Rinehart, and Winston, 1972), 99, cited in Harald E. L. Prins and John Bishop, "Edmund Carpenter: Explorations in Media and Anthropology," *Visual Anthropology Review* 17, no. 2 (Fall–Winter 2001–2002): 110.

44. Edmund Carpenter was engaged with radio and television broadcasts for Canadian Broadcasting Company (CBC-TV) from 1950 to 1957 (Prins and Bishop, "Edmund Carpenter," 114). In 1957, he left the University of Toronto to join the San Fernando Valley State University (117).

45. Victor J. Papanek, letter to Mrs Tenkku, including résumé, November 18, 1966, personal collection of the late Liisa Tenkku, Finland. Permission courtesy of the owner.

46. Victor Papanek, "Student Project: An Angry Young Instructor Attacks Teaching Practices and Problems," *Industrial Design* (March 1959): 74–77.

47. Papanek, "Student Project," 74.

48. Papanek, 74.

49. Papanek, 74.

50. Papanek, 76.

51. Papanek,77.

52. Papanek, 74–77.

53. Papanek, 76.

54. See Prins and Bishop, "Edmund Carpenter," 119.

55. See *Explorations* promotional material, 1957, McLuhan Archive, Library and Archives Canada (LAC), vol. 146, file 5; and subscriptions, 1955, McLuhan Archive vol. 145, file 1.

56. H. R. Jolliffee, "The Colleges: Communication Arts," *Michigan State University Magazine*, January 1957, 10–16, in subscriptions, 1955, McLuhan Archive, Library and Archives Canada (LAC), vol. 145, file 1.

57. Jolliffee, "The Colleges," 14.

58. "Design Analysis Show to Commence on WNED," undated newspaper clipping. Victor J. Papanek Foundation, University of Applied Arts Vienna Austria.

59. Conferee list, International Design Conference, Aspen, Colorado, June 23–July 1, 1956, University of Illinois at Chicago Special Collections, box 1, fol. 12.

60. Certificate of marriage, March 27, 1962, Baltimore, Maryland.

61. Sue Fruchtbaum, "Design Professor's Home a Place of Individuality," newspaper clipping, *Buffalo Evening News*, October 1959, Ideas for Better Living, 31. Victor J. Papanek Foundation, University of Applied Arts Vienna.

62. Disassembled and reassembled in the course of Papanek's world travels, a version of this portable interior culminated in the 1970s with his coauthored work *Nomadic Furniture 1* (1973).

63. Letter of Acceptance, Victor Papanek's Resignation, from State University of New York at Buffalo, May 16, 1962. Victor J. Papanek Foundation, University of Applied Arts Vienna.

Chapter 5

1. Gui Bonsiepe, recorded interview with the author, La Plata, Argentina, October 5, 2017.

2. The Italian translation of *Design for the Real World* was published in 1973 under the title *Progettare per il mondo reale* (Milan: Arnoldo Mondadori, 1973).

3. Carnegie Institute of Technology (after 1967 known as Carnegie Mellon University) was the first US institution to issue a formal bachelor's degree in industrial design, beginning in 1934. See Jim Lesko, "Industrial Design at Carnegie Institute of Technology, 1934–1967," *Journal of Design History* 10, no. 3 (1997): 269. As a Fulbright scholar, Huff had himself been trained by Maldonado at the Ulm School of Design between 1956 and 1957.

4. An exhibition, *Basic Design. Von Ulm in die USA und zurück* exploring the dialogue between US and Ulm design pedagogy and based on the archive of William S. Huff, took place at the Ulm Museum, Germany, in the winter of 2013–2014. Lesko observes that the Bauhaus and Ulm influence originated in the architecture department of CIT, and through Huff went on to influence the industrial design program: "Within the architecture department Professor William S. Huff taught a Bauhaus-based foundation course. But Huff's course was not available to ID students because of schedule conflicts and an apparent bureaucratic unwillingness to change. Huff, while never a student of Albers, was influenced by him and taught the same foundation course at the Hochscule für Gestaltung at Ulm, Germany." Lesko, "Industrial Design at Carnegie Institute of Technology," 285–286.

5. Cited in Lesko, "Industrial Design at Carnegie Institute of Technology," 287.

6. Lesko, 287.

7. Bonsiepe interview.

8. "Education of a Designer I," *Industrial Design* 10, no. 11 (November 1963): 36.

9. "Education of a Designer I," 37.

10. Victor Papanek, cited in R. M., "Education of a Designer I," 36.

11. "Education of a Designer I," 63.

12. "Education of a Designer I," 39.

13. "Education of a Designer I," 36.

14. See Roger H. Clark, *School of Design: The Kamphoefner Years, 1948–1973: Reflections and Recollections* (Raleigh: NC State University College of Design, 2007), 30.

15. Alison J. Clarke, "Design and Development: ICSID and UNIDO: The Anthropological Turn in 1970s Design," *Journal of Design History* 29, no.1 (2016): 43–57.

16. See Russell Flinchum, "North Carolina Design: How the Tar Heel State's Modernist Heritage Endures Today," *Metropolis Design Journal*, March 14, 2018, https://www.metropolismag.com/design/north-carolina-modernism-design.

17. See Clark, *School of Design, 30.*

18. Eva Diaz, *The Experimenters: Chance and Design at Black Mountain College* (Chicago: University of Chicago Press, 2015), 1.

19. Carey Hedlund, "Black Mountain College and Penland School of Crafts: Neighbors Without Introduction," in *Appalachian Journal* 44, nos. 3–4, and 45, nos. 1–2 (2017–2018): 527.

20. From its inception in 1929 until 1962, Penland School of Crafts was known as Penland School of Handicrafts. In November 1998 the school's name was legally changed. The author kindly thanks Penland archivist Carey Hedlund for this information.

21. Hedlund, "Black Mountain College," 529.

22. Diaz, *Experimenters*, 1.

23. See Hedlund, "Black Mountain College," 526.

24. Martin Duberman, *Black Mountain: An Exploration in Community* (Evanston, IL: Northwestern University Press, [1972] 2009), 38.

25. The author wishes to thank Carey Hedlund, archivist, Jane Kessler Memorial Archives, Penland School of Craft, Penland, North Carolina, for her archival assistance and invaluable intellectual insight into the regional politics of segregation during the period of the operational overlap of Black Mountain College and Penland School of Crafts.

26. See Julie Levin Caro and Jeff Arnal, *Between Form and Content: Perspectives on Jacob Lawrence and Black Mountain College* (Durham, NC: Duke University Press, 2019).

27. Micah Wilford Wilkins, "Black Mountain College: An Experiment in Integration: The Recruitment of African American Student and Faculty Members at a Southern Arts School in the 1940" (seminar paper, 2013), 1, https://www.academia.edu/36733047/Black_Mountain_College_An_Experiment_in_Integration_The_Recruitment_of_African_American_Students_and_Faculty_Members_at_a_Southern_Arts_School_in_the_1940s. See Camille Clark, "Black Mountain College: A Pioneer in Southern Racial Integration," *The Journal of Blacks in Higher Education* 54 (2006): 46–48, http://www.jstor.org/stable/25073587.

28. See Micah Wilkins, "Social Justice at BMC before the Civil Rights Age: Desegregation, Racial Inclusion, and Racial Equality at BMC," *Black Moutain Studies Journal* 6, http://www.blackmountainstudiesjournal.org/volume6/6-17-micah-wilkins/, accessed June 29, 2020.

29. Duberman, *Black Mountain*, 174.

30. Hampton Institute became an African American university—Hampton University.

31. Lucy Morgan (founder of Penland School of Crafts), "Facts and Figures about Penland School of Handicrafts," c. 1946, transcription of archival records given to the author in personal communication by Carey Hedlund, archivist, Jane Kessler Memorial Archives, Penland School of Crafts, February 8, 2019.

32. Morgan, "Facts and Figures."

33. Board of trustees' records, acting director's report (Bonnie Willis Ford), December 1959. Transcription of records kindly provided to the author by Carey Hedlund, archivist, Jane Kessler Memorial Archives, Penland School of Crafts, personal communication, February 8, 2019.

34. Board of trustees' records.

35. Board of trustees' records.

36. Board of trustees' records.

37. Board of trustees' records.

38. Board of trustees' records.

39. Board of trustees' records.

40. Board of trustees' records.

41. Given the institutional history of the Penland School of Crafts, and its checkered relationship to racial segregation, its seems all the more incongruous that Papanek, a self-proclaimed socially responsibly designer, enrolled there as a visiting tutor from 1963 through to 1967. But Papanek combined his position of associate professor at NSCU School of Design in Raleigh with summers spent in the rural environs of North Carolina (often accompanied by his young daughter Nicolette) working with traditional makers and craftspeople. His bond with Penland proved strong enough to ensure his continued visits even after quitting NCSU to join the design faculty of Purdue University in the fall of 1964.

42. During the period of Papanek's involvement there, the school had irrevocably changed and was overseen by the new directorship of sculptor William "Bill" Brown, who took over in August 1962. His first act was to demand the removal of a formal ban on applications from African American students. When the first officially permitted African American students arrived at Penland in July 1963, their presence was met with hostility by at least one instructor; further members of the faculty socially shunned the new students during communal supper time. These incidents, redolent of the external racial and political climate of rural South more generally, were gradually quashed under the new directorship. See oral history interview with the former director William J. Brown, January 19–March 2, 1991, Smithsonian Institution Archives of American Art, Washington, DC, https://www.aaa.si.edu/collections/interviews/oral-history-interview-william-j-brown-13122.

43. See for example Papanek, *Design for the Real World: Human Ecology and Social Change* (New York: Pantheon, 1971, 266.

44. Papanek, 266.

45. Harriet Simpson, "'Pop Culture' Host Inspires Interest in World of Design," unidentified newspaper clipping, Victor J. Papanek Foundation, University of Applied Arts Vienna.

46. Simpson, "'Pop Culture' Host."

47. Simpson, "'Pop Culture' Host."

48. Simpson, "'Pop Culture' Host."

49. Paul Jacques Grillo, *What Is Design?* (Chicago: Paul Theobald and Company, 1960). Paul J. Grillo inscribed copy, Victor J. Papanek Foundation, University of Applied Arts Vienna.

50. Inner flap, Grillo, *What Is Design?*

51. Vance Packard, *The Waste Makers* (Longmans: London, [1960] 1961), 125.

52. Daniel Horowitz, "The Birth of a Salesman: Ernest Dichter and the Objects of Desire," 1986, available as an unpublished paper at Hagley Museum, https://www.hagley.org/sites/default/files/HOROWITZ_DICHTER.pdf; Stefan Schwarzkopf, "Ernest Dichter, Motivation Research and the 'Century of the Consumer,'" in *Ernest Dichter and Motivation Research: New Perspectives on the Making of Post-war Consumer Culture*, ed. Stefan Schwarzkopf and Rainer Gries (New York: Palgrave Macmillan, 2010).

53. Ernest Dichter, *The Strategy of Desire* (Reprint, Eastford, CT: Martino Fine Books, [1960] 2012), 62.

54. Dichter, *The Strategy of Desire,* 67.

55. Daniel Horowitz, "The Émigré as Celebrant of American Consumer Culture," in *Getting and Spending: European and American Consumer Societies in the Twentieth Century*, ed. Susan Strasser, Charles McGovern, and Matthias Judt (Cambridge: Cambridge University Press, 1998), 149–166.

56. Review of *GOD'S OWN JUNKYARD: The Planned Deterioration of America's Landscape*, by Peter Blake, *New York Times Book Review*, January 12, 1964.

57. Bernard Rudofsky, *Architecture without Architects* (New York: Museum of Modern Art, 1964).

58. For a broader discussion of the emergence of design for development, see Alison J. Clarke, "Design for Development, ICSID and UNIDO: The Anthropological Turn in 1970s Design," in "Design Dispersed: Anthropology and Design," special issue of *Journal of Design History* 29, no. 1 (2016): 43–57.

59. Subtitle cited from R. M., "Education of a Designer I," 35.

60. See Andrew Blauvelt, ed., *Hippie Modernism: The Struggle for Utopia* (Walker Art Center: Minneapolis, 2015).

61. The author wishes to thank the archivists at the Special Collections Research Center, NCSU Libraries, for confirming that the first African American student to graduate from the College of Design was Arthur J. Clement, who received his degree in Architecture in 1971. Only 2.8 percent of students at NCSU in 1962 were women, none of whom were enrolled in product design. For general gender enrollment statistics at NSCU, see https://oirp.ncsu.edu/fall-enrollment-by-gender-1950-present.

62. "New Professor's Approach: Designing for Need, Instead of Want Makes Products More Usable," unknown publication, North Carolina, 1962, newspaper clipping, unpaginated, Victor J. Papanek Foundation, University of Applied Arts Vienna.

63. "New Professor's Approach."

64. "New Professor's Approach."

65. "New Professor's Approach."

66. "U Haul It in 'Mini Haul,'" newspaper clipping, unpaginated, North Carolina c. 1962–1963. The design team included students Bert Olivari, Bill Phifer, George Heeden, Bruce Auld, William Huntley, and Don Peeler.

67. R. M., "Education of a Designer I," 63.

68. R. M., 35.

69. R. M., 35.

70. Felicity D. Scott, "Talking Teacher": Radio, Television, and the Oral Channel," in *Victor Papanek: The Politics of Design*, ed. Mateo Kries, Amelie Klein, and Alison J. Clarke (Weil am Rhein: Vitra Design Museum, 2018), 53.

71. See William H. Chafe, "The Sit-Ins Begin," in *Civilities and Civil Rights: Greensboro, North Carolina and the Black Struggle for Freedom* (New York: Oxford University Press, 1980), 71–102.

72. Flora Bryant Brown, "African American Civil Rights in North Carolina," reprinted with permission from the *Tar Heel Junior Historian*, fall 2004, Tar Heel Junior Historian Association, NC Museum of History, https://www.ncpedia.org/history/20th-Century/african-american-civil-rights.

73. Penny M. Eschen, *Race against the Empire: Black Americans and Anticolonialism, 1937–1957* (Ithaca, NY: Cornell University Press), 168.

74. Barbara J. Beeching, "Paul Robeson and the Black Press: The 1950 Passport Controversy," *The Journal of African American History* 87, no. 3 (2002): 339–354.

75. "Afro-Asian Conference," *Baltimore Afro-American,* April 9, 1950; cited in Eschen, *Race against the Empire*, 168–169.

76. See Penny M. Eschen, "No Exit: From Bandung to Ghana," chap. 7 in Eschen, *Race against the Empire*, 167–184.

77. Arturo Escobar, "Notes on the Ontology of Design," unpublished manuscript circulated for the panel "Design for the Real World: But Which World?," American Anthropological Association, San Francisco, November 13–18, 2012, 2.

78. "Education of a Designer I," 59.

79. "Education of a Designer I," 59.

80. Papanek, *Design for the Real World*, 189–190.

81. Author's correspondence with Archive Stadt Ulm, November 17, 2014. Bonsiepe, the obvious bridge from the US designer to the Ulm School of Design, did not personally attend the lecture, and it is uncertain how Papanek found his way to giving a lecture there other than traveling in northern Europe during this period making an unsolicited offer to guest lecture.

82. Papanek, *Design for the Real World*, 164.

83. Interview with Dieter Raffler, former Ulm School of Design student, on the occasion of the Bauhaus Fundamental School Centenary event, Bauhaus Dessau, Germany, March 20, 2019.

84. "Industrial Design Will Speak at Dow," *Midland Daily News*, October 25, 1963. Victor J. Papanek Foundation, University of Applied Arts Vienna.

85. Creative Problem Solving Workshop brochure, July 21–23, 1964, North Carolina State University at Raleigh.

86. Al Gowan, *Victor Papanek: Path of a Design Prophet* (Cambridge, MA; Merrimack Media, 2015), 38–39. Gowan does not attribute any of his sources in citing Papanek, but the quoted passage is most likely derived from the recording made for Papanek's uncompleted autobiography, "Second Only to Donald Duck."

87. Gowan, *Victor Papanek*, 38–39.

88. Gowan, 39.

89. Gowan, 40.

90. Regarding rejection from IDSA, personal communication with archivist, IDSA Collection, Special Collections Research Center, Syracuse University Libraries, Syracuse, New York.

91. On July 8, 1964, Diana Jean Ann Kleefeld entered into a deed of separation from Papanek.

92. *I.D.* magazine (November 1963).

Chapter 6

1. Thomas Willis, "Design and the New Environment," *Chicago Tribune Magazine*, April 12, 1970, 62.

2. Willis, "Design and the New Environment," 62.

3. Willis, 64.

4. Felicity D. Scott, *Outlaw Territories: Environment of Insecurity/Architecture of Counterinsurgency* (New York: Zone Books, 2016), 247.

5. Dedicated copy of R. Buckminster Fuller, *Nine-Chains to the Moon: An Adventure Story of Thought* (1938), Papanek's personal library collection, Victor J. Papanek Foundation, University of Applied Arts Vienna.

6. Al Gowan, *Victor Papanek: Path of a Design Prophet* (Cambridge, MA: Merrimack Media, 2015), 48.

7. Victor Papanek, "The Tree of Knowledge: Bionics," chap. 5 in *Design for the Real World*. Seed-related experiments from this period are recounted in detail.

8. Nancy Pelaez, Trevor R. Anderson, Samuel N. Postlethwait, "A Vision for Change in Bioscience Education: Building on Knowledge from the Past," *BioScience* 65, no 1 (January 2015): 90–100, https://doi .org/10.1093/biosci/biu188.

9. Al Gowan, personal email communication with author, October 28, 2010.

10. *Bio Graphics* formed a part of the Papanek's presentation at the Jyväskylä Arts Festival 1967, Finland. Program notes to accompany the film showing were available in Finnish. Jyväskylä University Library Archive (JULA), Jyväskylä, Finland (translated).

11. John W. Dawson, letter to Victor Papanek, US Army Research Office, Durham, North Carolina, January 9, 1968, Victor J. Papanek Foundation, University of Applied Arts Vienna.

12. Gowan, *Path of a Design Prophet*, 54.

13. See Tara Abraham, *Rebel Genius: Warren S. McCulloch's Transdisciplinary Life in Science* (Cambridge, MA: MIT Press, 2016).

14. See Abraham, *Rebel Genius*.

15. Gowan, *Path of a Design Prophet*, 54.

16. Gowan, 54.

17. Warren F. Seibert and Richard E. Snow, "Cine-Psychometry," *AV Communcations Review* 13, no. 2 (1965): 150–158; 140.

18. See Felicity D. Scott, "'Talking Teacher': Radio, Television, and the Oral Channel," in *Victor Papanek: The Politics of Design*, ed. Mateo Kries, Amelie Klein, and Alison J. Clarke (Weil am Rhein: Vitra Design Museum, 2018), 48–63; 60.

19. Pamela M. Lee, *Think Tank Aesthetics: Midentury Modernism, the Cold War, and the Neoliberal Present* (Cambridge, MA: MIT Press, 2020), 41.

20. Papanek, *Design for the Real World*, 193–194.

21. Papanek, 186.

22. Papanek, 186.

23. Alison J. Clarke, ed., *Design Anthropology: Object Cultures in Transition* (London: Bloomsbury, 2017).

24. Alison J. Clarke, "Design for Development, ICSID and UNIDO: The Anthropological Turn in 1970s Design," *Journal of Design History* 29, no. 1 (2016): 43–57.

25. Papanek, *Design for the Real World*, 187.

26. Papanek, 187.

27. Papanek, 192.

28. Papanek, 188; emphasis in original.

29. Papanek, 194.

30. Papanek, 201.

31. Gowan, *Path of a Design Prophet*, 52.

32. Papanek, *Design for the Real World,* 192.

33. Papanek, 192.

34. Gowan, *Path of a Design Prophet*, 54. Although Gowan recounts this project as being devised and overseen by Papanek, a local paper report of 1986 celebrating twenty years of the Women's Clinic in Lafayette makes no mention of Papanek or the Dow Chemical Company and attributes the design instead to a young local architect named E. H. Brenner who is quoted as saying he "wanted [the design] to express the joy of a woman experiencing pregnancy." *Journal and Courier* (Lafayette, IN), May 10, 1986. The author wishes to thank Dr. Jennifer Kaufmann for sourcing this information.

35. Marriage certificate, Victor Joseph Papanek and Harlanne Herdmann, Wicomico County, Maryland, July 29, 1966.

36. Certificate of Death, Helene Papanek, Lafayette, Indiana, October 21, 1967.

37. "A number of things pleasant and unpleasant have happened to me: I had somewhat of a nervous breakdown, needed to relax and catch up with myself in the Southern Appalachian mountains and I also got married to a very lovely former students of mine who is a weaver!" Excerpt of letter to Mrs. Tenkuu, Jyväskylä, Finland from Victor Papanek, Purdue University, Indiana, November 18, 1966, kindly donated by Liisa Tenkuu to the Victor J. Papanek Foundation, University of Applied Arts Vienna.

38. Malcolm Brookes, "American Views on Design Education: A Report on the First Educators' Conference Held by the Industrial Design Society of America," *Design Journal* 216 (May 1966): 19–23; 22.

39. Brookes, "American Views," 19–23; 22.

40. Brookes, 19–23; 22.

41. Brookes, 19–23; 22.

42. Brookes, 19–23; 23.

43. Parsons School of Design Prospectus, 1966–1967, Industrial Design Program, 60, the New School Archives and Special Collections, New York City, New York.

44. New School for Social Research (New York, 1919–1997), *New School Bulletin* 26, no. 5 (Spring 1969), 82, New School course catalog collection, New School Archives and Special Collections Digital Archive.

45. New School, *Invitation to the Opening of the Human Relations Center,* 1969, Human Relations Center records, New School Archives and Special Collections Digital Archive.

46. *Radical Shifts: Reshaping the Interior at Parsons, 1955–1985*. The exhibition ran from March 23 to April 8, 2011, in the Arnold and Sheila Aronson Galleries of the Sheila C. Johnson Design Center, Parsons The New School for Design, 66 Fifth Avenue, New York City.

47. Gowan, *Path of a Design Prophet*, 50.

48. *Design Course* 1, no. 1 (Spring 1969): 2. The issue was edited by Al Gowan and published by Purdue University. Victor Papanek was a member of the editorial advisory board of *Design Course*, and its reviews editor.

49. Stephen Frazier, "Me, Design and Niggertown Slums," *Design Course*, 1, no. 3 (Fall 1969): 23–28.

50. Frazier, "Me, Design and Niggertown Slums," 23.

51. Frazier, 23.

52. Frazier, 24.

53. Frazier, 24.

54. Frazier, 24,

55. Frazier, 26.

56. Frazier's activism was not confined to his employment with CCA. He regularly utilized his music as a backdrop to political demonstrations within Illinois black communities. According to Mary Barr, in 1970, for example, the music of Steve Frazier and the Soul Experience accompanied a street demonstration initiated by Concerned Black Parents in Evanston, Illinois. Barr, *Friends Disappear: The Battle for Racial Equality in Evanston* (Chicago: University of Chicago Press, 2014), 136.

57. Frazier, "Me, Design and Niggertown Slums," 26.

58. Frazier, 26,

59. Frazier, 28.

60. Frazier, 28,

61. Swedish Public Television (SVT), Interview with Victor J. Papanek, July, 23, 1968. Archival footage, SVT (Sveriges Television AB).

62. "Letters to the Editor," *Design Course* 1, no. 1 (Spring 1969): 14–19.

63. Gowan, *Path of a Design Prophet*, 51.

64. Gowan, 17.

65. Gowan, 17.

66. Per Johansson, "Design and Youth Revolution," *Design Course* 1, no. 2 (Summer 1969): 23–26.

67. Johansson, "Design and Youth Revolution," 23.

Chapter 7

1. Pekka Korvenmaa, "From Policies to Politics: Finnish Design on the Ideological Battlefield in the 1960s and 1970s," in *Scandinavian Design: Alternative Histories*, ed. Kjetil Fallan (Oxford: Berg, 2012), 222–235.

2. Antti Nurmesniemi, report on design seminar, Jyväskylä Arts Festival Design Seminar, July 3–16, 1966, Jyväskylä University Library Archive.

3. Päivö Oksala and Seppo Numi, program brochure introduction, Jyväskylä Arts Festival, July 3–16, 1966. Translation of Finnish original.

4. Oksala and Numi, program brochure introduction.

5. Oksala and Numi.

6. Oral history conducted by the author with Finnish designer Vuokko Eskolin-Nurmesniemi, Helsinki, February 19, 2015.

7. Eskolin-Nurmesniemi oral history.

8. Victor Papanek, letter to director of Art and Design, Swedish University of Turku (Akademic), Turku, Finland, March 29, 1966, Jyväskylä Regional Archive, Finland.

9. Papanek, letter to director of Art and Design.

10. Papanek, letter to director of Art and Design.

11. Tappio Periäinen (director of Visual Arts section of the Jyväskylä Arts Festival), letter to Mr. Victor J. Papanek, Purdue University, Department of Art, Lafayette, Indiana, April 13, 1966, Jyväskylä Regional Archive, Finland.

12. Victor Papanek, letter to Mrs. Tenkuu, Jyväskylä, Finland, November 18, 1966. Kindly donated by Liisa Tenkuu to the Victor J. Papanek Foundation, University of Applied Arts Vienna.

13. Victor Papanek, letter to Olli Alho, Secretary General, Jyväskylän Kulttuuripäivät, Jyväskylä, Finland, October 10, 1968, Jyväskylä Regional Archive, Finland.

14. Papanek, letter to Alho, October 10, 1968.

15. Victor Papanek, letter to Olli Alho, Secretary General, Jyväskylän Kulttuuripäivät, Jyväskylä, Finland, June 26, 1967, Jyväskylä Regional Archive, Finland.

16. Victor Papanek, "A Bridge in Time: An Attempt at Neo-Euclidean Aesthetics," in Marshall McLuhan, *Verbi-Voco-Visual: Explorations* (New York: Something Else Press, 1967), 1–10. First published as the last edition of Marshall McLuhan and Edmund Carpenter's edited journal *Explorations: Studies in Culture and Communication* 8 (January 1957).

17. For an extensive discussion of this first meeting, and the crossovers between Fuller and McLuhan's ideas, see Mark Wigley, "Network Fever," *Grey Room* 4 (Summer 2001): 82–122.

18. Wigley, "Network Fever," 86.

19. Wigley, 87.

20. Marshall McLuhan, letter to Buckminster Fuller, September 17, 1964, correspondence 1935–1982, MG 31 D vol. 24, files 1–71, National Archives of Canada.

21. Buckminster Fuller, letter to John Ragdale, November 7, 1966. correspondence 1935–1982, MG 31 D vol. 24, files 1–71, National Archives of Canada.

22. Jyväskylä Arts Festival Program, July 2–16, 1967, 13.

23. Olli Alho, Secretary General, Jyväskylän Kulttuuripäivät, Jyväskylä, Finland, letter to Professor Victor Papanek, December 9, 1967, Jyväskylä Regional Archive, Finland.

24. Jertta Roos, "Kolmas Vaihtoehto," *Me Naiset* 39 (September 27, 1967): 76. The work of Hungarian-British émigré Arthur Koestler influenced Papanek enormously from his first reading of the author's anti-totalitarian novel *Darkness at Noon* (1940) onward, and he kept a set of Koestler's complete works in his personal library.

25. Roos, "Kolmas Vaihtoehto," 76.

26. Roos, 78.

27. Roos, 78.

28. Roos, 78.

29. Lauri Perkki, "Press Coverage of the Programme Policy and Central Themes of the Jyväskylä Arts Festival 1964–69" (in Finnish, 1967), stencil in *Keskisuomalaine*, ed. M. Pyhälä for the Jyväskylä Arts Festival, October 13, 1969 (Jyväskylän kulttuuripäivät), 19.

30. Personal communication between Yrjö Sotamaa and author.

31. Victor Papanek, "Northern Lights," *Industrial Design* 14, no. 8 (1967): 29.

32. See Gay McDonald, "The Modern American Home as Soft Power: Finland, MoMA and the American Home 1953 Exhibition," *Journal of Design History*, 23, no. 4 (2010): 387–408; Jukka Rislakki, "Without Mercy: US Strategic Intelligence and Finland in the Cold War," *Economist* (Dec. 1, 2011), http://www.economist.com/blogs/easternapproaches/2011/12/finland-and-american-intelligence.

33. Papanek, "Northern Lights," 29.

34. Papanek, 29.

35. Papanek, 29–30.

36. Papanek, 29–30; emphasis in original.

37. Papanek, 30.

38. Papanek, 29.

39. Korvenmaa, "From Policies to Politics," 222.

40. Jørn Guldberg, "'Scandinavian Design' as Discourse: The Exhibition Design in Scandinavia, 1954–57," *Design Issues* 27, no. 2 (2011): 41–58.

41. Korvenmaa, "From Policies to Politics," 225.

42. Don Willcox, *Finnish Design: Facts and Fancy* (Finland: Werner Söderström Osakeyhtiö, 1973), 47.

43. Willcox, *Finnish Design*.

44. In 1973, Willcox forwarded a signed copy of his book to Papanek, claiming its synchronicity with *Design for the Real World* and pointing out that the manuscript for his *Finnish Design: Facts and Fancy* had actually been completed in 1970, but that the press had taken a tardy three years to release it. Papanek and the author had, according to Willcox's account in *Finnish Design*, directly corresponded regarding their conflicting claims, as North Americans, to an authentic insight into their host culture.

45. Victor Papanek, "Do-It-Yourself Murder," chapter in *Design for the Real World: Human Ecology and Social Change* (New York: Pantheon, 1971), 66.

46. Papanek, "Do-It-Yourself Murder."

47. Papanek, *Design for the Real World*, 70.

48. Papanek, 70.

49. Papanek, 71.

50. Apuega R. Arikawei, "African Socialism in Tanzania: Lessons of a Community Development Strategy and Rural Transformation for Developing Countries," *Mediterranean Journal of Social Sciences*, 2nd ser., 6, no. 4 (2015): 540.

51. Victor Papanek, "Project Batta-Kōya," *Industrial Design* (July/August 1975): 56–57.

52. Papanek, *Design for the Real World*, 71.

53. See for example, the bestselling Teresa Hayter, *Aid as Imperialism* (Middlesex: Penguin, 1973).

54. Victor Papanek, "The Future of Design," editorial, *Industrial Design* 17, no. 3 (1970): 33.

55. Papanek, "The Future of Design."

56. Papanek, "Project Batta-Kōya," 57.

57. Translation of Gui Bonsiepe, "Design and Undevelopment" ("Design E Sottosviluppo"), *Casabella* 385 (January 1974): 42–44; 42.

58. Translation of Bonsiepe, "Design and Undevelopment," 42–44; 43.

59. Translation of Bonsiepe, 42–44; 43.

60. For an extended discussion of the cross-overs between the Italian radical designers and Papanek's work and ideas, see Alison J. Clarke, "The Indigenous and the Autochthon," in *Global Tools: When*

Education Coincides With Life, ed. Valerio Boronuovo and Silvia Francheschini (Nero: Rome, 2018), 135–139; "The Anthropological Object in Design: From Superstudio to Victor Papanek," in *Design Anthropology: Object Cultures in Transition*, ed. Alison J. Clarke (London: Bloomsbury, 2017), 37–52.

61. Franco Raggi, cited in Clarke, "The Indigenous and the Autochthon," 137.

62. Papanek, "Project Batta-Kōya," 57.

63. Franco Raggi describes Papanek's decolonizing design as "an autochthonous low-technology." Cited in Clarke, "The Indigenous and the Autochthon," 137.

64. Barbro Kulvik-Siltavuori, dust jacket, back cover, Papanek, *Design for the Real World*.

65. Fuller attended with the support of the Finnish National Fund for Research and Development.

66. F. D. Scott, "Fluid Geographies," in *New Views on R. Buckminster Fuller*, ed. H. Y. Chu and R. Trujillo (Palo Alto, CA: Stanford University Press, 2009), 161.

67. See J. Guldberg, "'Scandinavian Design' as Discourse: The Exhibition Design in Scandinavia, 1954–57," *Design Issues* 27 (Spring 2011): 41–58.

68. Korvenmaa, "From Policies to Politics," 225.

69. G. McDonald, "The Modern American Home as Soft Power: Finland, MoMA and the 'American Home 1953' Exhibition," *Journal of Design History* 23, no. 4 (2010): 387–408.

70. Korvenmaa, "From Policies to Politics," 225.

71. The eight-day Suomenlinna design symposium took place in July 1968 as a two-part event spanning the month. The pan-Scandinavian Design Students' Organization began in Finland in 1965, was formally organized by 1967, and by 1969 began to face dissolution. See "SDO in Crisis," *FORM* 8 (1969): 372.

72. Stanford Anderson, Gail Fenske, and David Fixler, eds., *Aalto and America* (New Haven: Yale University Press, 2012).

73. "SDO in Crisis," 372, trans. by author.

74. "SDO in Crisis," 372.

75. Victor Papanek, "The Needs of the Underdevelopment and Backward Areas" (Suomenlinna lecture transcript, July 2, 1968), Victor J. Papanek Foundation, University of Applied Arts Vienna.

76. "SDO in Crisis," 372.

77. See "Kuntoutusvälineitä ja teurastamo Suomenlinnan seminaarin tuloksina" ("Rehabilitation Equipment and a Slaughterhouse: The Results of the Suomenlinna Workshop"), *Helsingin Sanomat*, July 18, 1968, 10.

78. "A Place Where Everyone Can Play," *Apu*, no. 31 (1968): 22–25, trans. by author.

79. "Rehabilitation Equipment," 10.

80. "Rehabilitation Equipment," 10.

81. "Rehabilitation Equipment," 10.

82. "Rehabilitation Equipment," 41.

83. "Rehabilitation Equipment," 41.

84. Nick Rutter, "Look Left Look Right: Internationalisms at the 1968 World Youth Festival," in *The Socialist Sixties: Crossing Borders in the Second World War*, ed. A. E. Gorsuch and D. P. Koenker (Bloomington: Indiana University Press, 2012), 194.

85. Rutter, "Look Left Look Right," 194.

86. See J. Krekola and S. Mikkonen, "Backlash of the Free World: The U.S. Presence at the World Youth Festival in Helsinki, 1962," *Journal of Scandinavian History* 36, no. 2 (2011): 230–255.

87. Howard Smith, interviews with the author, Fiskars, Finland, August 2012 and August 2014.

88. See Matt Novak, "We Got Buckminster Fuller's FBI File," May 20, 2015, Gizmodo Paleofuture, https://paleofuture.gizmodo.com/we-got-buckminster-fullers-fbi-file-1704777475.

89. Novak, "We Got Buckminster Fuller's FBI File."

90. See P. Nicolin, *Castelli di carte: la XIV Triennale di Milano 1968* (Macerata, Italy: Quodlibet, 2011).

91. Victor Papanek, "Actions Speak Louder," *Industrial Design* 15, no. 9 (1968): 40–45.

92. See Krekola and Mikkonen, "Backlash of the Free World."

93. From author's English translation of "Prophets of Time: Victor Papanek's Broadcast," April 20, 1969, Finnish Broadcasting Company.

94. At the time of taking part in the radio broadcast, Juhani Pallasmaa openly opposed the work and legacy of Alvar Aalto as the archetypal Finnish architect, and was mainly focused on the concepts of rationalization and prefabrication represented in his Moduli 255 (1969–1971), an industrially produced summerhouse.

95. From author's English translation of "Prophets of Time."

96. Translation of "Prophets of Time."

97. Translation of "Prophets of Time."

98. See Clarke, *Design Anthropology*.

Chapter 8

1. See Elizabeth Guffey, "Designing for Disability, Historical and Contemporary Perspectives," in *Victor Papanek: The Politics of Design*, ed. M. Kries, A. Klein, and A. Clarke (Weil am Rhein: Vitra Design Museum, 2018), 198.

2. Victor Papanek, *Design for the Real World: Human Ecology and Social Change* (New York: Pantheon, 1971), 277.

3. Papanek, *Design for the Real World*, 277.

4. Papanek, 278.

5. Papanek, 278.

6. Pull-out flowchart from Papanek, *Design for the Real World*, between 276–277.

7. The Copenhagen flowchart designates "retarded, handicapped and disadvantaged children" as a specific subsection of those dispossessed by mainstream design; Papanek, *Design for the Real World*, insert between 276–277.

8. Gunilla Lundahl, design journalist and former editor *FORM* journal, interview with author, Stockholm, Sweden, November 2013.

9. Sam Binkley, *Getting Loose: Lifestyle Consumption in the 1970s* (Durham, NC: Duke University Press, 2007), 21.

10. Binkley, *Getting Loose*, 21.

11. Information arising from oral histories conducted with six former Danish students of Victor Papanek between 1969 and 1973 in Copenhagen conducted with the author, November 2013.

12. Victor Papanek, biographic data, revised September 1970, CalArts Special Collections & Institute Archives, Valencia, California.

13. Binkley, *Getting Loose*, 21.

14. Pull-out flowchart, Papanek, *Design for the Real World*, between 276 and 277.

15. Andrew Blauvelt, *Hippie Modernism: The Struggle for Utopia* (Minneapolis: Walker Art Center, 2015), 11.

16. Blauvelt, *Hippie Modernism, 11.*

17. Blauvelt, *11.*

18. Susan Sontag, "Posters: Advertisement, Art, Political Artifact, Commodity," in *Looking Closer 3: Classic Writings on Graphic Design*, ed. M. Bierut, J. Helfand, S. Heller, and R. Poynor (New York: Allworth, 1999), 214. Originally published in D. Stermer, *The Art of Revolution* (New York: McGraw-Hill, 1970), quoted in R. Poynor, "Utopian Images: Politics and Posters," *Design Observer*, October 3, 2013, http://designobserver.com/feature/utopian-image-politics-and-posters/37739.

19. "SDO I kris" (SDO in Crisis), *FORM* 8 (1969): 372.

20. &/*sdo Magazine*, Scandinavian Design Students Organization 1, no. 2 (1968).

21. Papanek, *Design for the Real World*, 282.

22. Papanek, 284.

23. Papanek, 285.

24. Bess Williamson, *Accessible America: A History of Disability and Design* (New York: New York University Press, 2019), 170.

25. For an expanded and historical contextualization of design for disability, see Williamson, *Accessible America;* and Elizabeth Guffey, *Designing Disability: Symbols, Space and Society* (Oxford: Bloomsbury, 2017).

26. Thomas Willis, "Design and the New Environment," *Chicago Tribune Magazine*, April 12, 1970, 62–64.

27. Papanek, *Design for the Real World*, 178–179.

28. Papanek, 178–179.

29. Papanek, 109.

30. Papanek, 52–53.

31. Papanek, 177–179. Regarding designs for rehabilitation vehicles, see also 285.

32. Lasse Brunnstörm, *Swedish Design: A History* (London: Bloomsbury Press, 2019), 165.

33. "UN Official Programme," published by ARse, London 1968. See Felicity D. Scott, "Fluid Geographies: Politics and the Revolution by Design," in *New Views on Buckminster Fuller*, ed. Hsiao-yun Chu and Roberto G. Trujillo (Stanford, CA: Stanford University Press, 2009), 164.

34. Victor Papanek (Pasadena, California) letter to R. Buckminster Fuller, Carbondale, Illinois, February 5, 1971, Stanford Archives.

35. Papanek letter to Fuller, 1.

36. Papanek letter to Fuller, 2.

37. Naomi E. Wallace, Administrative Secretary, Office of Buckminster Fuller, Carbondale, Illinois, letter to Verne Moberg, Pantheon Books, New York, New York, March 3, 1971, Stanford Archives.

38. R. Buckminster Fuller, introduction to Papanek, *Design for the Real World*, vii.

39. Fuller, introduction, xvi.

40. Fuller, xvi.

41. Fuller, xvi.

42. Fuller, xxvi.

43. Fuller, xxii.

44. R. Buckminster Fuller, letter to Victor Papanek, Chairman, Industrial and Environmental Design, Purdue University, Department of Creative Arts, Lafayette, Indiana, February 6, 1969, Victor J. Papanek Foundation, University of Applied Arts Vienna.

45. Victor Papanek, Purdue University, letter to Dean Richard E. Farson, School of Design California Institute of the Arts, Los Angeles, January 1970, 1.

46. Victor Papanek, Purdue University, letter to Dean Richard E. Farson, School of Design, California Institute of the Arts Los Angeles, February 26, 1970, 1.

47. Victor Papanek, Chair, Industrial and Environmental Design, Purdue University, letter to Richard Farson, Dean of Design, California Institute of the Arts, Los Angeles, March 25, 1970, CalArts Special Collections & Institute Archives, Series 1, administrative records, 1966–1976; Subseries 1.2 Deans.

48. Program, "Paradox," International Design Conference in Aspen, June 20–24, 1971.

49. See A. Twemlow, "I Can't Talk to You If You Say That: An Ideological Collision at the International Design Conference at Aspen, 1970," *Design and Culture* 1, no. 1 (2009): 23–50.

50. See R. Farson, "Aspen Project, Networks 1," *CalArts Journal* (1971–1972): 17.

51. Papanek, *Design for the Real World*, xxvi.

52. Richard Farson, "Aspen Project, Networks 1," 1971–1972, report from Aspen.

53. Papanek, *Design for the Real World*, 51.

54. Memorandum to Robert W. Corrigan from Victor J. Papanek, "Restructuring the Design School," November 4, 1971, CalArts Special Collections & Institute Archives.

55. Memo to Corrigan from Papanek, 2.

56. Memo to Corrigan from Papanek, 2.

57. School of Design, CalArts Prospectus, 1970, designed by Sheila Levrant de Bretteville.

58. See transcript of Jenni Sorkin's oral history interview of Sheila Levrant de Bretteville, cofounder, Woman's Building, ArtSpaces Archives Project, June 22, 2010, Getty Research Institute.

59. Victor Papanek was a short-lived dean of CalArts in 1971 following the resignation of Richard Farson. Sheila Levrant de Bretteville recounts her struggle with Papanek over her ambition to create a separate women's design program, which, according to de Bretteville, Papanek controversially described—as someone of a Jewish émigré background to another—as a form of "ghettoization." Author's interviews with Shelia Levrant de Bretteville were conducted in Warsaw, Poland, in October 2013, and in Minneapolis, Minnesota, in October 2015.

60. Papanek, *Design for the Real World*, xxvi.

61. Memorandum, "Some Thoughts on Starting the Year," from Victor Papanek to Design School Faculty and TAs (cc: Student Organization and Other Deans), October 25, 1971, CalArts Special Collections & Institute Archives.

62. Memo, "Some Thoughts."

63. Memorandum, "Design School Structure," from Victor Papanek to Design School Faculty, Students, and TAs (cc: Student Organization and Other Deans), February 11, 1972, CalArts Special Collections & Institute Archives.

64. Memo, "Design School Structure."

65. Unknown author, "Controversial Design Teacher Seeks to Stir Imagination," *Product Engineering*, June 20, 1966, 98–99.

66. Course Catalogue, California Institute of the Arts, Term II, January 10, 1972, 34.

67. Course Catalogue, 12–13.

68. Alan Cartnal, "Dean's Aim to Clean Up L.A.," *Los Angeles Times*, March 5, 1972.

69. Cartnal, "Dean's Aim."

70. Unknown author, "Down with Designers?," *Time*, February 21, 1972, 47.

Chapter 9

1. Victor Papanek, "What to Design and Why," *Mobilia* 193 (August 1971), n.p.

2. Papanek, "What to Design and Why," n.p.

3. Tobias Faber, Rector of the Royal Danish Academy of Fine Arts, Copenhagen, letter to Director Mr. A. Brøndum, The National Bank of Denmark, Copenhagen, February 25, 1972. Kind permission of The National Bank of Denmark.

4. See Kirsten Birk Hansen, "The Danish Project: Developing Greenland into a Modern Welfare Society," in *Forming Welfare*, ed. Katrine Lotz et al. (Copenhagen: The Danish Architectural Press, 2018), 92–109.

5. Hansen, "The Danish Project," 103–104.

6. Paul Erik Skriver, "Grønland I Focus," *Arkitekten* 23 (1970): 541–542. Cited in Hansen, "The Danish Project," 104.

7. Skriver, "Grønland I Focus," in Hansen, 104.

8. Andreas Dalgaard, "Get Rid of Flying Alarm Clocks," *Berlingske Weekendavisen*, September 22, 1972. Author's translation, unpaginated.

9. See "Project Lolita," *Miljön och miljonerna: design som tjänst eller förtjänst* (Stockholm: Bonniers, 1970), 63–65; and "The Lolita Project," in Victor Papanek, *Design for the Real World: Human Ecology and Social Change* (New York: Pantheon, 1971), 88–90.

10. Ingrid Hjreting, "Few Profession[s] Are as Damaging as the Designer's," *Børsen*, December 28, 1972. Author's translation, unpaginated.

11. Unknown author, "Problem Solver and Missionary," *Ingeniørens Ugeblad* (*The Engineer's Weekly Paper*), November 10, 1972. Author's translation, unpaginated. Kind permission of the National Bank of Denmark.

12. George Day and Simon Croxton, "Appropriate Technology, Participatory Technology Design and the Environment," *Journal of Design History* 6, no. 3 (1993): 179–183.

13. Jørgen Lind, "The Professor's Biggest Success," *MORGENAVISEN Jyllands-Posten*, October 8, 1972. Author's translation, unpaginated. Kind permission of the National Bank of Denmark.

14. Steen Juhler, oral history interview with the author, Denmark, November 2013.

15. Beatriz Colomina, Esther Choi, Ignacio Gonzalez, and Anna-Maria Meister, "Radical Pedagogies in Architectural Education," *Architectural Review*, September 26, 2012, https://www.architectural-review .com/today/radical-pedagogies-in-architectural-education/8636066.article.

16. Mette Bovin, "PORTRÆT: Victor Papaneks ideer og antropologien," *Dagbladet Information*, September 11, 1973, n.p.

17. Bovin, "PORTRÆT," n.p.

18. Bovin, n.p.

19. Victor Papanek, Copenhagen, Denmark, letter to William S. Lund, President, CalArts Special Collections & Institute Archives, February 3, 1973.

20. See Victoria Gerrard, "What Did You Teach My Students?!," a narrated interview with Pak Imam Buchori, in *Victor Papanek: The Politics of Design*, ed. Mateo Kries, Amelie Klein, and Alison J. Clarke (Weil am Rhein: Vitra Design Museum, 2018), 172–179; 177.

21. Gerrard, "What Did You Teach My Students?!," 177.

22. See Alison J. Clarke, *Design Anthropology: Recolonizing and Decolonizing the Material World* (MIT Press, forthcoming).

23. Victor Papanek and James Hennessey, *Nomadic Furniture 1* (New York: Pantheon Books, 1973), back cover.

24. Papanek, letter to Lund, 1–3; 2.

25. Ken Isaacs, interviewed by Susan Snodgrass, "An Interview with Ken Isaacs," in *Hippie Modernism: The Struggle for Utopia,* ed. Andrew Blauvelt (Minneapolis: Walker Art Center, 2015), 370.

26. Peter de Bretteville, Toby Cowan, and Jack Reineck, *Dorm Book One*, 1971, CalArts Special Collections & Institute Archives.

27. Christina Cogdell and Simon Sadler, "Fast Forward and Rewind," *Volume* 2 (2010): 51.

28. Silvia Bottinelli, "The Discourse of Modern Nomadism: The Tent in Italian Art and Architecture of the 1960s and 1970s," *Art Journal* (2015): 63.

29. Jean Baudrillard, *The System of Objects* (London: Verso, [1968] 1996), 15.

30. Baudrillard, *System of Objects*, 17–18.

31. Raoul Vaneigem, *The Revolution of Everyday Life* (Oakland: PM Press, [1967] 2012), 50.

32. Vaneigem, *Revolution of Everyday Life*, 55.

33. See Alice Twemlow, "I Can't Talk to You If You Say That: An Ideological Collision at the International Design Conference at Aspen, 1970," *Design and Culture* 1, no. 1 (2009): 23–50.

34. Victor Papanek, "Actions Speak Louder," *Industrial Design* 15, no. 9 (1968): 40–45.

35. Greg Castillo, "Hippie Modernism," *Places Journal*, October 2015, https://doi.org/10.22269/151026, accessed June 8, 2018.

36. Greg Castillo, "Counter Culture Terroir: California's Hippie Enterprise Zone," in *Hippie Modernism: The Struggle for Utopia*, ed. Andrew Blauvelt (Walker Art Center: Minneapolis, 2015), 87–101; 96.

37. See Peter Lang and William Menking, eds., *Superstudio: Life without Objects* (Milan: Skira, 2003).

38. Petra Eisele, "Do-It-Yourself-Design: Die IKEA-Regale IVAR und BILLY," *Zeithistorische Forschungen / Studies in Contemporary History* 3 (2006): 439–448, http://www.zeithistorische-forschungen.de/3-2006/id=4458.

39. Niels Krygers, *DanskForm,* no. 10 (December 1973): 38–39. Victor Papanek responds to Niels Krygers, Letters to the Editor, *DanskForm,* no. 3 (March 1974): 3.

40. Sten (pseudonym), "Promises of More Bullets against 'Lazy Students,'" *Politiken*, August 8, 1973, n.p. Kind permission of the National Bank of Denmark.

41. Joe [pseudonym], *Ingeniørens Ugeblad* (*The Engineer's Weekly Paper*), August 3, 1973, n.p. Kind permission of the National Bank of Denmark.

42. Joe, *Ingeniørens Ugeblad.*

43. Joe, *Ingeniørens Ugeblad.*

44. Anonymous, *Information*, "The Head of the Design School Answers to the Guest Professor's Criticism," August 4–5, 1973, n.p. Kind permission of the National Bank of Denmark.

45. David Jenkins, "Professor for the Real World," *NOVA*, August 1974, 50–55.

46. Jenkins, "Professor for the Real World," 50. The same article reports that 125,000 copies of *Nomadic Furniture* had sold, and that the publisher had informed Papanek that it was the book most often stolen from bookstores.

47. Jenkins, "Professor for the Real World," 50.

48. Jenkins, 55.

49. Jenkins, 50.

50. Jenkins, 52.

51. Mark Brutton, "Goodbye Papanek: Manchester Poly Loses Its Most Celebrated Staff Member in All-Department Vintage Year," *Design* 323 (November 1975): 44.

52. Unknown, "New York Taxi: Five New Prototypes at MoMA," *Domus* 560 (July 1976): 42.

53. Julian Bicknell and Julia McQuiston, eds., *Design for Need: The Social Contribution of Design* (Oxford: ICSID, Pergamon Press), 1976.

54. Mark Brutton, "Third World Cautions," *Design* 322 (October 1975): 56. For an expanded insight into ICSID as an organization, see Tania Messell, "Design Across Borders: The Establishment of the International Council of Societies of Industrial Design (ICSID) 1957–1963," in *Making Trans/National Contemporary Design History,* ed. Wendy Siuy Wong, Yuko Kikuchi, and Tingyi S. Lin (São Paulo: Blucher, 2016), 131–137.

55. ICSID, minutes of executive board meeting, Leningrad, June 12–14, 1970, 16, ICD/04/1. Brighton University Design Centre Archive, UK. The author wishes to thank Tania Messell for assistance with this archival research.

56. ICSID, minutes of executive board meeting, Berlin (GDR), April 25–29, 1974, 9, ICD/04/2, Brighton University Design Centre Archive, UK.

57. Josine des Cressonnières, letter to Sudhakar Nadkarni, July 18, 1977, ICD/11/1/1, Brighton University Design Centre Archive, UK.

58. Victor Papanek, "For the Southern Half of the Globe," *Design Studies* 4, no. 1 (1983): 61–64.

59. Victor Papanek, letter to Carl Auböck, January 12, 1978, ICD/06/4/4, University of Brighton Design Archive, UK.

60. Victor Papanek, letter to Hélène de Callataÿ, ICSID's Chief Executive, March 23, 1981, ICD/11/3, University of Brighton Design Archive, UK.

61. The lecture "Victor Papanek USA" took place in the course of the congress, on January 18, 6:30 pm, at the Tagore Hall auditorium, Paldi, Ahmedabad, India.

62. Arthur J. Pulos, "The Profession of Industrial Design," Professors' Seminar at the Congress and Assembly of the International Council of Societies of Industrial Design Mexico City, October 14–19, 1979.

63. A. Escobar, "Notes on the Ontology of Design," draft circulated for the panel "Design for the Real World: But Which World?," American Anthropological Association, San Francisco, November 13–18, 2012.

64. Arturo Escobar, *Encountering Development: The Making and Unmaking of the Third World* (Princeton, NJ: Princeton University Press, [1995] 2012).

65. Escobar, *Encountering Development*, 5. For a thorough outline of what became known as the Basic Needs Approach (BNA), see K. Willis, *Theories and Practices of Development* (Abingdon: Routledge, 2005), 103–105.

66. See, for example, A. Chatterjee, "Design in India: The Experience of Transition," *Design Issues* 21, no. 4 (2003): 4–10; S. Balaram, "Design in India: The Importance of the Ahmedabad Declaration," *Design Issues* 25, no. 4 (2009): 54–79. Both authors analyze the Ahmedabad Declaration and Congress through incorporation of their perspectives as firsthand experiences: as an attendee, in the case of Balaram, and as leading delegate and the executive director of NID, in the case of Chatterjee.

67. "Design in India: The Importance of the 'Ahmedabad Declaration' Meeting for Promotion of Industrial Design in Developing Countries," aide-memoire, UNIDO Archive, Vienna, id.78–5076, 3.

68. UNIDO aide-memoire, 3; Balaram, "Design in India."

69. Chatterjee, "Design in India."

70. John Reid, *The State of Industrial Design in Developing Countries 1978: A Report of the Pilot Mission to India, Pakistan, Egypt and Turkey*, UNIDO Archive, Vienna, id.78–7582, preface; emphasis in original.

71. Charles Eames and Ray Eames, *The India Report*, National Institute of Design, Ahmedabad, [April 1958] 1997, sourced from the National Institute of Design.

72. Reid, *The State of Industrial Design in Developing Countries 1978*, 13.

73. Art historian Saloni Mathur describes the 1970s design magazines that the Eameses had in their study collection as indicative of Nehru's notion of design as a catalyst for change, newness, and creativity, and a means "to communicate the problem-solving spirit of the Nehruvian era." See Saloni Mathur, "Charles and Ray Eames in India," *Art Journal Open*, May 29, 2011, https://artjournal.collegeart .org/?p=1735. However, this genre of hybridized "design for development" that entered mainstream design discourse in the late 1970s was arguably a post-Nehruvian response to broader policy initiatives, including the Green Revolution.

74. Escobar, "Notes," 5.

75. Papanek held the position of professor and chair, Department of Design, Kansas City Art Institute from 1976 to 1981. The radio show *By Accident or Design* aired on KCUR-FM, Kansas City, between 1978 and 1988.

76. Victor Papanek and James Hennessey, "No Roast Tonight—the Lights on My Carving Knife Need Realignment," chap. 6 of *How Things Don't Work* (New York: Pantheon Books, 1977), 79–92.

77. Papanek and Hennessey, "No Roast Tonight," 111.

78. Papanek and Hennessey, 111.

79. Papanek and Hennessey, *How Things Don't Work*, back cover.

80. Robert McCrum, foreword to Barbara Wood, *Alias Papa: A Life of Fritz Schumacher* (Totnes, UK: Green Books, [1984] 2011), 8.

81. Nigel Cross, David Elliott, and Robin Roy, eds., *Man-Made Futures: Readings in Society, Technology, and Design* (London: Hutchinson Education, 1974).

82. The work of Danish economist Ester Boserup, author of *Woman's Role in Economic Development* (1970), had led to a rethinking of classic Malthusian theories of economy and scarcity. Ideas of decentralization and microcredit were initiated in these same movements. Ivan Illich, a devout Catholic, took on the issue of gender in his 1982 book *Gender* (New York: Pantheon Books), largely dismissing decades of feminist and women's studies that viewed the eradication of gendered roles, bolstered by feminism, as a key driver of capitalism rather than a condition of social equality. See book review by Ninna Nyberg-Sørensen, *Acta Sociologica* 30, no. 2 (1987): 219–223.

83. J. Roger Guilfoyle, "Papanek's Latest Book: Still Naive But More Reasonable and Responsible," *Industrial Design* 6, no. 24 (November–December 1977): 39.

84. Guilfoyle, "Papenek's Latest Book," 39.

85. Guilfoyle, 39.

86. Guilfoyle, 39.

Chapter 10

1. Stephen Bayley, ed., *The Conran Dictionary of Design* (London: Octopus Conran, 1985), 202.

2. John R. Short, "Yuppies, Yuffies and the New Urban Order," *Transactions of the Institute of British Geographers* 14, no. 2 (1989): 173–188.

3. Stephen Bayley, "On Green Design," *Design* (April 1991): 52.

4. Victor Papanek, *Design for the Real World* (rev. ed.; New York: Van Nostrand Reinhold Co., 1984), 352; cited in Nigel Whiteley, *Design for Society* (London: Reaktion Books), 114.

5. Margaret Thatcher, "Speech Opening Design Museum," July 5, 1989, Margaret Thatcher Foundation, https://www.margaretthatcher.org/document/107722.

6. Stephen Bayley, "The Car Maker That Took Off," *Design* 382 (October 1980): 46–49; Victor Papanek, "Polemic: The Name Game," *Design* 382 (October 1980): 54–55; 54.

7. Papanek, "Polemic," 54. See Vance Packard, *The Waste Makers* (New York: David McKay, 1960).

8. Papanek, "Polemic," 55.

9. See Victor Papanek, "Five Ways to Bring Design to the People," in *Design* 387 (March 1981): 58–59.

10. See Guy Julier, *Economies of Design* (London: Sage, 2017), 25.

11. Victor Papanek, "The Cultural Object," *MIMAR: Architecture in Development* 12 (1984): 47–53; 50. For "McDonaldization," see George Ritzer, *The McDonaldization of Society: An Investigation into the Changing Character of Contemporary Life* (Thousand Oaks, CA: Pine Forge Press, 1993).

12. *The Earth We Inherit* (*Jorden vi ärvde*), Sveriges Utbildningsradio, 1981, television documentary, 60 min.

13. Transcript, *By Accident or Design with Victor Papanek*, KCUR-FM (PBS) program no. 36, c. 1980–1981, Victor J. Papanek Foundation, University of Applied Arts Vienna. Papanek continued to adhere to British English spellings, as opposed to US English, throughout his life, possibly on account of his (unsubstantiated) claim to have attended boarding school in the United Kingdom prior to his arrival as a refugee in the United States in 1939, and due to his long residence in, and professional affiliation with Canada.

14. Transcript, *By Accident or Design with Victor Papanek.*

15. Transcript, *By Accident or Design with Victor Papanek.*

16. Papanek, "The Cultural Object," 47–53.

17. Papanek, 47–53.

18. Papanek, 49.

19. Papanek, 50.

20. Victor Papanek, *Design for Human Scale* (New York: Van Nostrand Reinhold Company, 1983), reverse external dust jacket.

21. Papanek, *Design for Human Scale.*

22. Pauline Madge, "Design, Ecology, Technology: A Historiographical Review," *Journal of Design History* 6, no. 3 (1993): 154.

23. Victor Papanek, "The Future Isn't What It Used to Be," *Design Issues* 5, no. 1 (1988): 4.

24. Papanek, "The Future Isn't What It Used to Be," 4.

25. Victor Papanek, "Design: The View from Now," *Mobilia* 296/297 (1980): 107.

26. Victor Papanek, "Various Correspondence, 1983–1998," Victor J. Papanek Foundation, University of Applied Arts Vienna.

27. Anonymous, "Victor Papanek," *Amerika* 319 (June 1983): 30–33, author's translations.

28. Anonymous, "Victor Papanek," 31.

29. Victor Papanek, "Ecological Design: Quest, Results," *Tekhnicheskaya Estetika* 10 (1988): 16–17, author's translations; emphasis in original.

30. Victor Papanek, résumé, June 1997, Victor J. Papanek Foundation, University of Applied Arts Vienna.

31. Ken Isaacs, review of *The Green Imperative*, by Victor Papanek, *Design Issues* 18, no. 2 (1997): 78–79.

32. The author wishes to thank Satish Kumar, founder of Schumacher College, for assistance in researching Papanek's involvement with the institution.

33. Victor Papanek, "Sensing Dwelling," *Resurgence* 152 (May/June 1992): 15.

34. Related to the author in personal communication by Nicolette Papanek.

35. The review was written by self-described design forecaster James Woudhuysen, a libertarian, anti-environmentalist campaigner, pitted against the interventionist approach Papanek and his supporters represented. James Woudhuysen, "Victor Papanek: The Pioneer of Patronising Design," review of Victor Papanek, *Design for the Real World: Human Ecology and Social Change*, Thames and Hudson, second edition, 1985, https://www.woudhuysen.com/victor-papanek-pioneer-of-patronising-design/.

Index